S. J. BENKOVIC

A Specialist Periodical Report

Carbohydrate Chemistry

Volume 4

A Review of the Literature Published during 1970

Senior Reporter
J. S. Brimacombe, *Chemistry Department, University of Dundee*

Reporters
R. J. Ferrier, *Victoria University of Wellington, New Zealand*

R. D. Guthrie, *University of Sussex*

T. D. Inch, *C.D.E.E., Porton, Wilts*

J. F. Kennedy, *University of Birmingham*

SBN: 85156 032 X

The Chemical Society
Burlington House, London, W1V 0BN

Organic formulae composed by Wright's Symbolset method

PRINTED IN GREAT BRITAIN BY JOHN WRIGHT AND SONS LTD., AT THE STONEBRIDGE PRESS, BRISTOL

Preface

The compilation of this, the fourth volume of the Series, has met with a number of difficulties. In the first instance, the preparation of Part I was seriously disrupted by a postal strike in the United Kingdom, as a consequence of which the later issues of a number of overseas Journals, which would normally be covered by this Report, were unavailable at the time of writing. Coverage of these will be delayed until the next Report.

Most regrettable of all is that it has been necessary to go to press without four of the chapters (General Methods, Plant Polysaccharides, Microbial Polysaccharides, and Physicochemical Properties) which were promised for Part II. These chapters had not been received by the end of August, when it was necessary to choose between the unpalatable alternatives of a further delay of uncertain duration or of publishing the material, much of it in page proof, that was already to hand. While the loss of continuity of the series is to be deplored, further delays could only seriously diminish the topicality of the Report, which is one of its primary objectives. In the circumstances, a notable expansion of Dr. Kennedy's chapters for Part II has been much appreciated.

As in previous years, *Abstracts of the American Chemical Society Meetings*, *Dissertation Abstracts*, and the patent literature have not been abstracted. The abbreviation 'Bn' to denote the benzyl group does not appear to have met with disapproval and is again used throughout.

We are indebited to Professor N. K. Kochetkov for providing the English abstracts of a large number of Russian papers, and to Dr. L. C. N. Tucker for reading and commenting on the whole of Part I.

This year Dr. R. D. Guthrie has relinquished his position as Senior Reporter although, happily, he continues as a Reporter. I am sure that readers of the Specialist Periodical Reports on Carbohydrate Chemistry would wish me to thank him for his expert guidance of these Reports during their formative years and for the high standard of reportage he has set.

J. S. B.

August 1971.

Contents

Part I Mono-, Di-, and Tri-saccharides and their Derivatives

1 Introduction . . . 3

2 Free Sugars . . . 5
Isolation and Synthesis . . . 5
Physical Measurements . . . 7
Reactions . . . 8

3 Glycosides . . . 13
O-Glycosides . . . 13
Synthesis . . . 13
Hydrolysis . . . 18
Other Reactions and Features of Glycosides . . . 18
Natural Products . . . 19
S-Glycosides . . . 19
C-Glycosides . . . 20

4 Ethers and Anhydro-sugars . . . 22
Ethers . . . 22
Methyl Ethers . . . 22
Substituted Alkyl Ethers . . . 25
Silyl Ethers . . . 27
Intramolecular Ethers (Anhydro-sugars) . . . 27
Epoxides . . . 27
Other Anhydrides . . . 28

5 Acetals . . . 31
Reactions and Properties . . . 31
Synthesis . . . 32
From Diol Groups on Acyclic Carbohydrates . . . 32
From Diol Groups on Cyclic Carbohydrates . . . 33
(i) Free Sugars . . . 33
(ii) Glycosides *etc.* . . . 34
From Single Hydroxy-groups . . . 34

6 Esters . . . 35
Acetates . . . 35
Substituted Acetate and Other Non-aromatic Carboxylates . . . 37

Benzoates . . . 39
Carboxylic Orthoesters . . . 39
Carbonates . . . 40
Phosphates . . . 41
Carbohydrate Phosphates . . . 41
Nucleoside Phosphates . . . 42
Sulphates . . . 43
Sulphonates . . . 44
Displacement Reactions without Participation . . . 44
Displacements with Participation of Oxygen Functions . 47
Displacement Accompanied by Rearrangement . . 48

7 Halogenated Sugars . . . 50
Glycosyl Halides . . . 50
Other Halogenated Derivatives . . . 53

8 Amino-sugars . . . 57
Natural Products . . . 57
Synthesis . . . 57
Reactions . . . 63
Di- and Poly-amino-sugars . . . 64

9 Hydrazones, Osazones, and Formazans . . . 67

10 Miscellaneous Nitrogen-containing Compounds . . . 70
Glycosylamines and Related Compounds . . . 70
Azides . . . 71
Nitro-compounds . . . 72
Epimines . . . 74
Other Heterocyclic Derivatives . . . 75
Other Nitrogen-containing Compounds . . . 77

11 Thio-sugars . . . 80

12 Derivatives with Sulphur or Nitrogen in the Sugar Ring . . 83
Nitrogen Derivatives . . . 83
Sulphur Derivatives . . . 87
Phosphorus Derivatives . . . 88

13 Deoxy-sugars . . . 89

14 Unsaturated Derivatives . . . 93
Glycals . . . 93
Other Unsaturated Compounds . . . 98

15 Branched-chain Sugars . . . 104
Compounds with an R^1—C—OR^2 Branch . . . 104
Compounds with an R—C—H Branch . . . 106

16 Aldoses, Dialduloses, and Diuloses 110

17 Sugar Acids and Lactones 113
Aldonic Acids 113
Ulosonic Acids 114
Uronic Acids 116
Aldaric Acids 119
Ascorbic Acid 120
Muramic Acid 120

18 Inorganic Derivatives 121

19 Cyclitols 123
Nitrogen-containing Derivatives 128

20 Antibiotics 130

21 Nucleosides 135
Synthesis 135
Nucleosides containing Unsaturated Carbohydrate Components 141
Nucleosides containing Branched-chain Sugar Components 142
Physical Measurements 143

22 Oxidation and Reduction 145
Periodate Oxidation 145
DMSO-based Oxidations 146
Other Oxidations 147
Reduction 148

23 N.M.R. Spectroscopy and Conformational Features of Carbohydrates 149
Pyranoid Systems 149
General Observations on Model Compounds . . . 149
General Observations on Pyranoid Compounds . . 150
Specific Pyranoid Compounds 151
Furanoid Systems 153
Acyclic Systems 155
Heteronuclear N.M.R. Studies 155
^{13}C N.M.R. Spectroscopy 155
^{19}F N.M.R. 157

24 Other Physical Methods 158
Infrared Spectroscopy 158
Mass Spectrometry 158
X-Ray Crystallography 160

25 Polarimetry 164

26 Separatory and Analytical Methods 167
Chromatographic Methods 167
Gas–Liquid Chromatography 167
Column and Ion-exchange Chromatography . . . 168
Paper Chromatography and Electrophoresis . . . 169
Thin-layer Chromatography 169
Other Analytical Methods 169

27 Alditols 171

Part II Macromolecules

1 Introduction 177

2 Glycoproteins, Glycopeptides, and Animal Polysaccharides . . 178
Microbial Glycoproteins 180
Plant Glycoproteins 180
Blood-group Substances 182
Collagens 187
Glycogens 189
Glycosaminoglycuronans, Glycosaminoglycans, and their Protein and Peptide Derivatives 190
Occurrence, Isolation, Measurement, and Structure . . 190
Clinical Conditions 199
Biosynthesis and Metabolism 201
Hormonal Glycoproteins 206
Milk Glycoproteins 209
Serum, Plasma, and Blood Cellular Element Glycoproteins, and Immunoglobulins 211
Serum and Plasma Glycoproteins 211
Blood Cellular Element Glycoproteins 213
Immunoglobulins 214
Urinary Glycoproteins 215
Miscellaneous Glycoproteins 216

3 Enzymes 222
Acetamidodeoxyhexosidases 222
β-Fructofuranosidases 223
Fucosidases 223
Galactosidases. 224
Glucosidases 226

Glucuronidases 227
Mannosidases 228
Sialidases 228
α-Amylases 229
β-Amylases 231
Glucoamylases 232
Dextrin-1,6-glucosidases 233
Lysozymes 234
Other Polysaccharidases and Oligosaccharidases 237
Cellulases 237
Chitinases 238
Dextranases 238
Galactanases 238
Heparin Lyases and Heparan Sulphate Lyases 238
Hyaluronidases and Hyaluronate Lyases 239
Laminarinases 239
Oligo-1,6-glucosidases 239
Polygalacturonases and Polygalacturonate Lyases . . . 239
Pullulanases 240
Trehalases 240
Xylanases 240
Carbohydrate Oxidases 240
Glucose Oxidases 240
Galactose Oxidases 241
Peptidases 241
Ribo- and Deoxyribo-nucleases 242
Miscellaneous Enzymes 243
Index of Enzymes Referred to in Chapter 3 246

4 Glycolipids and Gangliosides 248

5 Chemical Synthesis and Modification of Polysaccharides *etc.* . 255
Synthesis of Polysaccharides, Oligosaccharides, and Glycopeptides 255
Modification of Polysaccharides and Uses of Modified Polysaccharides 256
Modification of Glycoproteins and Enzymes and Uses of Modified Glycoproteins and Enzymes 264
Author Index 267

Abbreviations

The following abbreviations have been used

ATP	adenosine triphosphate
Bn	benzyl
c.d.	circular dichroism
CDP	cytidine diphosphate
CMP	cytidine monophosphate
DCC	dicyclohexylcarbodi-imide
DEAE	diethylaminoethyl
DMF	*NN*-dimethylformamide
DMSO	dimethyl sulphoxide
e.s.r.	electron spin resonance
g.l.c.	gas–liquid chromatography
HMPT	hexamethylphosphortriamide
i.r.	infrared
NBS	*N*-bromosuccinimide
n.m.r.	nuclear magnetic resonance
o.r.d.	optical rotatory dispersion
py	pyridine
THF	tetrahydrofuran
t.l.c.	thin-layer chromatography
TMS	trimethylsilyl
UDP	uridine diphosphate

Part I

MONO-, DI-, AND TRI-SACCHARIDES AND THEIR DERIVATIVES

By

J. S. Brimacombe
R. J. Ferrier
R. D. Guthrie
T. D. Inch

1
Introduction

The general terms of reference remain those set out in the Introduction to Volume 1 (vol. 1, p. 3), and the arrangement of subject matter follows previous Reports.

There have been no major new developments in the area of synthetic carbohydrate chemistry, although unsaturated and dicarbonyl sugars continue to be used increasingly as precursors for synthetic work. There has been the expected expansion in the use of o.r.d., mass spectrometry, and *X*-ray diffraction analysis. A notable paper by Kim and Jeffrey[1] on the crystal structures of nine alditols has related conformation to configuration in the solid state; it is interesting that the results parallel those previously obtained for acyclic sugars in solution.[2] The reagent europium tris-(dipivaloylmethane) appears to be of potential value for simplifying complex n.m.r. spectra of carbohydrate derivatives.[3] Several distinguished papers have appeared on ^{13}C and ^{19}F n.m.r. spectroscopy, and the importance of their application in stereochemical analysis is discussed in Chapter 23. Empirical rules for estimating the molar rotation of saturated pyranoid carbohydrates from their geometric structure have been devised and applied by Lemieux and Martin.[4] The effects of solvation and hydrogen bonding on the rotation were also interpreted in terms of these empirical rules.

The International Union of Pure and Applied Chemistry (IUPAC) have produced tentative rules dealing with the main principles of Fundamental Stereochemistry[5a] and, jointly with the International Union of Biochemistry (IUB), have issued Information Bulletin No. 7 containing tentative proposals for the nomenclature of acyclic and cyclic forms of monosaccharides and their simple derivatives.[5b]

A book of general interest has appeared.[6] We omitted to note in last year's Report that the expanded title 'Advances in Carbohydrate Chemistry

[1] G. A. Jeffrey and H. S. Kim, *Carbohydrate Res.*, 1970, **14**, 207.
[2] D. Horton and J. D. Wander, *Carbohydrate Res.*, 1969, **10**, 279.
[3] I. Armitage and L. D. Hall, *Chem. and Ind.*, 1970, 1537.
[4] R. U. Lemieux and J. C. Martin, *Carbohydrate Res.*, 1970, **13**, 139.
[5a] *J. Org. Chem.*, 1970, **35**, 2849.
[5b] Information Bulletin, No. 7, Carbohydrate Nomenclature, 1970; see also *Arch. Biochem. Biophys.*, 1970, **136**, 1.
[6] C. Prevost, 'Traité de Chimie Organique, Tome 2: Principaux Types de Réactions, Glucides,' Dunod, Paris; 410 pp.

and Biochemistry' has been introduced for this authoritative series commencing with Volume 24 (1969). Two special issues of *Carbohydrate Research* were published in 1970; the April issue[7] was dedicated to the memory of the late Professor M. L. Wolfrom, and the December issue[8] was dedicated to Professor F. Micheel in celebration of his seventieth birthday.

[7] *Carbohydrate Res.*, vol. 13, April 1970.
[8] *Carbohydrate Res.*, vol. 15, December, 1970.

2
Free Sugars

The use of stereonumbers for designating stereochemistry (*cf.* vol. 3, p. 5) has now been extended to the aldohexopyranoses.[9] A convention for the description of sugar conformations based on the Reeves' system has been described.[10] The *C*1(D) conformation becomes $^{4}C_{1}$ and 1*C*(D) becomes $_{1}C^{4}$; boat conformations may be described similarly. It was suggested that if axial substituents only were designated then for aldopyranoses $^{4}C_{1}$(1,2)-5-CH_2OH would represent α-D-mannopyranose with the ring in the *C*1 conformation.

A proposal has been put forward to account for the requirement of a free C-2 hydroxy-group in hexoses related to D-glucose in active intestinal transport.[11] It was suggested that sugars with this stereochemistry are covalently bound to a carrier by an ester linkage. The transport of a wide range of sugars, sugar derivatives, and related compounds was studied.

Isolation and Synthesis

D-Glucose, D-mannose, and a new disaccharide [3-*O*-(β-D-glucopyranosyl)-D-mannopyranose] have been extracted from the roots of *Asparagus racemosus*.[12] D-*manno*-Heptulose and the closely related D-*glycero*-D-*galacto*-heptitol (perseitol), D-*glycero*-D-*manno*-octulose, D-arabinitol, D-mannitol, galactitol, *myo*-inositol, D-xylose, D-galactose, and primeverose have been isolated from aqueous extracts of Pichi tops.[13] Among the carbohydrate components identified from the spores of *Nosema apis*, a protozoan parasite of the honey bee, were D-glucose, D-glucitol, α,α-trehalose, D-*glycero*-D-*gluco*-heptitol, and D-fructose.[14] Arabinose, galactose, xylose, and rhamnose have been identified in hydrolysates of the leaves of the Indian fig plant (*Opuntia ficus indica* Mill).[15] The two carbohydrate residues attached to the aglycone in panaxoside B^{I} are glucose and 2-*O*-glucopyranosyl-glucose. In panaxoside C, the two residues are either

[9] S. Neelakantan, *Current Sci.*, 1970, **39**, 85.
[10] J. Szejtli, *Acta Chim.* (*Budapest*), 1969, **61**, 57 (*Chem. Abs.*, 1970, **72**, 32157*j*).
[11] J. E. G. Barnett, A. Ralph, and K. A. Munday, *Biochem. J.*, 1970, **116**, 537, 843.
[12] A. B. Landge and J. L. Bose, *Indian J. Chem.*, 1970, **8**, 588.
[13] N. Richtmyer, *Carbohydrate Res.*, 1970, **12**, 233.
[14] P. J. Wood, I. R. Siddiqui, J. W. Vandermeer, and T. A. Gochnauer, *Carbohydrate Res.*, 1970, **15**, 154.
[15] El S. Amin, O. M. Awad, and M. M. El-Sayed, *Carbohydrate Res.*, 1970, **15**, 159.

rhamnose and 2-*O*-glucopyranosyl-glucose or glucose and 2-*O*-rhamnopyranosyl-glucose.[16] The sugar components of pharbitic acid, from the purgative 'resin glycoside' pharbitin, have been identified as D-quinovose, D-glucose, and L-rhamnose.[17]

DL-Glucose has been synthesised by the route outlined for the D-enantiomer (Scheme 1) in which the starting material was derived from acrolein dimer.[18] Tetroses (10%), pentoses (30%), and hexoses (50%) have

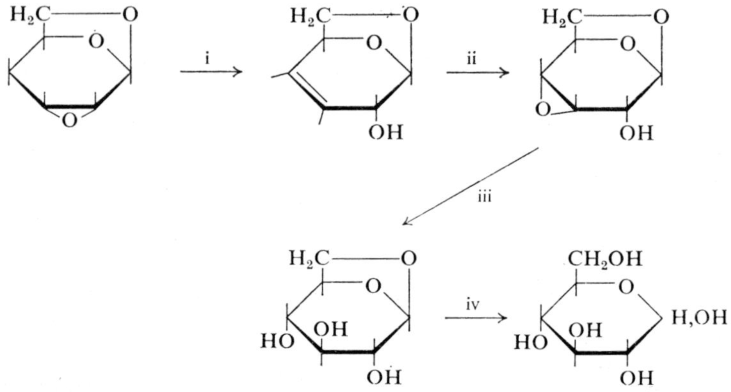

Reagents: i, BuLi; ii, $ClC_6H_4CO_3H$; iii, $Ba(OH)_2$; iv, HCl

Scheme 1

been obtained from formaldehyde and calcium hydroxide in a continuously-stirred tank reactor,[19] and detailed studies of this reaction have been reported.[20] The preparation of L-sorbose from glucitol has been investigated.[21]

D-Galactose and D-lyxose, and not the expected D-galactononitrile, were obtained on photolysis of D-galactose azine,[21a] and the mechanism illustrated in Scheme 2 was suggested to account for the observed products. D-Lyxose was the major product of photolysis of D-galactose oxime.[22]

3-*O*-Benzyl-D-*glycero*-L-*manno*-heptose has been prepared by nitromethane extension of 2-*O*-benzyl-D-galactose.[23] Surprisingly, only 3-*O*-benzyl-1-deoxy-1-nitro-D-*glycero*-L-*manno*-heptitol was formed and the C-2

[16] G. I. Shaposhnikova, N. A. Ferens, N. I. Uvarova, and G. B. Elyakov, *Carbohydrate Res.*, 1970, **15**, 319.
[17] H. Okabe and T. Kawasaki, *Tetrahedron Letters*, 1970, 3123.
[18] U. P. Singh and R. K. Brown, *Canad. J. Chem.*, 1970, **48**, 1791.
[19] A. H. Weiss and J. Shapira, *Hydrocarbon Process*, 1970, **49**, 119.
[20] A. H. Weiss, R. B. Lapierre, and J. Shapira, *J. Catalysis*, 1970, **16**, 332.
[21] N. S. Zolotarev, Z. G. Tkhorevskaya, M. D. Moskvin, K. D. Vasilenko, and I. V. Barsegyan, *Khim.-Farm. Zhur.*, 1970, **4**, 28.
[21a] R. W. Binkley and W. W. Binkley, *Carbohydrate Res.*, 1970, **13**, 163.
[22] W. W. Binkley and R. W. Binkley, *Tetrahedron Letters*, 1970, 3439.
[23] P. P. Singh and G. A. Adams, *Carbohydrate Res.*, 1970, **12**, 261.

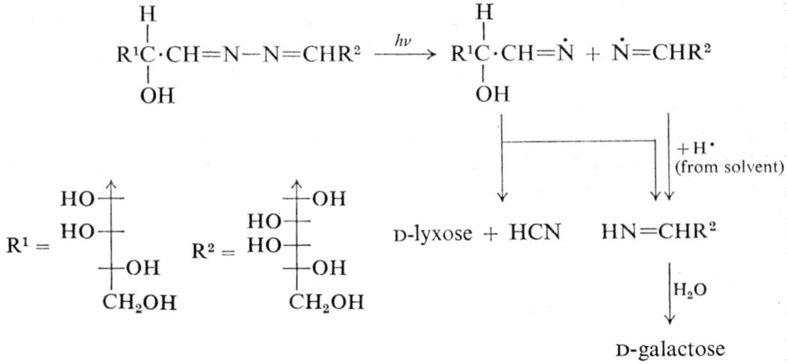

Scheme 2

epimer could not be detected. It is unusual for such stereospecificity to be encountered in a reaction of this type.

D-Galactose and D-talose (ratio 4 : 1) were obtained on hydroxylation of D-galactal in the presence of osmium tetroxide, whereas in the presence of either molybdenum trioxide or tungsten trioxide, D-talose was obtained in 90% yield (see also Chapter 14).[24]

An improved synthesis of L-idose from D-glucose has been reported.[25] Several modifications were discussed and the recommended procedure is illustrated in Scheme 3 (see also p. 47). A new route to L-allose is described in Chapter 6.

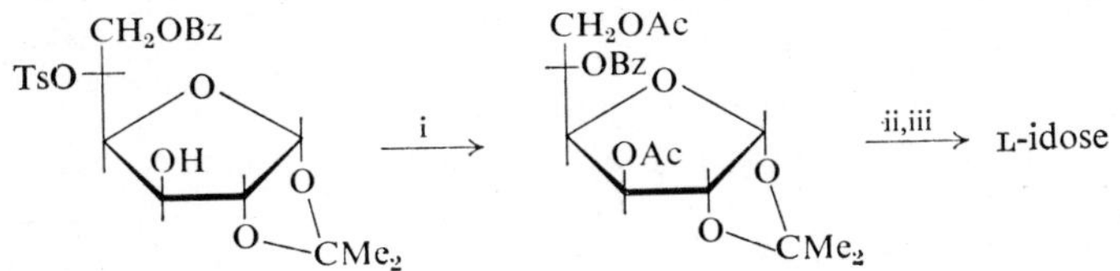

Reagents: i, Dowex resin (^-OAc), Ac_2O; ii, MeONa; iii, H_3O^+

Scheme 3

Physical Measurements

Coriose (D-*altro*-3-heptulose) exists in the crystal in the furanose ring form.[26]

The kinetics of the hydrogenolysis of D-glucose with a nickel catalyst in basic media[27] and those of the oxidation of eleven free sugars by Cu^{2+} in basic media[28] have been discussed. For the latter, reaction rates were

[24] V. Bílik and Š. Kucar, *Carbohydrate Res.*, 1970, **13**, 311.
[25] J. Kovář, *Canad. J. Chem.*, 1970, **48**, 2383.
[26] T. Tagu, K. Osaki, and T. Okuda, *Acta Cryst.*, 1970, **B26**, 991.
[27] I. D. Rozhdestvenskaya, T. N. Fadeeva, N. B. Titova, and L. V. Shileiko, *Kineticai Kataliz*, 1970, **11**, 696.
[28] S. V. Singh, O. C. Saxena, and M. P. Singh, *J. Amer. Chem. Soc.*, 1970, **92**, 537.

independent of the copper ion concentration and were first-order with respect to the reducing sugar and base. Enolisation of the sugar appears to be the rate-controlling step.

Absorption spectroscopy and circular dichroism were used to demonstrate that D-fructose exists to the extent of 2% in the acyclic form in equilibrated aqueous solution, whereas the percentages of the acyclic forms of D-fructose-1-phosphate, D-fructose-6-phosphate, and D-fructose-1,6-diphosphate are somewhat larger. There is an appreciable increase in the acyclic form of D-fructose present in solutions of DMSO.[29]

E.s.r. spectroscopy has been used to study the radicals produced on γ-irradiation of frozen concentrated solutions of sucrose and D-glucose at 77 K.[30] Heats of combustion and enthalpies of formation of D-ribose, D-arabinose, and L-ascorbic acid have been determined.[31] Proton dissociations from several monosaccharides have been measured at 10 and 40 °C.[32]

Reactions

[^{14}C]-D-Mannoheptulose was not metabolised by rats.[33] D-Glucose has been used as starting material for a synthesis of (+)-desthiobiotin (1) (Scheme 4) thereby providing chemical confirmation of the absolute configuration of (+)-biotin.[34]

The acid-catalysed conversion of D-xylose and D-glucuronic acid into 2-furaldehyde has been shown to be analogous to the acid-catalysed conversion of hexoses into 5-(hydroxymethyl)-2-furaldehyde.[35] Specifically, the conversions of D-[2-^{2}H]glucose into 5-(hydroxymethyl)-2-furaldehyde (under acid conditions) and metasaccharinic acid (under basic conditions) have indicated that the former occurs by the irreversible sequence (2) → (3) → (5) → (7), whereas the latter proceeds by the sequence (2) → (3) → (4) → (6).[36] Anet has published a note on isotope exchanges occurring during the acid-catalysed conversion of hexoses to 5-(hydroxymethyl)-2-furaldehyde in deuterium oxide,[37] and the observed isotope effects were considered to reduce objections raised to an earlier proposal, which invoked 3-deoxyhexos-2-uloses as intermediates.

Thirteen products have been isolated from the degradation of D-glucose with methylamine and acetic acid.[38] The two compounds responsible for the caramel odour detected during this reaction were identified as acetylformoin and 4-hydroxy-2,5-dimethyl-3(2*H*)-furanone. The isomerisation of

[29] G. Avigad, S. Englard, and I. Listowsky, *Carbohydrate Res.*, 1970, **14**, 365.
[30] P. J. Baugh, K. Kershaw, and G. O. Phillips, *J. Chem. Soc.* (*B*), 1970, 1482.
[31] P. Desai, and R. C. Wilhoit, *Thermochimica Acta*, 1970, **1**, 61.
[32] J. J. Christensen, J. H. Rytting, and R. M. Izatt, *J. Chem. Soc.* (*B*), 1970, 1646.
[33] W. Nelkin and E. Simon, *Biochem. Biophys. Res. Comm.*, 1970, **41**, 864.
[34] H. Kuzuhara, H. Ohrui, and S. Emoto, *Tetrahedron Letters*, 1970, 1185.
[35] M. S. Feather, *Tetrahedron Letters*, 1970, 4143.
[36] M. S. Feather and J. F. Harris, *Carbohydrate Res.*, 1970, **15**, 304.
[37] E. F. L. J. Anet, *Austral. J. Chem.*, 1970, **23**, 2383.
[38] G. R. Jurch, jun., and J. H. Tatum, *Carbohydrate Res.*, 1970, **15**, 233.

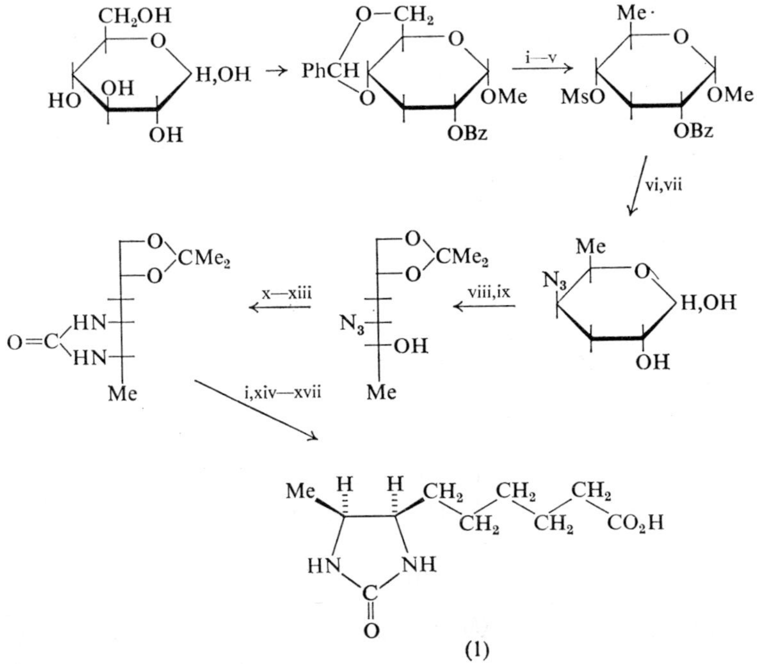

Reagents: i, H^+; ii, TsCl, py; iii, NaI; iv, H_2; v, MsCl, py; vi, NaN_3; vii, H^+; viii, $NaBH_4$; ix, Me_2CO, H^+; x, MsCl, py; xi, NaN_3; xii, H_2–Ni; xiii, $COCl_2$; xiv, H^+; xv, $NaIO_4$; xvi, Wittig reagent; xvii, H_2

Scheme 4

lactose to lactulose by aqueous calcium hydroxide has also been studied.[39] The reactions shown in Scheme 5 were proposed to account for some of the products formed on thermal degradation of 1-deoxy-1-piperidino-D-fructose.[40]

A detailed report on the thermal decomposition of sugars has appeared.[41] Studies with [^{14}C]-labelled D-glucose have shown that 65% of the formaldehyde produced by thermal degradation arose from C-6, whereas C-1 and C-2 contributed 15% and 5%, respectively.[42]

Deoxy and deoxycarbonyl compounds are formed on γ-radiolysis of mono- and di-saccharides, glycosides, and certain polysaccharides;[43] the formation of deoxy-sugars on irradiation of aqueous solutions of carbohydrates appears to be a general process. The production of formaldehyde

39 S. Zagrodzki and B. Krol, *Rocz. Technol. Chem. Zyarn.*, 1969, **15**, 189 (*Chem. Abs.*, 1970, **73**, 4102q).
40 F. D. Mills, B. G. Baker, and J. E. Hodge, *Carbohydrate Res.*, 1970, **15**, 205.
41 M. Popa-Luchian, Gh. Rozmarin, and I. A. Schneider, *J. Therm. Anal.*, 1969, **1**, 211.
42 Y. Houminer and S. Hoz, *Israel J. Chem.*, 1970, **8**, 97.
43 N. K. Kochetkov, L. I. Kudryashov, S. M. Yarovaya, and S. V. Voznesenskaya, *Bull. Acad. Sci., U.S.S.R.*, 1970, 211.

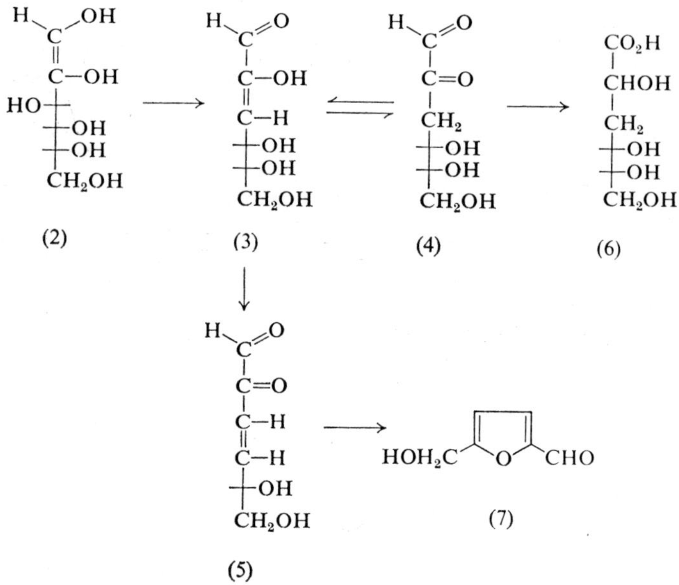

(2) (3) (4) (6) (5) (7)

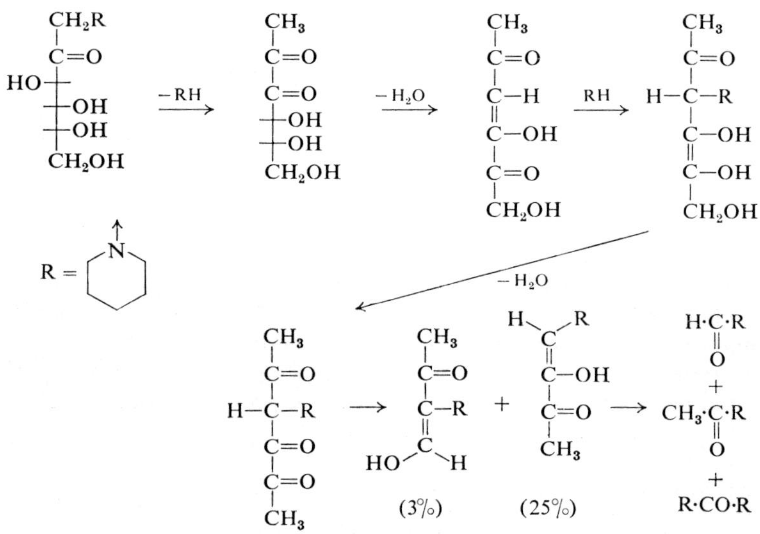

Scheme 5

during the γ-irradiation of fructose has been studied.[44] Malonaldehyde has been identified as a product of the u.v. photolysis of aqueous solutions of such carbohydrates as D-glucose, D-fructose, D-mannitol, and D-glucono-1,5-lactone.[45] The pathway shown in Scheme 6 was proposed to explain the formation of both malonaldehyde and deoxy-sugar in this process.

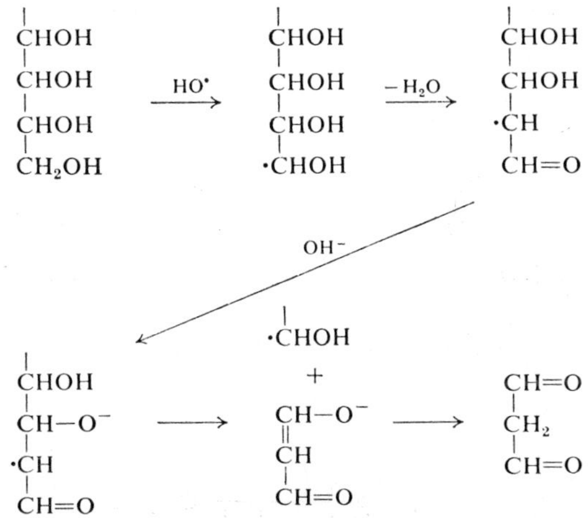

Scheme 6

The mechanism of formation of (8) from the reaction of D-glucose with liquid hydrogen fluoride in toluene has been discussed.[46] Treatment of

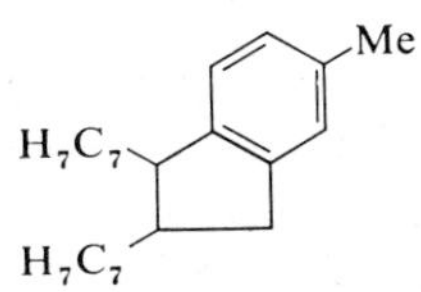

(8)

D-glucose with anisole in liquid hydrogen fluoride afforded 1-deoxy-1,1-bis-(p-methoxyphenyl)-D-glucitol (70%, after 1 h) and 2,5-anhydro-1-deoxy-1,1-bis-(p-methoxyphenyl)-D-glucitol (18%, after 3 h), whereas after 40 h the reaction mixture contained about twenty products.[47] The crystalline pigments (9)—(12) have been isolated from the reaction of either D-glucose or D-fructose with p-toluidine.[48]

[44] G. Lofroth and C. Kim, *Acta Chem. Scand.*, 1970, **24**, 749.
[45] H. Scherz, *Carbohydrate Res.*, 1970, **14**, 417.
[46] F. Micheel and H. Sobitzkat, *Tetrahedron Letters*, 1970, 1605.
[47] F. Micheel and J. Stanek jun., *Tetrahedron Letters*, 1970, 1609.
[48] T. Ozawa and N. Kinae, *Chem. and Pharm. Bull.* (*Japan*), 1970, **18**, 1293.

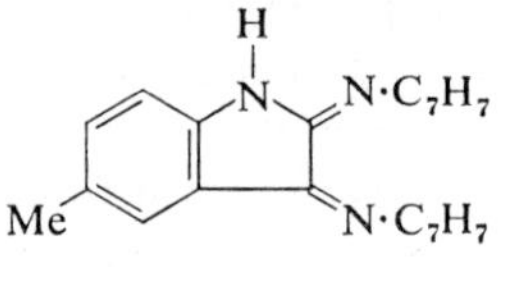

(9)

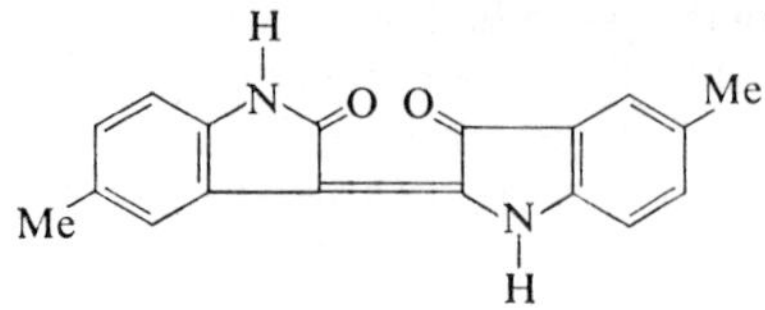

(10)

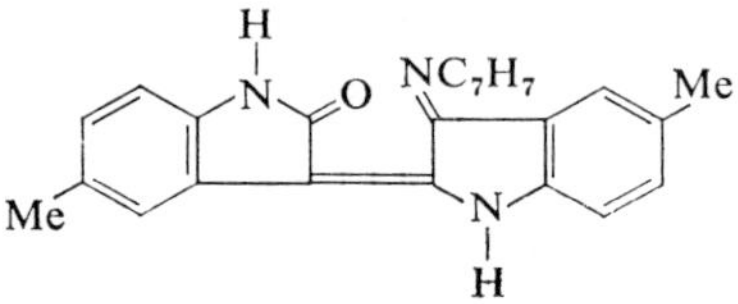

(11)

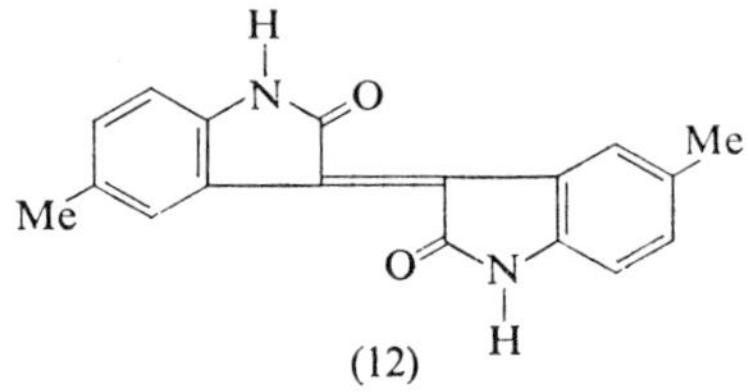

(12)

The reactions of D-glucose and D-galactose with 2-aminothiophenol and 2-amino-4-chlorothiophenol gave products of the type (13).[48a] Acetylation afforded hexa-acetyl derivatives which could also be prepared by reaction

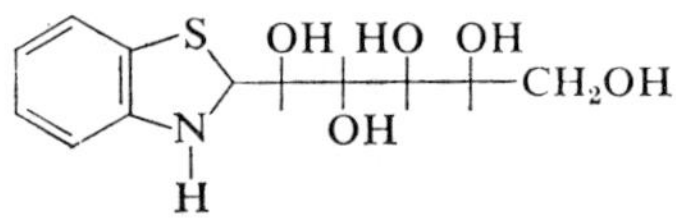

(13)

of the thiophenols with the *aldehydo*-sugar penta-acetate followed by *N*-acetylation.

The reactions of 2,4:3,5-di-*O*-benzylidene-*aldehydo*-D-ribose and penta-*O*-acetyl-*aldehydo*-D-glucose with seven resonance-stabilised ylides have been studied.[49] Whereas all the ylides reacted with the D-ribose derivative in high yield, the reactions of the ylides with the D-glucose derivative generally gave lower yields and, in some cases, the expected product was not isolated. It was concluded that steric hindrance by substituents on the carbohydrate and ylide probably affects the overall rate of the Wittig reaction.

N.m.r. studies of the anomeric proton signals have shown that enzymically-catalysed isomerisations of D-fructose and D-xylulose give the α-pyranoses preferentially and that isomerisation occurs by mechanisms involving C-1 → C-2 hydrogen transfer.[50]

[48a] R. Bognár, Z. Kolodynska, L. Somogyi, Z. Györgydeák, L. Szilágyi, and É. N. Nemes, *Acta Chim. Acad. Sci. Hung.*, 1969, **62**, 65.

[49] R. E. Harmon, G. Wellman, and S. K. Gupta, *Carbohydrate Res.*, 1970, **14**, 123.

[50] M. S. Feather, V. Deshpande, and M. J. Lybyer, *Biochem. Biophys. Res. Comm.*, 1970, **38**, 859.

3
Glycosides

O-Glycosides

Synthesis.—A new modification of the Koenigs–Knorr reaction involved carrying out the condensations in the presence of silver salts of 2-, 3-, or 4-hydroxyalkanoic acids, or of 1,3- or 1,4-dicarboxylic acids. Acetobromoglucose in ether afforded good yields of β-glycosides, particularly when silver 4-hydroxyvalerate is used. A mechanism involving the step illustrated in Scheme 7 was envisaged for the reaction.[51] Other workers

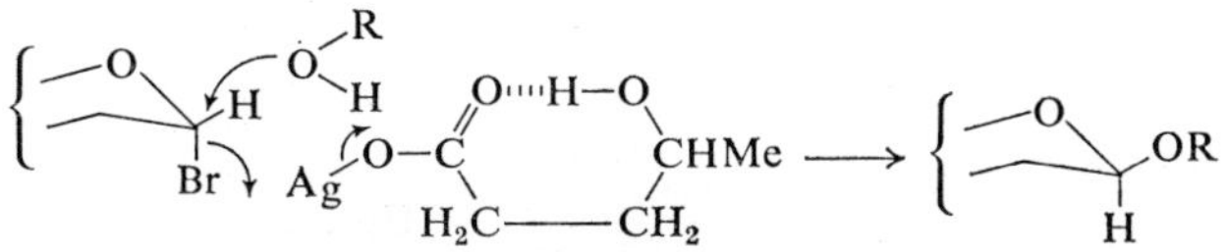

Scheme 7

have discovered that silver succinate can be effective in facilitating glycoside synthesis.[52]

An improved synthesis of methyl β-D-glucopyranoside involves methylation of the free sugar with dimethyl sulphate at low temperature.[53] Methyl 4-*O*-methyl-α-DL-arabinopyranoside has been synthesised (Scheme 8) by

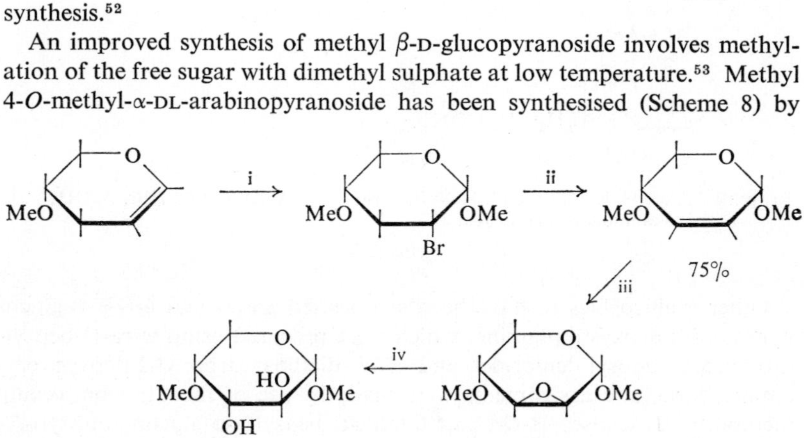

Reagents: i, Br_2, MeOH; ii, KOH, MeOH; iii, $Cl\cdot C_6H_4\cdot CO_3H$; iv, KOH

Scheme 8

[51] G. Wulff, G. Röhle, and W. Krüger, *Angew. Chem. Internat. Edn.*, 1970, **9**, 455.
[52] B. Helferich and W. M. Müller, *Chem. Ber.*, 1970, **103**, 3350.
[53] D. M. Hall and O. A. Stamm, *Carbohydrate Res.*, 1970, **12**, 421.

Brown's group in a continuation of their studies on the formation of specific carbohydrate derivatives from non-carbohydrate precursors.[54]

α-D-Glucopyranosides have been prepared by the reaction of 3,4,6-tri-*O*-acetyl-β-D-glucopyranosyl chloride in benzene with alcohols in the presence of zinc oxide or silver or mercury salts. Best results were obtained using mercury succinate.[55] An application of the nitrosyl chloride method of α-glycoside synthesis to the preparation of a penicillin derivative is shown in Scheme 9.[56]

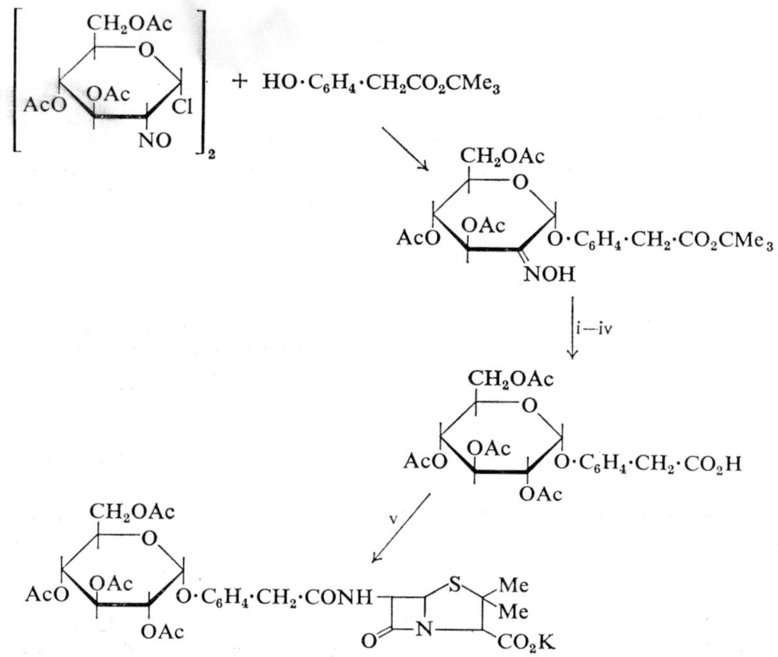

Reagents: i, AcOH, levulinic acid, HCl; ii, $NaBH_4$; iii, Ac_2O, py; iv, HBr, AcOH; v, 6-aminopenicillanic acid

Scheme 9

Other α-glucosides to have been synthesised are 4- and 6-*O*-α-D-glucopyranosyl-2-deoxystreptamine, which were prepared using tetra-*O*-benzyl-α-D-glucopyranosyl chloride and *NN'*-diethoxycarbonyl-2-deoxystreptamine. Structural assignments were made by use of the Cu^{II}-complexing method.[57] Likewise, 3-*O*-(2,3,4,6-tetra-*O*-benzyl-α-D-glucopyranosyl)-*N*-(2,4-dinitrophenyl)-L-serine and -L-threonine methyl esters have been

[54] R. M. Srivastava and R. K. Brown, *Canad. J. Chem.*, 1970, **48**, 2341.
[55] B. Helferich and W. M. Müller, *Naturwiss.*, 1970, **57**, 496.
[56] T. L. Nagabhushan and C. Chin, *Canad. J. Chem.*, 1970, **48**, 3097.
[57] Y. Nishimura, T. Tsuchiya, and S. Umezawa, *Bull. Chem. Soc. Japan*, 1970, **43**, 2960.

prepared. The kinetics of the base-catalysed eliminative degradation of these compounds were found to be first order, and the significance of the results in relation to studies of glycopeptides was stressed.[58] β-Glycosides of *N*-benzyloxycarbonyl-L-serine and -L-threonine methyl esters (and simple alcohols) were otherwise prepared by condensing the corresponding t-butyl ethers with acetohalogeno-sugars in toluene in the presence of silver carbonate, toluenesulphonic acid, and pyridine perchlorate.[59] Another report of the synthesis of serine *O*-glycosides describes the derivatives of 2-acetamido-2-deoxy-D-glucose and -D-galactose.[60]

Appreciable additional attention has been given to the synthesis of glycosides of 2-amino-2-deoxy-sugars: the oxazoline method has thus afforded 2-*O*-(2-acetamido-2-deoxy-β-D-glucopyranosyl)glycerol and 2-acetamido-3-*O*-(2-acetamido-2-deoxy-β-D-glucopyranosyl)-2-deoxy-D-glucose.[61] Other results were reported which indicate that oxazoline derivatives of mono- and oligo-saccharides are effective glycosylating agents for the preparation of 1,2-*trans*-2-acetamido-2-deoxyglycosides including oligosaccharides.[62] Such compounds can, of course, be prepared by the use of peracetylglycosyl chlorides, as demonstrated by the preparation of 2′,3′-epoxypropyl 2-acetamido-2-deoxy-β-D-glucopyranoside and corresponding di- and tri-saccharide glycosides. The procedure involved Koenigs–Knorr condensation with allyl alcohol followed by epoxidation of the initial products and deacetylation.[63]

The use of the trifluoroacetamido-group in glycosyl halide condensations affords a means of obtaining 2-amino-2-deoxy-β-D-glucopyranosides, whereas the 2,4-dinitrophenyl group leads to products with the α-configuration, presumably because the amino-function protected in this way does not take part in the displacement processes.[64] The dichloroacetyl group has also been used in both di- (Scheme 10)[65] and tri-saccharide (Scheme 11)[66] syntheses. These transformations were carried out using standard procedures, the dichloroacetyl group being reduced to an acetyl group with hydrogen over a palladium catalyst. 2-Amino-2-deoxy-β-D-glucopyranosides of digitoxigenin, strophanthidin, and pregnenolone have been synthesised by the Koenigs–Knorr procedure,[67] and 4-*O*-(2-acetamido-

[58] J. R. Vercellotti, N. Nienaber, and C. J. Chang, *Carbohydrate Res.*, 1970, **13**, 63.

[59] N. K. Kochetkov, E. M. Klimov, and V. A. Derevitskaya, *Doklady Acad. Nauk S.S.S.R.*, 1970, **192**, 376.

[60] E. Werries and E. Buddecke, *Z. physiol. Chem.*, 1970, **351**, 1089.

[61] T. S. Antonenko, S. E. Zurabyan, and A. Ya. Khorlin, *Izvest. Akad. Nauk S.S.S.R., Ser. khim*, 1970, 1153.

[62] S. E. Zurabyan, T. S. Antonenko, and A. Ya. Khorlin, *Carbohydrate Res.*, 1970, **15**, 21.

[63] E. W. Thomas, *Carbohydrate Res.*, 1970, **13**, 225.

[64] W. Meyer Zu Reckendorf and N. Wassiliadou-Micheli, *Chem. Ber.*, 1970, **103**, 1792.

[65] D. Shapiro and A. J. Acher, *J. Org. Chem.*, 1970, **35**, 229.

[66] A. J. Acher, Y. Rabinsohn, E. S. Rachaman, and D. Shapiro, *J. Org. Chem.*, 1970, **35**, 2436.

[67] W. Meyer Zu Reckendorf, N. Wassiliadou-Micheli, and H. Machleidt, *Arch. Pharm.*, 1970, **303**, 17.

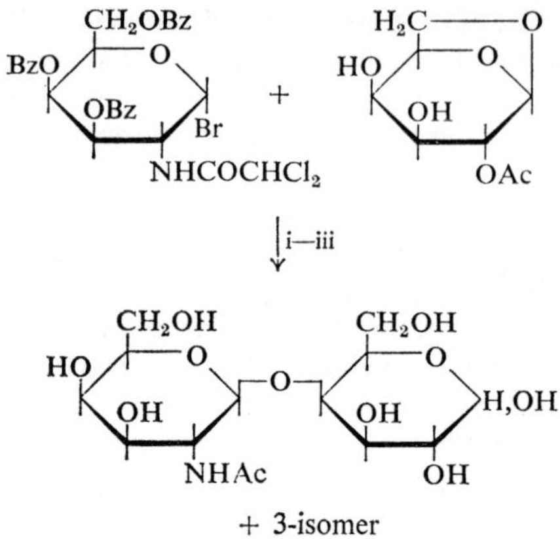

Reagents: i, $C_2H_4Cl_2$, $Hg(CN)_2$; ii, $Ba(OMe)_2$; iii, Pd–C, H_2

Scheme 10

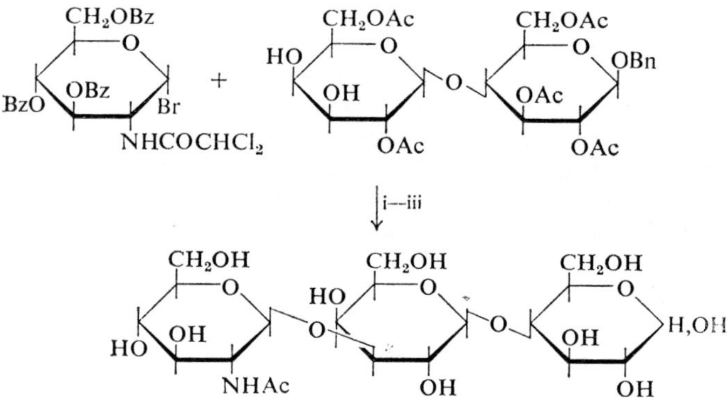

Reagents: i, $C_2H_4Cl_2$, $Hg(CN)_2$; ii, $Ba(OMe)_2$; iii, Pd–C, H_2

Scheme 11

2-deoxy-β-D-glucopyranosyl)-D-glucopyranose has been prepared by glycosylation of 2,3-di-*O*-acetyl-1,6-anhydro-β-D-glucopyranose.[68]

Not surprisingly, fusion of 2-amino-2-deoxy-D-mannose penta-acetate with phenol in the presence of acid catalysts afforded mainly the α-phenyl glycoside, but a small proportion of the β-anomer was also isolable. Nitration of phenyl 2-acetamido-3,4,6-tri-*O*-acetyl-2-deoxy-α-D-galactopyranoside, followed by deacetylation and resin chromatography, afforded *o*- and *p*-nitrophenyl 2-acetamido-2-deoxy-α-D-galactopyranoside, which are useful substrates for α-acetylgalactosaminidase.[69]

[68] D. Shapiro, Y. Rabinsohn, A. J. Acher, and A. Diver-Haber, *J. Org. Chem.*, 1970, **35**, 1464.

[69] B. Weissmann, *J. Org. Chem.*, 1970, **35**, 1690.

Other reports have appeared on phenyl glycosides. Stannic-chloride-catalysed condensations of phenols and nitrophenols with glycose peracetates having 1,2-*trans*-related groups give mixtures of anomers rather than the β-products expected of reactions which involve anchimeric assistance. Presumably, anomerisations were also catalysed by the chloride.[70] A large number of substituted phenyl α-D-mannopyranosides have been prepared by standard procedures using zinc chloride as catalyst.[71] Likewise, bromo-, chloro-, and iodo-phenyl β-cellobiosides, β-lactosides, and β-maltosides were synthesised by use of acetobromo-sugars in acetone and potassium phenates in water. The yields for the bromo- and chloro-compounds decrease in the order $p > m > o$, and the conversion efficiencies also fall off in the order cellobioside > lactoside > maltoside.[72]

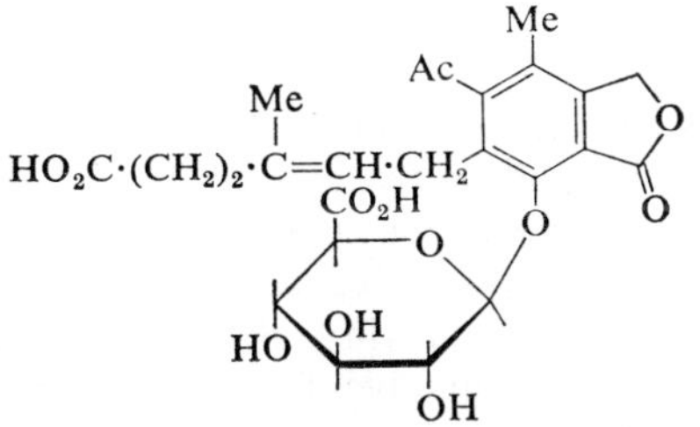

(14)

More specifically, mycophenolic acid β-D-glucuronide (14) has been prepared and found to have significant anti-tumour activity,[73] and twenty-eight steroidal tri-*O*-acetyl-β-D-glucopyranuronoside methyl esters have been synthesised and anomerised to the α-compounds with titanium tetrachloride. The optical rotations of the glycosides were examined with particular reference to Hudson's rules, and the anomers were distinguishable by i.r. spectroscopy — an absorption in the region 1140—1146 cm^{-1} was displayed only by the α-anomers — but not by mass spectrometry.[74] Flavonoid glucuronides have been isolated from natural sources and synthesised.[75, 76]

Two reports have appeared on the synthesis of non-reducing disaccharide derivatives. Reaction of trimethylsilyl 2,3,4,6-tetra-*O*-methyl-α-D-glucopyranoside with 2,3,4,6-tetra-*O*-methyl-α-D-glucopyranosyl chloride in the presence of silver perchlorate gave octa-*O*-methyl-αα-trehalose; the

[70] T. R. Ingle and J. L. Bose, *Carbohydrate Res.*, 1970, **12**, 459.
[71] A. Vervoort and C. K. De Bruyne, *Carbohydrate Res.*, 1970, **12**, 277.
[72] I. C. M. Dea, *Carbohydrate Res.*, 1970, **12**, 297.
[73] K. Ando, S. Suzuki, and M. Arita, *J. Antibiotics*, 1970, **23**, 408.
[74] J. J. Schneider, *Carbohydrate Res.*, 1970, **12**, 369.
[75] H. Wagner, H. Danninger, O. Seligmann, M. Nogradi, L. Farkas, and N. Farnsworth, *Chem. Ber.*, 1970, **103**, 3678.
[76] H. Wagner, H. Danninger, O. Seligmann, and L. Farkas, *Chem. Ber.*, 1970, **103**, 3674.

unsubstituted disaccharide and its octa-acetate were obtained by use of the corresponding benzyl ethers.[77] Condensation of tetra-*O*-methyl-D-glucose and -D-fructose in the presence of zinc chloride gave octa-*O*-methyl-α-D-glucopyranosyl α-D-fructofuranoside in 55—60% yield.[78] The disaccharide (15) was one of the products obtained on reaction of β-D-glucopyranose penta-acetate and silicon tetrachloride in propane-1,2-diol carbonate.[78a]

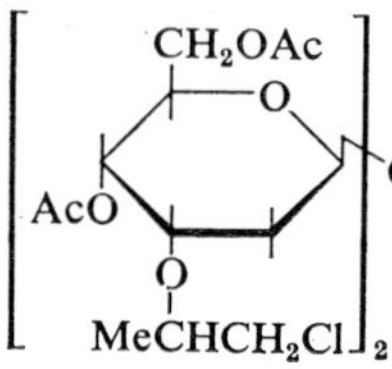

(15)

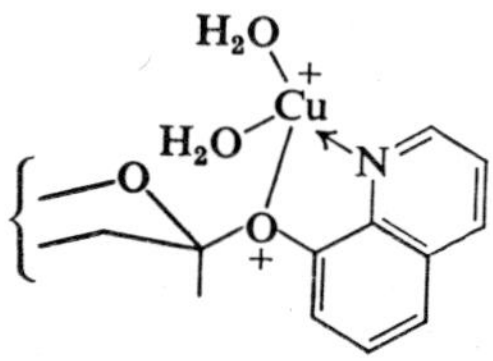

(16)

Hydrolysis.—It has been shown that the hydrolysis of 8-hydroxyquinoline β-D-glucopyranoside is significantly catalysed by Cu^{II} ions in the pH range 5·5—6·0, and it was presumed that the intermediate (16) was involved.[79] The influence of *para* and *meta* substituents on the enzymic hydrolysis o aryl β-D-xylopyranosides by a β-D-xylosidase has been investigated: electronic, steric, and hydrophobic factors were used to analyse the structure–activity relationships.[80]

Further information on the kinetics of the acid-catalysed hydrolysis of cellotriose has been reported. Cellobiose is hydrolysed 1·4 times more rapidly than the glucosidic bond at the non-reducing end of the trisaccharide.[81] The hydrolysis of raffinose and the separation of the products of hydrolysis on cation-exchange resins have also been described.[82]

Studies of the influence of an ultrasonic field on the hydrolysis of sucrose have been reported.[83]

Other Reactions and Features of Glycosides.—A study of the radiolysis of dilute aqueous solutions of methyl α-D-glucopyranoside, methyl α-D-galactopyranoside, and methyl β-D-arabinopyranoside has shown that the stereochemistry of the starting materials influences their stability towards radiation and also the ratios of the products formed. Compounds containing the methylene group were the main products of radiolysis but the formation of free sugars was not observed.[84]

[77] A. Klemer, E. Buhe, R. Kutz, S. Chahin, and L. Kleefeldt, *Annalen*, 1970, **739**, 185.
[78] A. Klemer and B. Dietzel, *Tetrahedron Letters*, 1970, 275.
[78a] A. Klemer and H. Gibmeier, *Carbohydrate Res.*, 1970, **15**, 411.
[79] C. R. Clark, R. W. Hay, and I. C. M. Dea, *Chem. Comm.*, 1970, 794.
[80] F. Van Wijnendaele and C. K. De Bruyne, *Carbohydrate Res.*, 1970, **14**, 189.
[81] A. Meller, *Carbohydrate Res.*, 1970, **13**, 455.
[82] D. Woermann, *Ber. Bunsengesellschaft. Phys. Chem.*, 1970, **74**, 441.
[83] R. Thomas, *Compt. rend.*, 1970, **270** *C*, 635.
[84] L. I. Kudrjashov, T. Ya. Liverstovskaya, S. V. Voznesenskaya, U. I. Kovalev, V. A. Sharpatyi, and N. K. Kochetkov, *Zhur. obshchei. Khim.*, 1970, **40**, 1133.

The migration of the aglycone of methyl glycosides is reported in Chapter 4 and the oxidation of acetylated derivatives is described in Chapter 17.

Natural Products.—Amongst several naturally occurring glucosides which have been investigated are a gibberellin derivative[85] and an α-glycoside obtained from estriol by a biochemical transfer from maltose.[86] Gynocardin has been shown to have structure (17) by a crystallographic study of its 6-brosylate.[87] *p*-Cresyl β-D-ribofuranoside is found as a

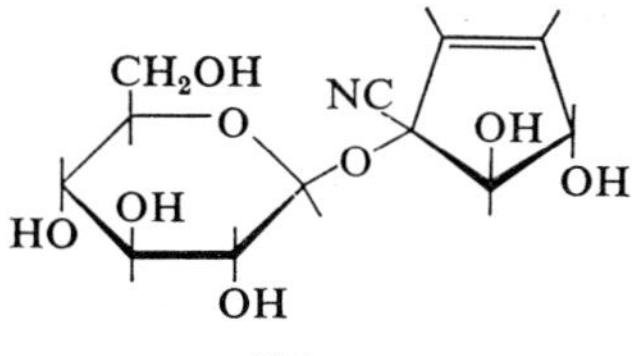

(17)

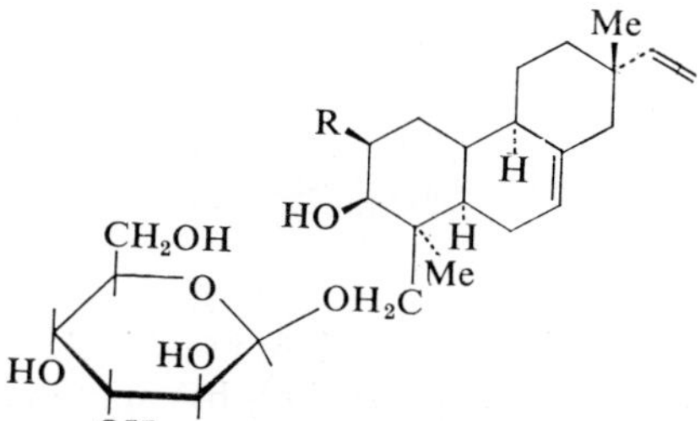

(18) R = OH
(19) R = H

component of a corrinoid factor isolated from sewage sludge,[88] and the discovery of the first naturally occurring D-altropyranosides, virescenosides A and B (18 and 19), is of considerable interest.[89]

S-Glycosides

The crystallographic analysis of the structure of sinigrin has been refined,[90] and three new glucosinolates [*S*-glycosides of general structure (20)] have been isolated from members of the Cruciferae.[91]

85 T. Yokota, N. Murofushi, and N. Takahashi, *Tetrahedron Letters*, 1970, 1489.
86 S. C. Pan, *Biochemistry*, 1970, **9**, 1833.
87 H. S. Khim, G. A. Jeffrey, D. Panke, R. C. Clapp, R. A. Coburn, and L. Long jun., *Chem. Comm.*, 1970, 381.
88 F. Dinglinger and I. Braun, *Z. physiol. Chem.*, 1970, **351**, 1157.
89 N. Cagnoli Bellavita, P. Ceccherelli, M. Ribaldi, J. Polonsky, and Z. Baskevitch, *Gazzetta*, 1969, **99**, 1354.
90 R. E. Marsh and J. Waser, *Acta Cryst.*, 1970, **B26**, 1030.
91 A. Kjaer and A. Schuster, *Acta Chem. Scand.*, 1970, **24**, 1631.

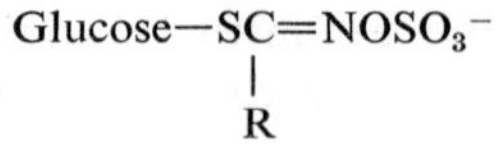

(20)

Benzyl 6-deoxy-1-thio-β-D-glucopyranoside has been prepared by the route outlined (Scheme 12).[92]

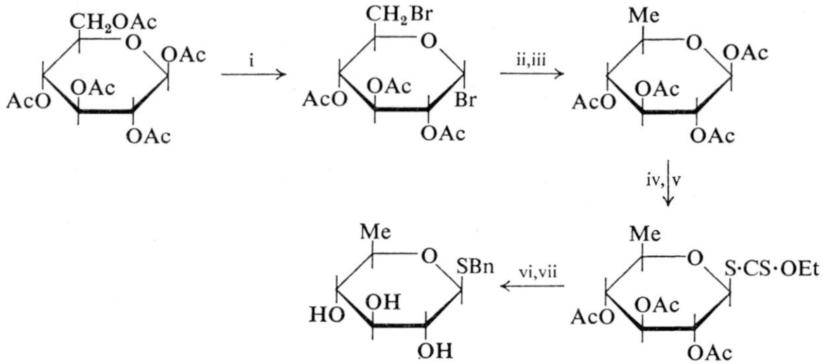

Reagents: i, HBr, $AlBr_3$; ii, Ac_2O, KOAc; iii, Pd–C, H_2; iv, HBr, Ac_2O; v, EtO·CS·SK; vi, NaOMe, MeOH; vii, BnBr

Scheme 12

C-Glycosides

Considerable interest continues to be shown in this class of compound and a review on natural and synthetic members has appeared in Russian.[93]

Vitexin, saponaretin, orientin, iso-orientin, and vicenin-1 have been isolated from the leaves of *Croton zambezicus*,[94] and 8-*C*-β-D-glucopyranosyl luteoline has been obtained from water nut.[95] Cratenacin has been assigned the structure (21).[96]

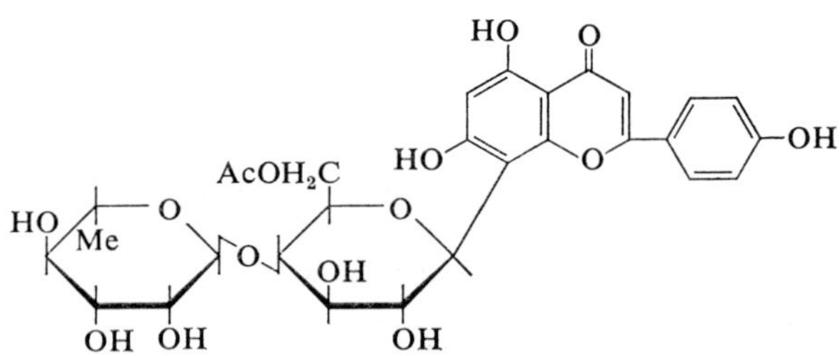

(21)

[92] W. Schüep and E. Hardegger, *Helv. Chim. Acta*, 1970, **53**, 1336.
[93] Yu. A. Zhdanov and L. A. Uzlova, *Nauch. Tr. Taskent, Gos. Univ.*, 1968, 15.
[94] H. Wagner, L. Hörhammer, and I. C. Kiraly, *Phytochemistry*, 1970, **9**, 897.
[95] Y. U. Okonenko and P. B. Krivchuk, *Khim. issled. Kiev 'Zdorov ya'*, 1970, 160.
[96] V. S. Batyuk, N. V. Chernobrovaya, and D. G. Kolesnikov, *Khim. prirod. Soedinenii*, 1969, **5**, 234.

Acetobromo-sugars have been used to synthesise 6-*C*-glucosyl- and 6-*C*-xylosyl-chrysine,[97] and several di-*C*-glucosyl flavones have been prepared by related reactions.[98]

The anomeric pseudouridines have attracted attention: the crystal structure of the α-anomer shows that the furanoid ring adopts the usual envelope conformation (p. 162), while an analysis of the 100 MHz n.m.r. spectrum of the β-anomer by computational methods suggests that the furanoid ring exists as an equilibrium mixture of various puckered conformations. The preferred rotamers about the 4′–5′ and 1′–5 bonds were also determined, and no change was observed in the spectrum over a 40 °C temperature range.[99]

Other compounds which might be considered to be related to *C*-glycosides are mentioned in Chapter 10, and the mass spectroscopy of these compounds is referred to in Chapter 24.

[97] J. Chopin, M.-L. Bouillant, and A. Durix, *Compt. rend.*, 1970, **270** *C*, 69.
[98] J. Chopin and M.-L. Bouillant, *Compt. rend.*, 1970, **270** *C*, 222, 331, 733.
[99] F. E. Hruska, A. A. Grey, and I. C. P. Smith, *J. Amer. Chem. Soc.*, 1970, **92**, 4088.

4
Ethers and Anhydro-sugars

Ethers

Methyl Ethers.—Attention continues to be given to the synthesis of specific methyl ethers of sugars, and several new preparations have been reported. 2-*O*-Methyl-L-ribose was synthesised by oxidation of benzyl 3,4-*O*-isopropylidene-β-L-arabinopyranoside to the uloside which, on reduction, gave the partially protected L-ribopyranoside required for methylation.[100] Methylation of methyl 2,3,4-tri-*O*-acetyl-α-D-mannopyranoside with diazomethane–boron trifluoride etherate, and subsequent removal of the protecting groups, has led to a ready synthesis of 6-*O*-methyl-D-mannose.[101] Two components of cardiac glycosides, 6-deoxy-3-*O*-methyl-D-gulose[102] and -L-mannose,[103] have been obtained from 1,2-*O*-isopropylidene-3-*O*-methyl-D-gulose as shown in Scheme 13.

6-Deoxy-3-*O*-methyl-L-mannose has been identified as a component of a Gram-negative bacterial polysaccharide,[104] and it has also been isolated from mycoside G, a specific glycolipid of *Mycobacterium marinum*.[105] In the latter case, the structure was deduced from the mass and n.m.r. spectra of the acetylated methyl glycoside, and was confirmed by oxidation of the sugar with periodate. The circular dichroism curve (−ve Cotton effect) of the acetylated mycoside was taken as evidence for the α-L-glycosidic link. 3-*O*-Methyl-D-galactose (madurose) has been isolated from whole-cell hydrolysates of some *Actinomycetes*.[106]

The dimethyl sugars 2,4- and 3,6-di-*O*-methyl-D-mannose have been synthesised (Scheme 14),[107] and 2,4-di-*O*-methyl-D-fucose, required for polysaccharide studies, was prepared from methyl 6-*O*-benzyl-α-D-galactopyranoside by acetonation, methylation, deacetonation, and debenzylation to give methyl 2-*O*-methyl-α-D-galactopyranoside. The latter triol was converted into the 4,6-*O*-benzylidene acetal which was

[100] A. H. Haines and P. R. Lundie, *J. Chem. Soc.* (*C*), 1970, 1691.
[101] E. G. Gros and E. M. Gruñeiro, *Carbohydrate Res.*, 1970, **14**, 409.
[102] J. S. Brimacombe, N. Robinson, and J. M. Webber, *Chem. and Ind.*, 1970, 655.
[103] J. S. Brimacombe, N. Robinson, and J. M. Webber, *Carbohydrate Res.*, 1970, **14**, 420.
[104] J. Weckesser, H. Mayer, and G. Drews, *European J. Biochem.*, 1970, **16**, 158.
[105] C. Villé and M. Gastambide-Odier, *Carbohydrate Res.*, 1970, **12**, 97.
[106] M. P. Lechevalier and N. N. Gerber, *Carbohydrate Res.*, 1970, **13**, 451.
[107] G. Alfredsson, P. J. Garegg, and B. Lindberg, *Acta Chem. Scand.*, 1970, **24**, 2671.

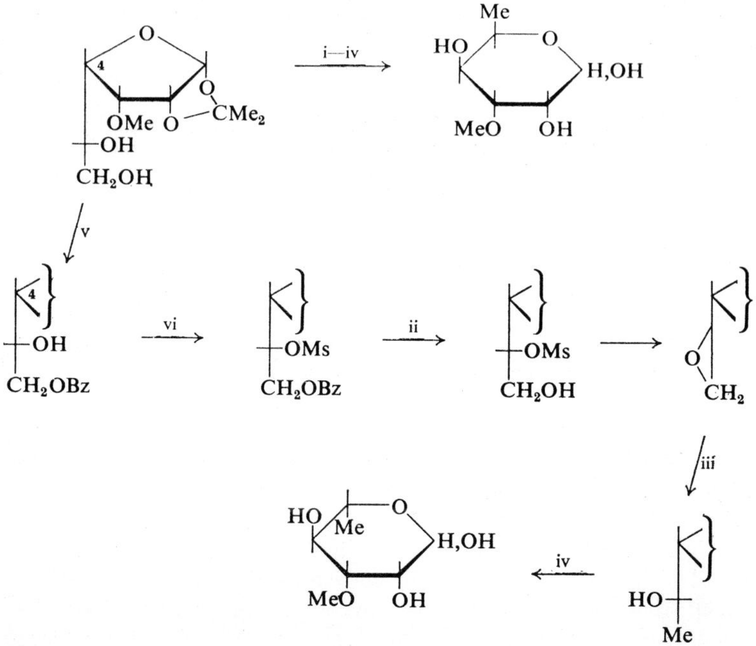

Reagents: i, TsCl, py; ii, MeO^-; iii, LAH; iv, H_3O^+; v, BzCl, py; vi, MsCl, py

Scheme 13

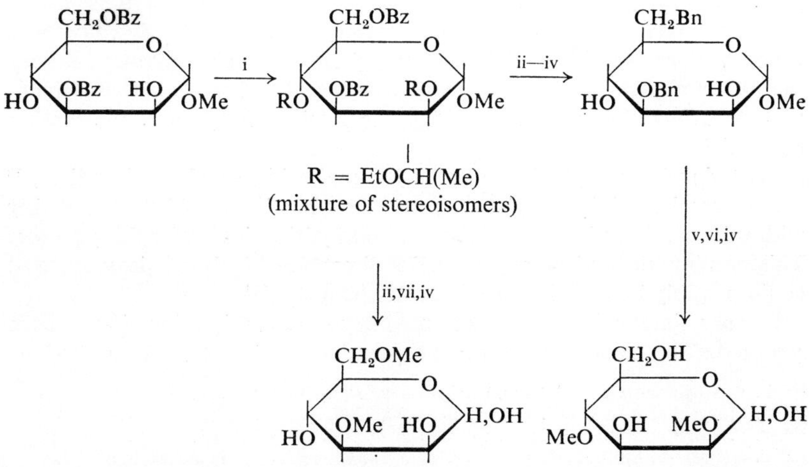

Reagents: i, $EtOCH{=}CH_2$; ii, BaO, MeOH; iii, BnCl, NaOH; iv, H_3O^+; v, MeI, NaOH; vi, Pd–C, H_2; vii, Me_2SO_4, NaOH

Scheme 14

benzylated, debenzylidenated, and tosylated selectively at position 6. The required ether was then obtained by standard reactions on the resulting methyl 3-*O*-benzyl-2-*O*-methyl-6-*O*-*p*-tolylsulphonyl-α-D-galactopyranoside.[108]

2,3,6-Tri-*O*-methyl-D-galactose was obtainable in 15% overall yield from methylated cellulose by oxidation and reduction of the derived methyl 2,3,6-tri-*O*-methyl-αβ-D-glucopyranoside (Scheme 15).[109] 2,3,4,6,7-Penta-

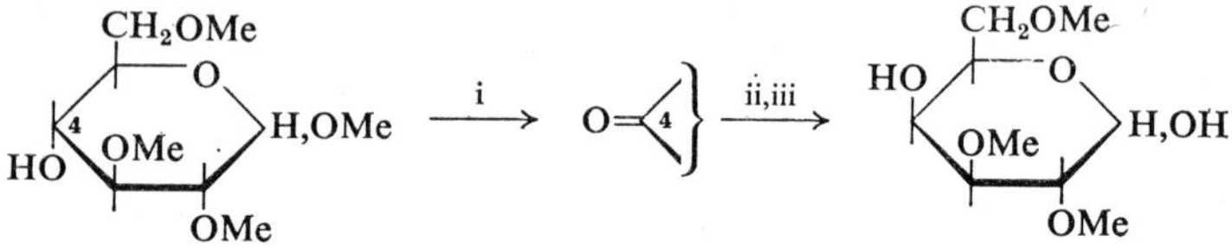

Reagents: i, Ac_2O, DMSO; ii, $NaBH_4$; iii, H_3O^+

Scheme 15

O-methyl-D-*glycero*-L-*manno*-heptose and the corresponding 2,4,6,7-tetramethyl ether have been prepared as reference compounds for structural studies on Gram-negative lipopolysaccharides. The ethers were obtained by methylation of the parent sugar and its 3-benzyl ether, respectively, and removal of the protecting groups by conventional methods.[110]

Other ethers have been prepared by non-specific methylations. Thus, treatment of phenyl 6-*O*-trityl-β-D-glucopyranoside with methyl iodide–silver oxide gave a mixture consisting of the 2,4-di- and 2,3,4-tri-*O*-methyl derivatives from which the free sugars were readily liberated.[111] In other work, partial methylation of *C*- and *O*-glucoside derivatives was studied using diazomethane in methanol or DMF–methanol in the presence of small amounts of stannous chloride. Methylation at O-3 was a favoured process. For example, methyl 4,6-*O*-benzylidene-α-D-glucopyranoside gave the 3-*O*-methyl derivative (93%) and a trace of the diether, whereas the β-anomer afforded the 3-ether (52%), the 2-ether (34%), and the diether (trace).[112]

Several monomethyl ethers of phenyl α-maltoside have been described.[113]

A discussion of demethylation procedures has been published, together with a description of the use of sodium borohydride–iodine for this purpose. This combination of reagents causes direct O—Me bond-cleavage, and could possibly be used in carbohydrate chemistry.[114]

Further interesting cases of methoxy-group migration have been reported. The reaction of aluminium trichloride with (22, β-anomer) gave

108 P. P. Singh and G. A. Adams, *Carbohydrate Res.*, 1970, **13**, 229.
109 S. Morgenlie, *Acta Chem. Scand.*, 1970, **24**, 2240.
110 P. P. Singh and G. A. Adams, *Carbohydrate Res.*, 1970, **12**, 261.
111 P. Nánási, A. Lipták, and G. Nagy, *Acta Chim. Acad. Sci. Hung.*, 1970, **65**, 97.
112 M. Aritomi and T. Kawasaki, *Chem. and Pharm. Bull.* (*Japan*), 1970, **18**, 677.
113 H. Arita, M. Isemura, T. Ikenaka, and Y. Matsushima, *Bull. Chem. Soc. Japan*, 1970, **43**, 818.
114 G. Odham and B. Samuelsen, *Acta Chem. Scand.*, 1970, **24**, 468.

mainly (22, α-anomer) and the rearranged product (23); the latter is presumably formed by way of the bridged oxonium ion (24). Similarly, methyl α-D-glucopyranoside tetraester gave (25) as the product of methoxy-group migration thus establishing, not unexpectedly, that a 1,6-oxonium ion is energetically preferable to the isomeric 1,4-oxonium ion.[114a]

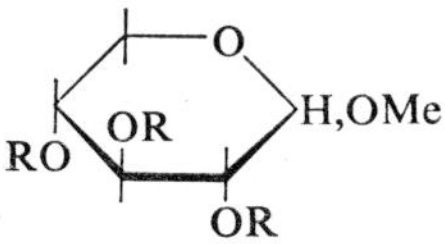

(22)

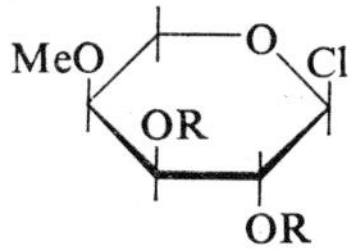

(23)

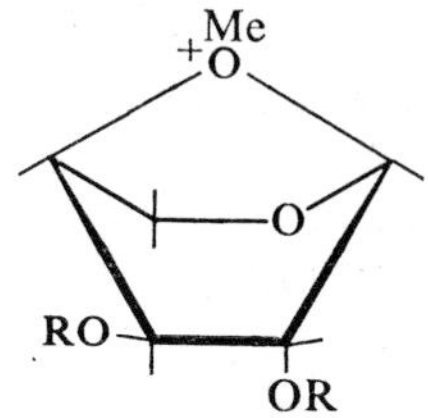

(24)

R = SO_2Cl

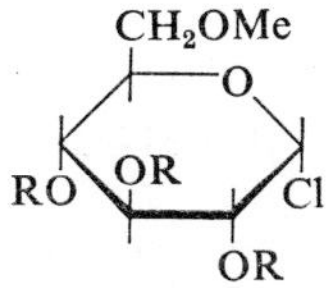

(25)

Substituted Alkyl Ethers.—3-*O*-(2′,3′-Epoxypropyl)-1,2:5,6-di-*O*-isopropylidene-α-D-glucose (26) was prepared as a precursor of possible analogues of starch derivatives by treating the sodio-derivative of (27) with 2,3-epoxypropyl chloride. The epithio analogue (28) of (26) was then synthesised by reaction of the epoxide with either thiourea or potassium thiocyanate.[115] Compound (26) was also converted into compounds (29)—(35) by a series of ring-opening reactions.[116] In an extension of this work, various methyl

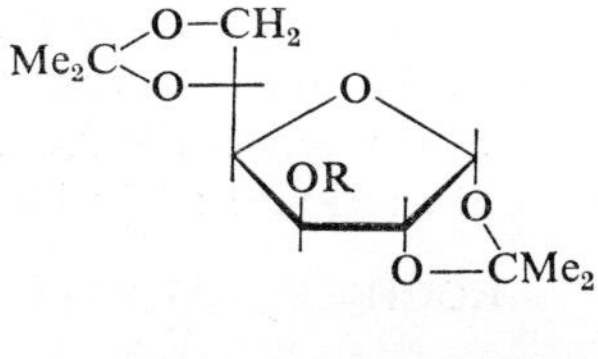

(26) R = $H_2CH\overset{O}{C—CH_2}$

(27) R = H

(28) R = $H_2CH\overset{S}{C—CH_2}$

(29) R = $CH_2CH(OH)CH_2OH$
(30) R = $CH_2CH(OH)CH_2OMe$
(31) R = $CH_2CH(OH)CH_2NH_2$
(32) R = $CH_2CH(OH)CH_2OPr$
(33) R = $CH_2CH(OH)CH_2NEt_2$
(34) R = $CH_2CH(OH)Me$
(35) R = $CH_2CH(OH)CH_2NHCH_2CH_2N{<}\begin{smallmatrix}CH_2\\|\\CH_2\end{smallmatrix}$

[114a] C. F. Gibbs and H. J. Jennings, *Canad. J. Chem.*, 1970, **48**, 2735.
[115] R. E. Wing, W. M. Doane, and C. E. Rist, *Carbohydrate Res.*, 1970, **12**, 285.
[116] R. E. Wing, W. M. Doane, and C. E. Rist, *Carbohydrate Res.*, 1970, **12**, 347.

α-D-glucopyranoside epoxypropyl ethers containing up to four ether groups were synthesised.[117]

Three methods of preparing 1,3,4,6-tetra-*O*-benzyl-D-fructofuranose have been devised. Two of these involved benzylation of methyl D-fructofuranoside and sucrose, respectively, followed by hydrolysis of the perbenzylated product. The third method involved acidic hydrolysis of the acetal group from 1,3,4,6-tetra-*O*-benzyl-2,5-*O*-methylene-D-mannitol followed by oxidation with acetic anhydride in DMSO.[118]

The reaction between β-D-glucopyranose penta-acetate and silicon tetrachloride in propane-1,2-diol carbonate afforded 2-deoxy-3-*O*-[1′-(chloromethyl)-ethyl] derivatives (see also p. 18); possible mechanisms for the formation of these ethers were discussed.[78a]

Russian workers have prepared 3-*O*- and 3,5,6-tri-*O*-allyl ethers and the corresponding vinyl ethers of D-glucose by way of the cyclohexylidene acetals and, following hydrogenation, also obtained the analogous alkyl ethers.[119] Derivatives of 3-*O*-(1-acyloxyethyl)-D-glucose were obtained by the same general approach.[120]

Further studies have been reported on 1,2-*O*-ethylene-D-glucose derivatives. The α-furanose and α- and β-pyranose compounds have been shown to form during the acid-catalysed hydrolysis of *O*-(2-hydroxyethyl) starch, and by intramolecular glycosidation of 2-*O*-(2-hydroxyethyl)-D-glucose so formed.[121]

Selective tritylation of sucrose with trityl chloride in pyridine yields a mixture of 6- and 6′-*O*-tritylsucrose which, on acetylation followed by detritylation of the products, gives 2,3,4,6,1′,3′,4′- and 2,3,4,1′,3′,4′,6′-hepta-*O*-acetylsucrose.[122]

The nature of the cross-linking in cellulose following treatment with *N*-methylbis-(2-chloroethyl)amine has been investigated by hydrolysing the products and comparing the simple substituted D-glucoses (36) and the cross-linked compounds (37) so produced with mono-substituted D-glucoses, prepared by unambiguous routes. The relative distribution of the 2-, 3-, and 6-monoethers was 2·3 : 0·28 : 1·0 whereas the 2,2′-, 2,6′-, and 6,6′-cross-linked products were found in the ratio 2·1 : 3·7 : 1·0.[123]

$ROCH_2CH_2N(Me)CH_2CH_2OH$ (36)

$ROCH_2CH_2N(Me)CH_2CH_2OR^1$ (37)

[117] R. E. Wing, W. M. Doane, and C. E. Rist, *Carbohydrate Res.*, 1970, **14**, 267.
[118] R. K. Ness, H. W. Diehl, and H. G. Fletcher jun., *Carbohydrate Res.*, 1970, **13**, 23.
[119] B. I. Mikhant'ev, V. L. Lapenko, and E. Yu. Ponomarenko, *Izvest. Vyssh. Ucheb. Zaved., Khim. Tekhnol.*, 1969, **12**, 1698 (*Chem. Abs.*, 1970, **73**, 4112*t*).
[120] B. I. Mikhant'ev, V. L. Lapenko, and E. Yu. Ponomarenko, *Zhur. obshchei Khim.*, 1970, **40**, 911.
[121] H. C. Srivastava, K. V. Ramalingam, N. M. Doshi, and A. S. Chaudhari, *Carbohydrate Res.*, 1970, **12**, 23.
[122] T. Otake, *Bull. Chem. Soc. Japan*, 1970, **43**, 3199.
[123] E. J. Roberts and S. P. Rowland, *Canad. J. Chem.*, 1970, **48**, 1383.

Silyl Ethers.—Evidence has been presented which shows that the trimethylsilyl ethers of glycolaldehyde, glyceraldehyde, and dihydroxyacetone are derivatives of the cyclic dimeric forms.[124]

Intramolecular Ethers (Anhydro-sugars)

Epoxides.—A variety of 2-alkoxy-5,6-dihydropyrans of general type (38), which serve as models for 2,3-unsaturated hexopyranosides, were

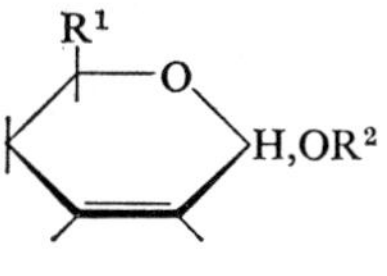

(38)

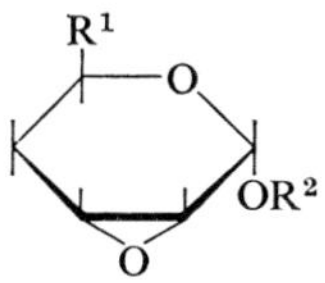

(39)

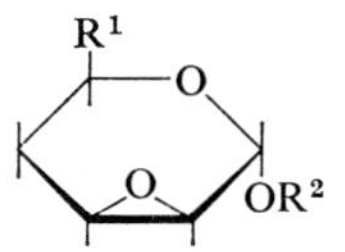

(40)

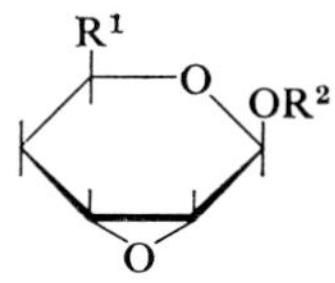

(41)

epoxidised with *m*-chloroperbenzoic acid. *trans*-Unsaturated compounds gave (39) and (40) with the latter predominating, whereas the '*ribo*' epoxides (41) were obtained exclusively from the *cis*-isomers.[125]

Syntheses of the remaining two epoxy-1,6-anhydro-β-D-hexopyranoses, namely 1,6:2,3- and 1,6:3,4-dianhydro-β-D-allopyranose (42) and (43), have

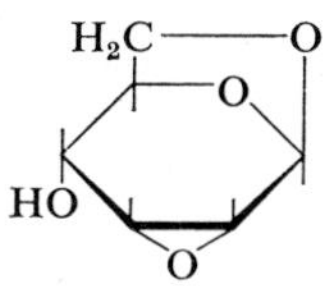

(42)

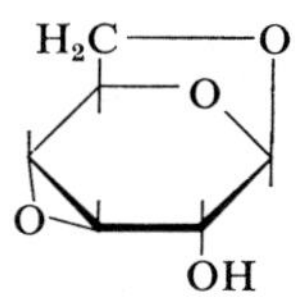

(43)

been described.[126] The 3,4-epoxide (43) was prepared from 1,6-anhydro-β-D-glucopyranose by dibenzyloxycarbonylation, mesylation (at position 3), and treatment with alkali, whereas the 2,3-epoxide (42) was prepared from 1,6-anhydro-4-*O*-benzyl-2-*O*-*p*-tolylsulphonyl-β-D-glucopyranose.

Methyl 2,3-anhydro-4,6-*O*-benzylidene-α-D-hexopyranosides were found to react with alkyl magnesium chlorides to give branched-chain sugars (see Chapter 15). Whereas methyl magnesium iodide reacted with methyl 2,3-anhydro-4,6-*O*-benzylidene-α-D-allopyranoside to give the 2-deoxy-2-iodo-α-D-altropyranoside derivative, the isomeric 3-deoxy-3-iodo-α-D-glucopyranoside derivative was formed when the reaction was performed in THF.[127]

The use of epoxides in the synthesis of amino-sugar nucleosides is referred to in Chapter 8.

[124] B. Arreguin and J. Taboada, *J. Chromatog. Sci.*, 1970, **8**, 187.
[125] A. Konowal, A. Zamojski, M. Masojidkova, and J. Kohoutova, *Roczniki Chem.*, 1970, **44**, 1741.
[126] M. Černý, T. Trnka, P. Beran, and J. Pacák, *Coll. Czech. Chem. Comm.*, 1969, **34**, 3377.
[127] T. D. Inch and G. J. Lewis, *Carbohydrate Res.*, 1970, **15**, 1.

Other Anhydrides.—2,3-Di-*O*-acetyl-1,6-anhydro-β-D-glucopyranose has been synthesised (Scheme 16) for use in Koenigs–Knorr preparations of 4-substituted D-glucose derivatives, since the reactivity of the 4-position of

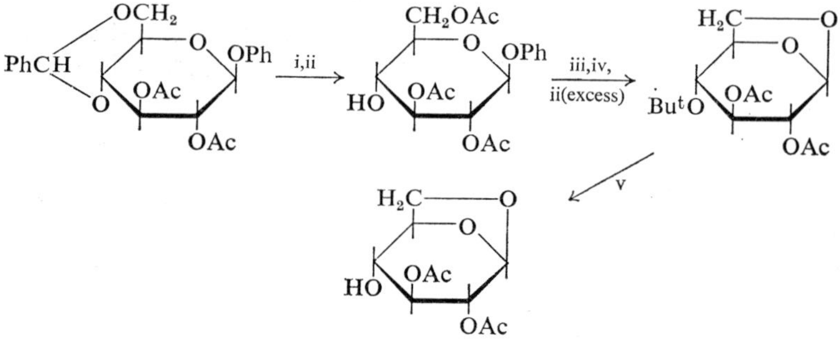

Reagents: i, H_3O^+; ii, Ac_2O(1 mol), py; iii, $Me_2C{=}CH_2$; iv, KOH; v, CF_3CO_2H

Scheme 16

normal glucopyranoses in the *C*1 conformation is so low. It was also obtained, in 13% yield, by direct acetylation of 1,6-anhydro-D-glucose with acetic anhydride (2·5 mol), and was then used in syntheses of lactose and 4-*O*-(2-acetamido-2-deoxy-β-D-glucopyranosyl)-D-glucopyranose.[68] Solid-phase transitions in 1,6-anhydro-β-D-glucopyranose have been examined by differential thermal analysis and by scanning calorimetry. Whereas carbohydrates generally exhibit an exothermic peak corresponding to the melting point, the anhydride behaves quite differently, showing an exothermic peak at 113 °C before melting at 180 °C.[128] Derivatives of 1,6-anhydro-hexoses are also mentioned in Chapters 7 and 25.

Barker's earlier work on the formation of THF derivatives from pentitols has been extended to the hexitols where variations of reaction rates have again been interpreted in terms of steric interactions in the transition states. The hexitols cyclise to 1,4-anhydrides more readily than do the pentitols, and the rates of reaction (relative to D-lyxitol = 1) are allitol (1700), talitol (930), iditol (670), glucitol (510), altritol (195), galactitol (173), gulitol (78), and mannitol (13). Where the reactions are slow, 2,5-anhydrides are also formed. The formation of 1,4:3,6-dianhydrides from 1,4-anhydrides was also studied kinetically.[129] In related work, the reaction of 1-amino-1-deoxypentitols with nitrous acid was found to give 1,4-anhydropentitols together with 2,5-anhydrides and alditols, and the results are compared with the formation of 1,4-anhydrides by treatment of pentitols

[128] F. Shafizadeh, G. D. McGinnis, C. W. Philpot, and R. A. Susott, *Carbohydrate Res.*, 1970, **13**, 184.

[129] R. Barker, *J. Org. Chem.*, 1970, **35**, 461.

with acid.[130] The dehydration of D-glucitol in the presence of acidic ion-exchange resins has been investigated by others.[131]

In experiments, the results of which correlate well with those of Barker's,[129, 130] the di-isobutyl dithioacetals of D-lyxose, D-ribose, and D-xylose have been shown to give 2,5-anhydrides on treatment with limited amounts of *p*-tolylsulphonyl chloride in pyridine at reduced temperatures, whereas the corresponding D-arabinose derivative (stereochemical analogue of D-lyxitol for this comparison) gives a 5-tosylate. The differences observed in these reactions have been rationalised in conformational terms by using the n.m.r. information that, whereas a planar zig-zag conformation is favoured for the D-arabinose thioacetal, sickle conformations are preferred by the other isomers. The 2,5-anhydro-ring obtained from pentose derivatives with the *arabino*-configuration is energetically the least preferred, so this too decreases the likelihood of its formation.[132]

In an interesting synthetic application of the above type of anhydro-ring closure, Japanese workers have prepared 2,5-anhydro-3-azido-3-deoxy-D-xylose dimethyl acetal (44), a key intermediate in the synthesis of

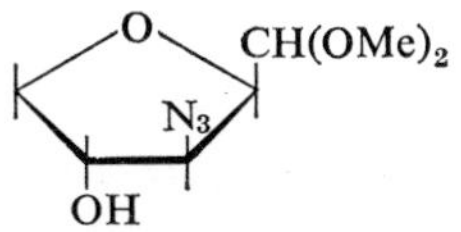

(44)

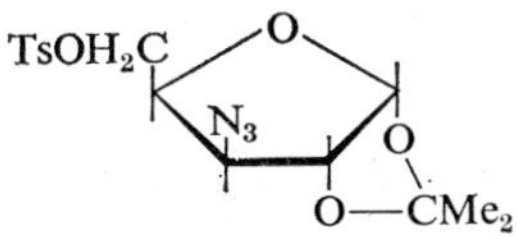

(45)

D-oxybiotin, by methanolysis of 3-azido-3-deoxy-1,2-*O*-isopropylidene-5-*O*-*p*-tolylsulphonyl-α-D-xylofuranose (45).[133]

A new trianhydrosucrose, namely 3,6-anhydro-α-D-glucopyranosyl 1,4:3,6-dianhydro-β-D-fructofuranoside (46), has been prepared and its structure established by *X*-ray diffraction.[134]

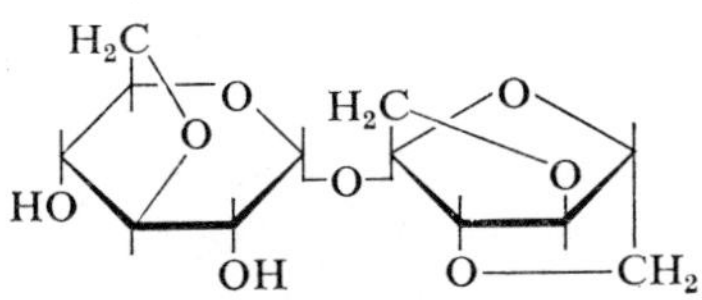

(46)

Two reports have appeared on intermolecular anhydrides of carbohydrates which resemble disaccharide derivatives but contain no glycosidic

[130] D. D. Heard, B. G. Hudson, and R. Barker, *J. Org. Chem.*, 1970, **35**, 464.
[131] S. Ropuszynski, H. Matyschok, and M. Rzepka, *Przem. Chem.*, 1969, **48**, 665 (*Chem. Abs.*, 1970, **72**, 79370p).
[132] J. Defaye and D. Horton, *Carbohydrate Res.*, 1970, **14**, 128.
[133] H. Ohrui, H. Kuzuhara, and S. Emoto, *Agric. and Biol. Chem.* (*Japan*), 1970, **34**, 375.
[134] N. W. Isaacs, C. H. L. Kennard, G. W. O'Donnell, and G. N. Richards, *Chem. Comm.*, 1970, 360.

bonds between the sugar moieties. A derivative (47) of the insecticidal exotoxin (48) has been synthesised (Scheme 17), the key step involving the

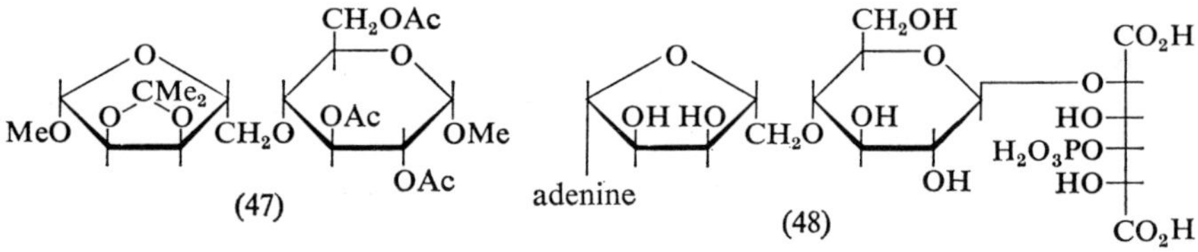

(47) (48)

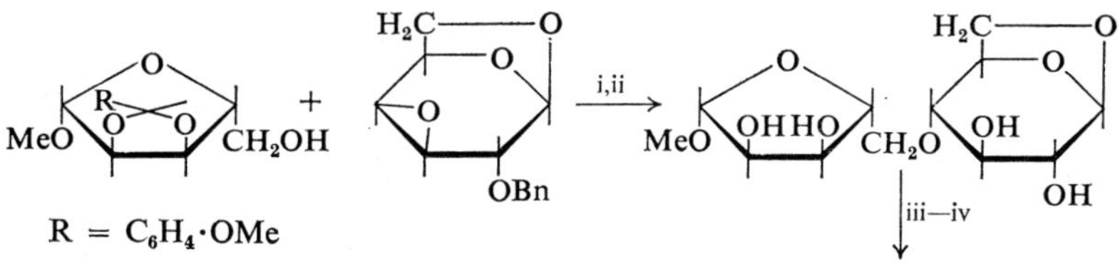

Reagents: i, $MeSOCH_2Na$, DMSO; ii, Pd–H_2; iii, MeOH, H^+; iv, Me_2CO, H^+; v, Ac_2O, py

Scheme 17

diaxial opening of the epoxide ring of 1,6:3,4-dianhydro-2-*O*-benzyl-β-D-galactopyranose by methyl 2,3-*p*-anisylidene-β-D-ribofuranoside.[135] The anhydro-compound (49) was obtained similarly from the reaction between

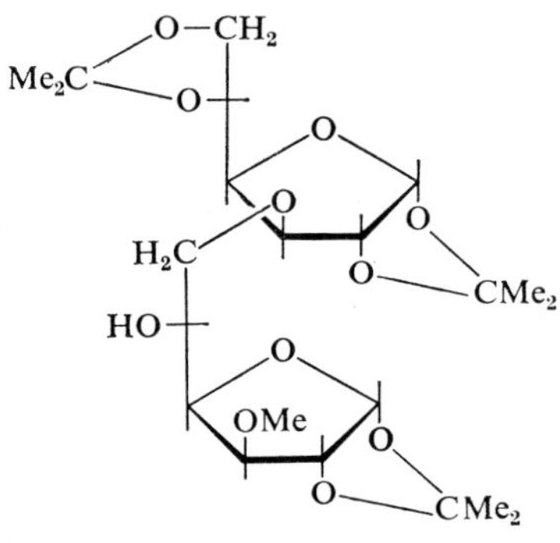

(49)

5,6-anhydro-1,2-*O*-isopropylidene-3-*O*-methyl-α-D-glucofuranose and the sodium salt of 1,2:5,6-di-*O*-isopropylidene-α-D-glucofuranose.[136]

6-Deoxy-6-iodo derivatives of cellulose were found to give anhydro and unsaturated sugars on treatment with base (see also Chapter 14).[137]

[135] M. Prystaš and F. Šorm, *Tetrahedron Letters*, 1970, 4097.
[136] A. Klemer and G. Uhlemann, *Carbohydrate Res.*, 1970, **13**, 331.
[137] Š. Bauer and K. Tihlárik, *Carbohydrate Res.*, 1970, **15**, 418.

5
Acetals

Reactions and Properties

Little additional information appears to have been published on this aspect of acetal chemistry, and this year the major effort has been devoted to synthetic aspects. However, the methanolysis of 1,2:5,6-di-*O*-isopropylidene-α-D-allofuranose has been examined and shown to give at least six products. Methyl α- and β-D-allopyranoside, methyl α- and β-D-allofuranoside, methyl 2,3-*O*-isopropylidene-β-D-allofuranoside, and methyl 2,3:5,6-di-*O*-isopropylidene-β-D-allofuranoside were recognised as products, the last two resulting from acetal migration.[138] The acetalation of D-allose is referred to later in this Section, and the formation of anhydro-derivatives from pentose acetals and thioacetals is noted in Chapter 4.

Several cyclic and acyclic carbohydrate orthoesters have been hydrogenolysed with a 1 : 1 mixture of $LiAlH_4$–aluminium trichloride and the mechanism of the reactions studied. Hydrogenolysis of a diastereoisomeric mixture of 1,2-*O*-(1′-methoxyethylidene)-3,4,6-tri-*O*-methyl-β-D-mannopyranose (50) and (51) gave the same 1,2-*O*-ethylidene derivative (52) having

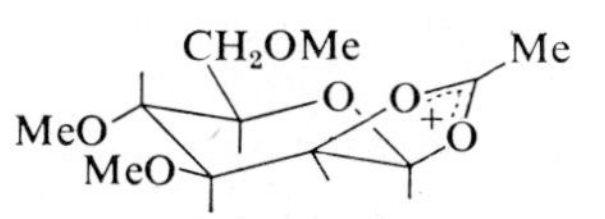

(50)

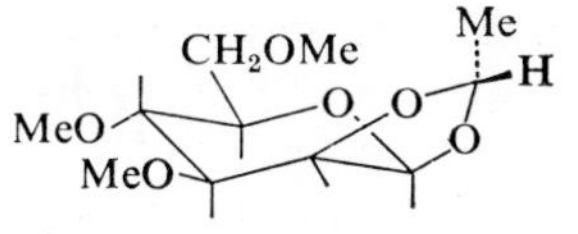

(51)

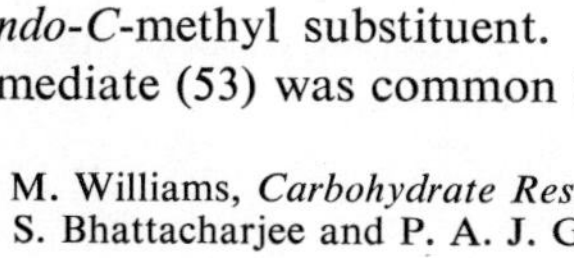

(53)

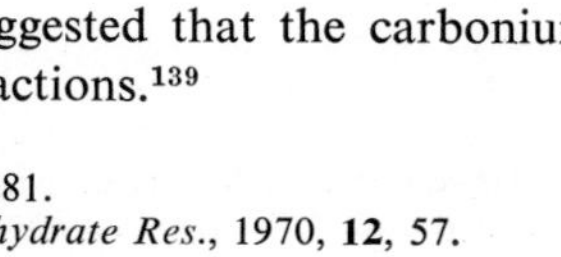

(52)

an *endo*-*C*-methyl substituent. It was suggested that the carbonium ion intermediate (53) was common to both reactions.[139]

[138] J. M. Williams, *Carbohydrate Res.*, 1970, **13**, 281.
[139] S. S. Bhattacharjee and P. A. J. Gorin, *Carbohydrate Res.*, 1970, **12**, 57.

Whereas cyclic orthoesters were cleaved *via* cyclic carbonium ions to give cyclic acetals, the cleavage of acyclic orthoesters was less specific. Thus, the orthoesters (55), (56), and (57), formed by treatment of (54) with

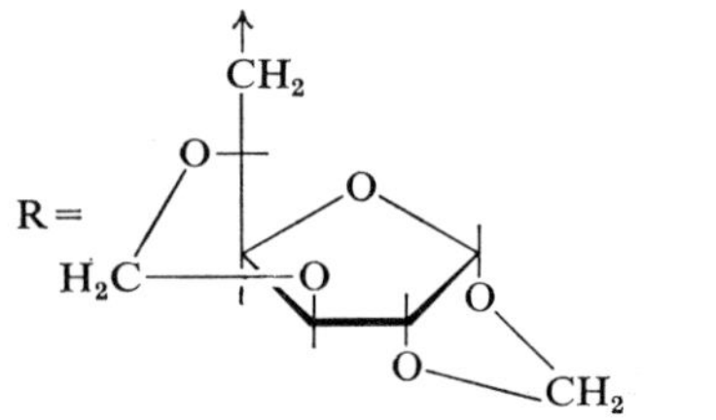

(54) ROH (55) $(RO)_2CHOMe$
(56) $ROCH(OMe)_2$ (57) $(RO)_3CH$
(58) $ROCH_2OMe$ (59) $ROCH_2OR$

methyl orthoformate, gave a mixture of (58) and (59) on reduction with $LiAlH_4$–aluminium trichloride.[139]

Oxidation of the acetal (60) with chromium trioxide in acetic acid afforded the hexulose derivative (61).[140]

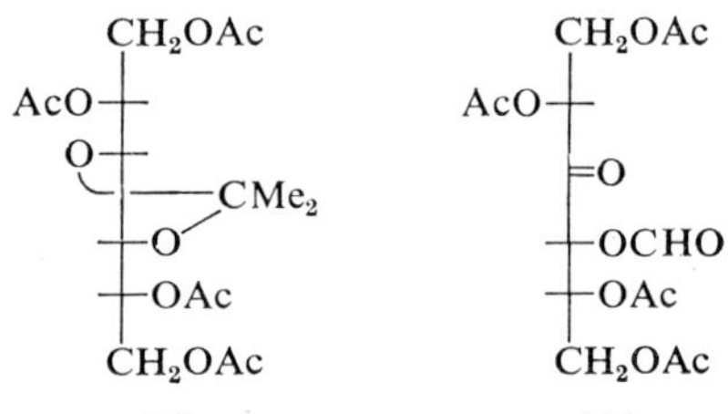

(60) (61)

Synthesis

From Diol Groups on Acyclic Carbohydrates.—D-Mannitol derivatives still attract considerable attention. The three possible 1,2:5,6-bis-*O*-(trifluoromethyl)ethylidene acetals have been isolated from the reaction between the hexitol and 1,1,1-trifluoroacetone and their structures were assigned by n.m.r. spectroscopy. Several 2,3-*O*-(trifluoromethyl)ethylideneglycerol derivatives were prepared and examined as model compounds to facilitate these assignments.[141] In a related study, further work has been carried out on the stereochemistry of 1,2-*O*-bromoethylidene-D-mannitol and on the corresponding 1,2:5,6-diacetal, which are formed from the hexitol with bromoacetaldehyde diethyl acetal in the presence of an acid catalyst. Mono-acetal formation was shown to be thermodynamically controlled, giving the *cis*- and *trans*-products in the ratio 65 : 35, whereas diacetal formation was kinetically controlled since the products precipitate from solution in the

[140] S. J. Angyal and K. James, *Chem. Comm.*, 1970, 320.
[141] T. D. Inch and R. V. Ley, *Carbohydrate Res.*, 1970, **14**, 95.

ratio *cis,cis* : *cis,trans* = 2 : 1; the *trans,trans*-isomer was not obtained. The influence of the first acetal ring on the formation of the second was discussed.[142] 1,3:4,6-Di-*O*-benzylidene-D-mannitol has been prepared by the acid-catalysed reaction of D-mannitol with benzaldehyde in DMSO.[143]

It had been established previously that acetalation of D-glucose phenylosotriazole with benzaldehyde in the presence of zinc chloride gave initially the diastereoisomeric 5,6-acetals which rearranged to a 4,6-isomer, but eventually dibenzylidene derivatives predominated (Vol. 3, p. 83). The diacetals have now been isolated and recognised as the four possible 3,4:5,6-tetrasubstituted derivatives.[144]

2,3:4,5-Di-*O*-isopropylidene acetals of several derivatives of xylonic and arabonic acids have been prepared.[145]

From Diol Groups on Cyclic Carbohydrates.—(*i*) *Free Sugars.* The principal product of acid-catalysed acetonation of D-allose is the 2,3:5,6-di-*O*-isopropylidene-β-D-furanose (62); it has been shown that the 1,2:5,6-diacetal rearranges to (62) under acidic conditions.[146] Japanese workers

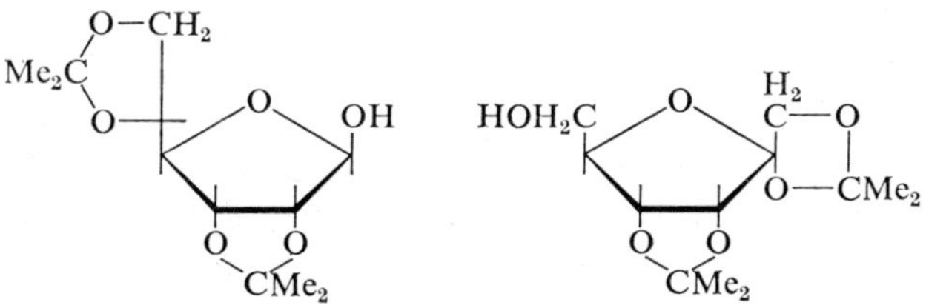

(62) (63)

have confirmed the former finding, and they have also shown that D-psicose affords mainly the 1,2:3,4-diacetal (63); the mass spectra of the products were reported.[147] Similar studies on an equilibrated mixture of D-altrose and 1,6-anhydro-β-D-altropyranose have shown that compounds (64), (65), and (66) are produced in 60%, 23%, and 17% yields, respectively.[148]

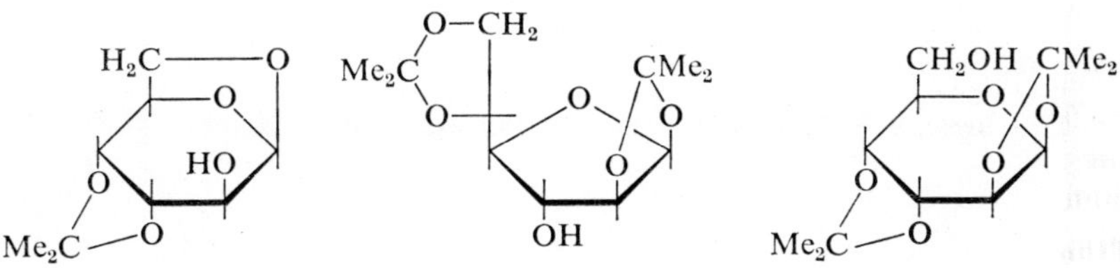

(64) (65) (66)

[142] H. B. Sinclair, *J. Org. Chem.*, 1969, **34**, 3845.
[143] H. B. Sinclair, *Carbohydrate Res.*, 1970, **12**, 151.
[144] R. M. Carman and J. D. Petty, *Austral. J. Chem.*, 1970, **23**, 801.
[145] H. Zinner and H. Nehring, *Z. Chem.*, 1970, **10**, 394.
[146] J. M. Ballard and B. E. Stacey, *Carbohydrate Res.*, 1970, **12**, 37.
[147] M. Haga, M. Takano, and S. Tejima, *Carbohydrate Res.*, 1970, **14**, 237.
[148] J. S. Brimacombe and P. A. Gent, *Carbohydrate Res.*, 1970, **12**, 1.

Condensation of D-fructose with acetone in the presence of sulphuric acid gives 1,2:4,5-di-*O*-isopropylidene-β-D-fructopyranose (67) as the kinetically-controlled product, but it is rapidly converted into an equilibrium mixture of products in which 2,3:4,5-di-*O*-isopropylidene-β-D-fructopyranose (68) predominates.[149]

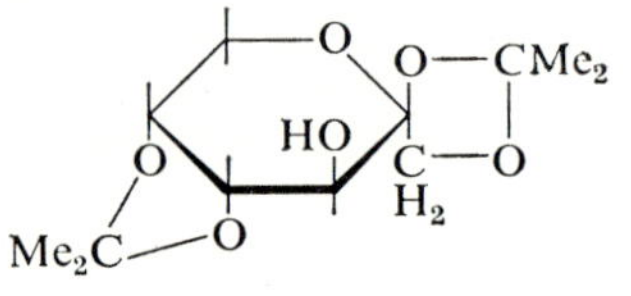

(67)

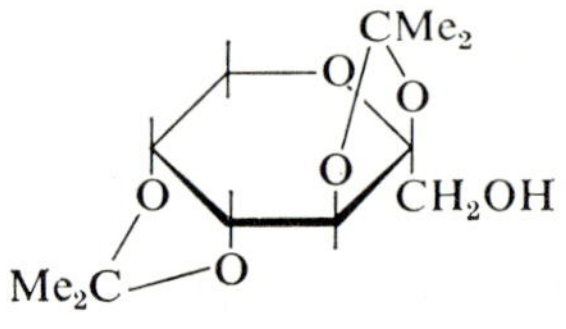

(68)

The crystal structure of 5-*O*-chloroacetyl-1,2:3,4-di-*O*-isopropylidene-α-D-glucoseptanose (Vol. 3, p. 38) has revealed that the large ring adopts a conformation between a chair and a twist-chair.[150]

(*ii*) *Glycosides*, etc. Addition of diethyl ether to the aldehydes used to prepare 4,6-*O*-alkylidene derivatives of D-glucose and its methyl pyranosides improved the yield of these acetals, apparently by reducing the yield of seven-membered-ring products formed by bridging the 2,3-diol groups.[53]

A 1 : 1 adduct of methyl 4,6-*O*-benzylidene-β-D-glucopyranoside and pyridinium chloride has been isolated. The anomers may be separated by means of the adduct, since the corresponding α-compound does not form a complex.[151]

From Single Hydroxy-groups.—A full paper has appeared on the use of the symmetrical 4-methoxytetrahydropyran-4-yl group for protection purposes in nucleoside chemistry (cf. Vol. 1, p. 51). The 2′-substituted derivative of uridine, for example, was prepared from the 3′,5′-diacetate. The acetal group is cleaved under acidic conditions at a suitable rate for its use as a protecting group in oligonucleotide synthesis.[152]

[149] R. F. Brady jun., *Carbohydrate Res.*, 1970, **15**, 35.
[150] J. Jackobs and M. Sandaralingam, *Chem. Comm.*, 1970, 157.
[151] J. Lehrfeld and J. C. Goodwin, *Carbohydrate Res.*, 1970, **14**, 412.
[152] C. B. Reese, R. Saffhill, and J. E. Sulston, *Tetrahedron*, 1970, **26**, 1023.

6

Esters

The use of the general sequence: protection of free hydroxy-groups as methylvinyl ethers, deacylation, *O*-methylation, and removal of the protecting groups, has been discussed as a method for locating acyl groups, and was applied to two partially acylated glycolipids. N.m.r. methods were also investigated as a means of establishing the positions of the acyl groups.[153]

Acetates

Partial acetylation of methyl 2,3-di-*O*-acetyl-α-D-glucopyranoside gave the 2,3,6-triester, a useful intermediate for the synthesis of 4-linked di- and oligo-saccharides.[154]

Similar studies on the partial acetylation of methyl 3-acetamido-3,6-dideoxy-α-D-gulopyranoside (69) with acetic anhydride in pyridine or with acetyl chloride in the same solvent gave the 2- and 4-esters, in a ratio of 85 : 15, and some diester. Deacetylation of the last compound with alkaline alumina gave the 4-ester and (69).[155] Studies on the L-*allo*-analogue (70) showed that acetic anhydride gave the 2- and 4-esters in a ratio of 15 : 85,

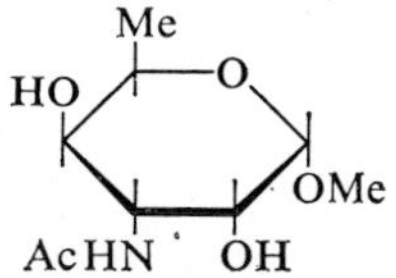

(69)

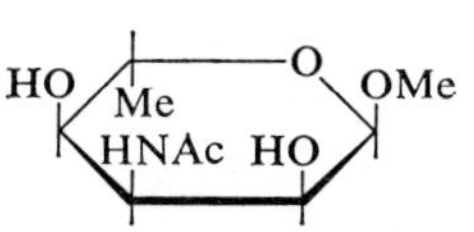

(70)

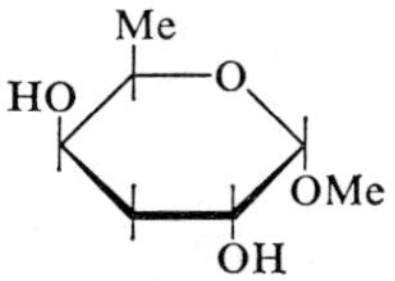

(71)

whereas for acetyl chloride it was 95 : 5. Deacetylation of the diester, as above, gave the 4-ester.[155] Partial acetylation of methyl 3,6-dideoxy-α-D-*xylo*-hexopyranoside (71) with acetic anhydride gave the 2- and 4-esters (in a ratio of 79 : 21) and some diester, whereas acetyl chloride gave the 2-acetate with only a trace of the 4-ester. Deacetylation of the diester, as

[153] S. S. Bhattacharjee, R. H. Haskins, and P. A. J. Gorin, *Carbohydrate Res.*, 1970, **13**. 235.

[154] A. Nowakowski and Z. Mroczkowski, *Roczniki Technol. Chem. Zywn.*, 1969, **15**, 37 (*Chem. Abs.*, 1970, **73**, 4147*h*).

[155] K. Čápek, J. Šteffková, and J. Jarý, *Coll. Czech. Chem. Comm.*, 1970, **35**, 107.

above, gave the 4-acetate.[156] The results of the above two papers further confirm (cf. Vol. 1, p. 54; Vol. 2, p. 47) that acetyl chloride in pyridine preferentially esterifies at C-2, and that in deacetylations the C-2 ester is the most easily hydrolysed.

aldehydo-Maltose deca-acetate has been prepared by way of maltose diethyl dithioacetal octa-acetate.[156a] The structures of the sucrose penta-acetates of McKeown (*Canad. J. Chem.*, 1957, **35**, 28) and of Bredereck (*Chem. Ber.*, 1958, **91**, 2824) have been firmly established by n.m.r. studies of deuteriated derivatives as the 2,3,6,3′,4′- and the 2,3,4,3′,4′-compounds, respectively.[157] A synthesis of 1,3,6,2′,3′,4′,6′-hepta-*O*-acetyl-β-maltose has been described. Maltose octa-acetate was treated with phosphorus penta-chloride to yield the hexa-*O*-acetyl-2-*O*-trichloroacetyl-maltosyl chloride which, without isolation, was converted into the required hepta-acetate.[158] Procedures for the synthesis of *p*-nitrophenyl glycosides of 2-*O*- and 6-*O*-acetyl-α-D-galactopyranose and of 2-*O*-acetyl-α-abequopyranose have been reported.[159]

Studies on the action of ammonia on acylated disaccharides have been extended to octa-*O*-acetyl-β-melibiose. The principal products were *N*-acetyl-6-*O*-(α-D-galactopyranosyl)-β-D-glucofuranosylamine and 1,1-bis-(acetamido)-1-deoxy-6-*O*-(α-D-galactopyranosyl)-D-glucitol.[160] The various products, *viz.* acyclic 1,1-bis(acetamido)-1-deoxyalditols and furanoid and pyranoid *N*-acetylaldosylamines, formed by the action of ammonia on acetylated aldopyranoses have also been investigated. However, the total yields of the products accounted for were $\leqslant$ 56% and the analytical procedures used in their determination were not described.[161]

Acetoxonium ion rearrangements in polyol and monosaccharide systems have been reviewed,[162] and further details of the D-glucose to D-idose inter-conversion given.[163] From penta-*O*-acetyl-D-glucopyranose and antimony pentachloride, the α-1,2-ion was first formed and this then equilibrated with the 2,3-α-D-*manno*, the 3,4-α-D-*altro*, and the 4,6-α-D-*ido* isomers. In nitromethane, the equilibrium mixture contained the four species in the ratio 60 : 12 : 7 : 21. The D-*ido* salt crystallised from methylene chloride as it was formed, thus offering the best method of obtaining D-idose derivatives. However, at − 10 °C in carbon tetrachloride the major isomer was precipitated. Treatment of the ion (72) with lithium bromide gave (73).

[156] K. Čapek, J. Capková-Šteffková, and J. Jarý, *Coll. Czech. Chem. Comm.*, 1970, **35**, 321.
[156a] A. Nowakowski and Z. Mroczkowski, *Roczniki Technol., Chem. Zywn.*, 1969, **15**, 17 (*Chem. Abs.*, 1970, **73**, 4119*a*).
[157] T. Suami, T. Otake, S. Ogawa, T. Shoji, and N. Kato, *Bull. Chem. Soc. Japan*, 1970, **43**, 1219.
[158] B. H. Koeppen, *Carbohydrate Res.*, 1970, **13**, 193.
[159] K. Stellner, O. Westphal, and H. Mayer, *Annalen*, 1970, **738**, 179.
[160] A. B. Zanlungo, J. O. Deferrari, and R. A. Cadenas, *J. Chem. Soc.* (*C*), 1970, 1908.
[161] A. S. Cerezo, A. B. Zanlungo, J. O. Deferrari, and R. A. Cadenas, *Chem. and Ind.*, 1970, 1051.
[162] H. Paulsen, *Chimia*, 1970, **24**, 290.
[163] H. Paulsen and C.-P. Herold, *Chem. Ber.*, 1970, **103**, 2450.

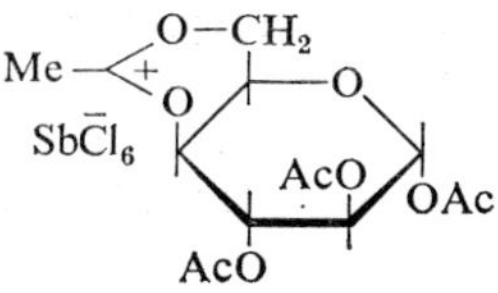

(72)

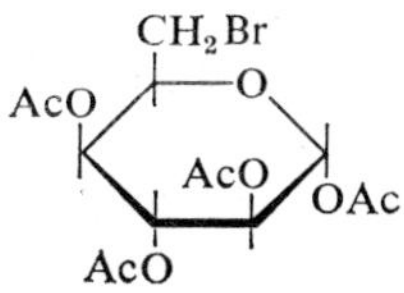

(73)

In a further paper,[164] related equilibria were reported. In nitromethane, the 1,2-α-D-*xylo*, 2,3-α-D-*lyxo*, and 3,4-α-D-*arabino* di-*O*-acetyl acetoxonium antimony hexachlorides were present in a ratio of 28 : 10 : 57. Pentofuranose equilibria were found always to contain preponderant amounts of the 1,2-acetoxonium salts. In the D-*galacto*, D-*talo* equilibrium the latter predominated, and in the D-*allo*, D-*altro* case the former was the major component.

The migration of the acetyl group in the conversion of (74) to (75) has been shown by i.r. and by n.m.r. spectroscopy to proceed through the

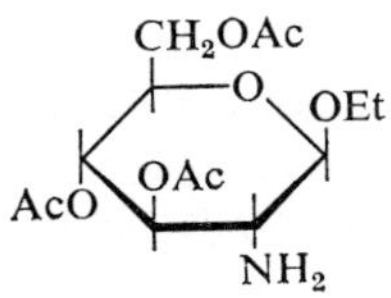

(74)

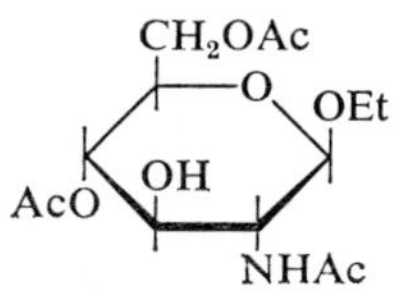

(75)

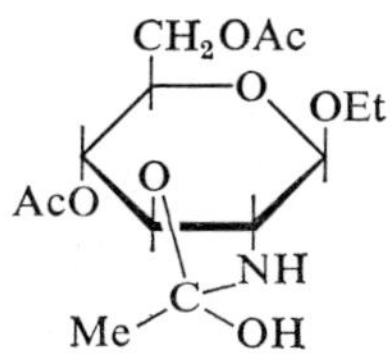

(76)

orthoamide (76) as expected.[165] The mechanism of isotopic exchange of acetoxy-substituents on some penta-*O*-acetyl-α- and -β-D-glycopyranoses has been studied in detail.[166]

Substituted Acetate and Other Non-aromatic Carboxylates

6-*O*-Acetyl-2,3,4-tri-*O*-[(+)-3-methylvaleryl]-β-D-glucopyranose has been isolated from Turkish tobacco.[167]

[164] H. Paulsen, C.-P. Herold, and F. G. Espinosa, *Chem. Ber.*, 1970, **103**, 2463.
[165] G. Fodor, F. Letourneau, and N. Mandava, *Canad. J. Chem.*, 1970, **48**, 1465.
[166] J. Strucinski, *Roczniki Chem.*, 1970, **44**, 355.
[167] J. N. Schumacher, *Carbohydrate Res.*, 1970, **13**, 1.

Chloroacetyl derivatives of nucleosides can be prepared using the acid anhydride in pyridine; the ester group can subsequently be removed under mild conditions, but is best removed with 2-mercaptoethylamine.[168] α-Sulphostearic and α-sulphopalmitic esters of D-mannitol, D-glucitol, and sucrose have been prepared.[169]

The chloroacetyl group has been shown to be a useful protecting group in carbohydrate chemistry. A variety of chloroacetates have been prepared and, after the required reaction, have been removed by treatment with thiourea in benzene–methanol at reflux temperature (Scheme 18).[170]

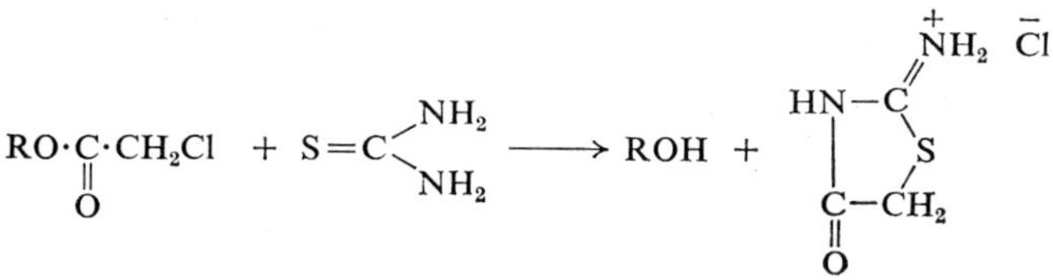

Scheme 18

A variety of methods for the formation of glycosyl and glucuronosyl esters of amino-acids of the general form (77) have been developed.[171] Syntheses of acetylated 1-*O*-(2-acyl-D- and -L-aminoacyl)-β-D-glucopyranoses have been accomplished either by condensing the silver salt of

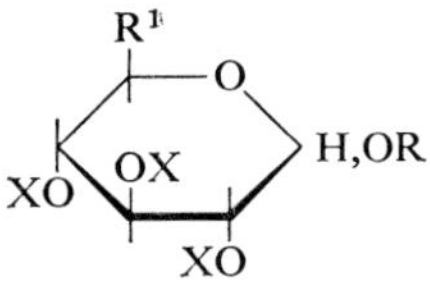

X = Bn or Ac
R^1 = CH_2OX or CO_2Me
R = Ac-L-Ala, Ac-L-Phe, or Ac-L-Met

(77)

the *N*-acylated amino-acid with the glycosyl halide, or by condensing the amino-acid and the sugar in the presence of DCC.[172] The former method proceeded with retention of configuration of the amino-acid irrespective of the nature of the *N*-blocking group. In the latter method, racemisation occurred when acetyl was the *N*-blocking group, whereas when Cbz or phthalyl were used retention was observed. Racemisation was postulated to occur *via* an intermediate oxazolone.

168 A. F. Cook and D. T. Maichuk, *J. Org. Chem.*, 1970, **35**, 1940.
169 R. G. Bistline, jun., F. D. Smith, J. K. Weil, and A. J. Stirton, *J. Amer. Oil Chemists' Soc.*, 1969, **46**, 549.
170 M. Bertolini and C. P. J. Glaudemans, *Carbohydrate Res.*, 1970, **15**, 263.
171 D. Keglević, A. Kornhauser, G. Roglić, and T. Kovač, *Tetrahedron Letters*, 1970, 2983.
172 A. Kornhauser and D. Keglević, *Carbohydrate Res.*, 1970, **13**, 433.

Bis-(2,3,4,6-tetra-*O*-acetyl-β-D-glucopyranosyl) derivatives of oxalic (78a), succinic (78b), maleic (78c), and malonic (78d) acids have been

RO·CO·CO·OR (78a)

RO·CO·X·CO·OR

(78b) X = $(CH_2)_2$
(78c) X = $-CH\underset{cis}{=}CH-$
(78d) X = CH_2

R = 2,3,4,6-tetra-*O*-acetyl-β-D-glucopyranosyl

prepared by condensing the di-silver salts with acetobromoglucose in benzene or ether. However, n-propyl 2,3,4,6-tetra-*O*-acetyl-β-D-glucopyranoside was formed when acetobromoglucose was reacted with silver succinate and propan-1-ol in ether.[52]

Benzoates

N-Benzoylimidazole has been shown to be a selective benzoylating agent.[173] Methyl 4,6-*O*-benzylidene-α-D-glucopyranoside gave the 2-ester in 78% yield. Methyl 3-azido-3-deoxy-α-D-xylopyranoside with one equivalent of benzoyl chloride in pyridine gave the 2-ester (70%), whereas the β-anomer gave low yields of both the 2- and 4-esters.[174]

The rearrangement of penta-*O*-benzoyl-D-glucopyranose into penta-*O*-benzoyl-α-D-idopyranose *via* benzoxonium ion rearrangements has been described.[163] Penta-*O*-benzoyl-β-D-glucopyranose reacted with titanium tetrachloride in chloroform to give a yellow complex, which was decomposed to the β-D-glucosyl chloride that then anomerised to the α-anomer in 15 minutes.[175]

Carboxylic Orthoesters

A useful survey of the hydrolysis of five-membered orthoester rings attached to six-membered rings, using both carbohydrate and other examples, has shown that products having axial ester groups and equatorial hydroxy-groups preponderate (Scheme 19).[176] For *exo*-isomers, it was proposed that for stereoselective reasons the conformations of the transition states leading to (79) and (80) are related to (81) and (82), respectively, and that for steric reasons (81) is favoured. For orthoformates, (*i.e.* when R = H) the selectivity was less pronounced.

Reactions of orthoesters with $LiAlH_4$ and aluminium trichloride are discussed on p. 31.

[173] F. A. Carey and K. O. Hodgson, *Carbohydrate Res.*, 1970, **12**, 463.
[174] T. Tsuchiya, K. Suo, and S. Umezawa, *Bull. Chem. Soc. Japan*, 1970, **43**, 531.
[175] Z. Csuros, G. Deak, and L. Fenichel, *Acta Chim.* (*Budapest*), 1966, **62**, 121 (*Chem. Abs.*, 1970, **72**, 44050*g*).
[176] J. F. King and A. D. Allbutt, *Canad. J. Chem.*, 1970, **48**, 1754.

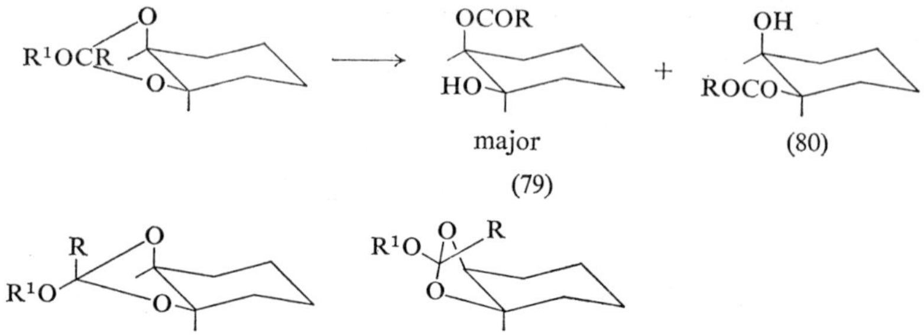

major (79) (80)

(81) (82)

Scheme 19

Carbonates

A six-membered, cyclic carbonate (83) has been prepared by reaction of methyl 2,3-di-*O*-methyl-α-D-glucopyranoside with ethyl chloroformate and triethylamine.[177] Compound (83) was converted into the acyclic carbonate

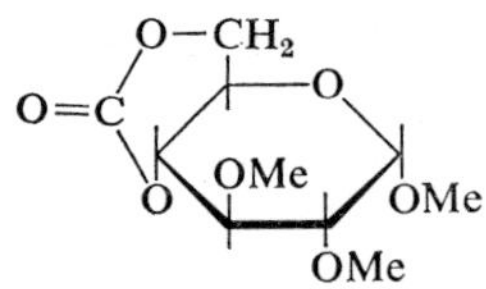

(83)

at twice the rate of a similar, five-membered, cyclic ester. Cyclic carbonates have been used as precursors for unsaturated sugars (see p. 98).

Application of the Corey and Winter procedure to the thionocarbonate (84) had previously been shown to yield the 5,6-olefin in high yield. The

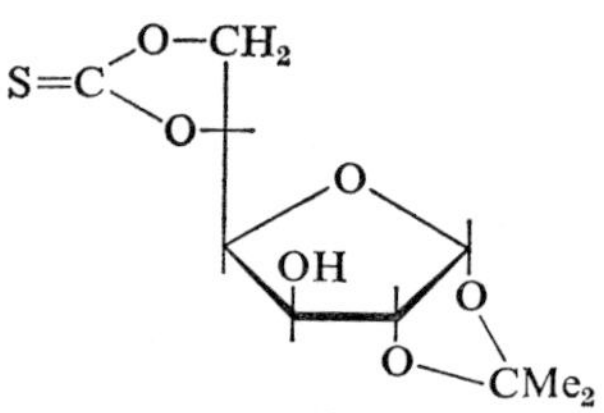

(84)

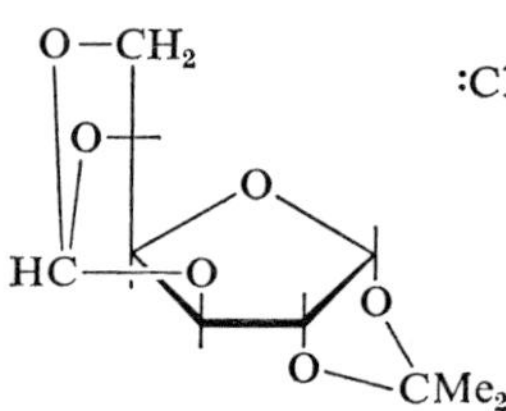

(85)

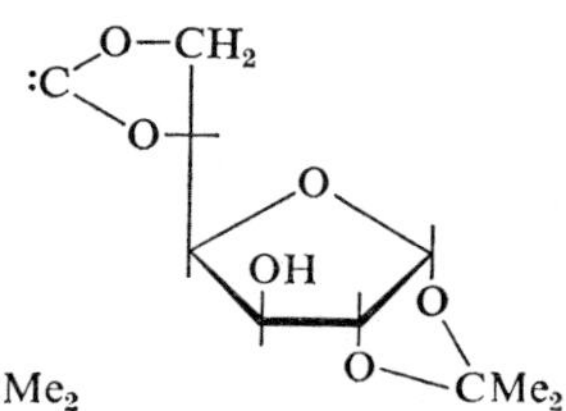

(86)

orthoester (85) has now been recognised as a by-product, and it was concluded that the elimination probably proceeds by way of the carbene (86).[178]

[177] D. Trimnell, W. M. Doane, C. R. Russell, and C. E. Rist, *Carbohydrate Res.*, 1970, **13**, 301.

[178] D. Horton and C. G. Tindall, *J. Org. Chem.*, 1970, **35**, 3558.

Phosphates

Carbohydrate Phosphates.—Full details of the synthesis of monophosphate esters from alcohols and *o*-phenylene phosphorochloridate (87) (Scheme 20)

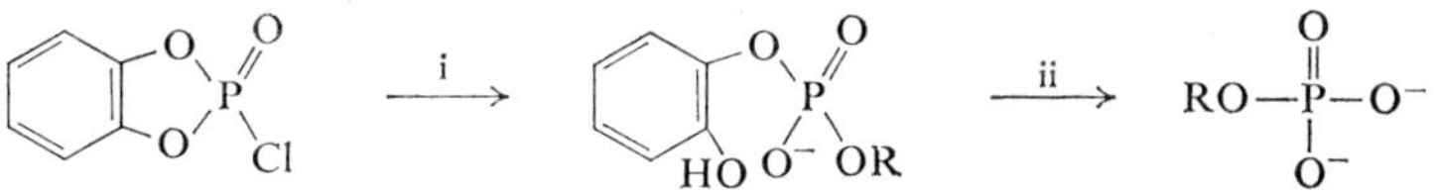

(87)

Reagents: i, ROX, excess base; ii, Br_2–aq. $Ba(OAc)_2$, or aq. HIO_4, or $Pb(OAc)_4$–dioxan

Scheme 20

have been given. Compound (87) was advanced as meeting all the criteria necessary for a general phosphorylating agent.[179]

A simple general procedure for the synthesis of 1,6-diphosphates of α-D-glucose, 2-acetamido-2-deoxy-α-D-glucose, and α-D-mannose, and also of α-D-ribose 1,5-phosphate has been described.[180] The fully acetylated sugar derivatives with a phosphate ester group on the primary hydroxy-group were phosphorylated at C-1 by reaction with crystalline phosphoric acid and the acetyl groups were then removed. 2,3,4,5-Tetra-*O*-stearoyl-D-glucitol and -D-mannitol have been phosphorylated in the 1- and 6-positions with *β*-bromoethyldichlorophosphate, and the products converted into lecithin analogues with trimethylamine.[181]

2-Acetamido-2-deoxy-α-D-mannopyranosyl phosphate has been prepared as shown in Scheme 21.[182] Similar reaction of the α-D-*gluco* analogue

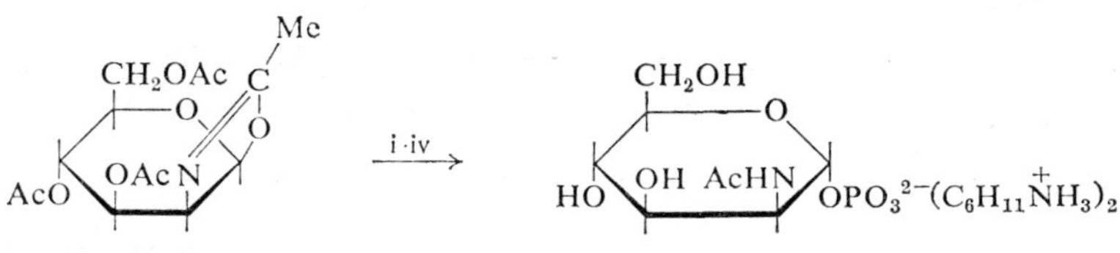

(88)

Reagents: i, $(BnO)_2P(O)OH$; ii, H_2–Pd; iii, NaOMe–MeOH; iv, $C_6H_{11}NH_2$

Scheme 21

of (88) led to the synthesis of 2-acetamido-2-deoxy-α-D-glucopyranosyl phosphate, isolated as its sodium salt.[183] The authors explained the nature of the product by an S_N1 mechanism of the glycosylation of hydrogen phosphates, in which phosphate anions attacked the intermediate cation in the *cis* position leading to the thermodynamically more stable α-anomer.

[179] T. A. Khwaja, C. B. Reese, and J. C. M. Stewart, *J. Chem. Soc.* (*C*), 1970, 2092.
[180] R. Hanna and J. Mendicino, *J. Biol. Chem.*, 1970, **245**, 4031.
[181] H. Eibl and O. Westphal, *Annalen*, 1970, **738**, 174.
[182] W. L. Salo and H. G. Fletcher, jun., *Biochemistry*, 1970, **9**, 878.
[183] A. Ya. Khorlin, S. E. Zurabyan, and T. S. Antonenko, *Tetrahedron Letters*, 1970, 4803.

Solutions in deuterium oxide or water of 5,6-dideoxy-D-*threo*-hexulose 1-phosphate, D-*erythro*-pentulose 1,5-diphosphate, 1,5-dihydroxy-2-pentanone 1,5-diphosphate, 1,3-dihydroxy-2-propanone diphosphate, D-fructose 1,6-diphosphate, and D-*glycero*-D-*altro*-octulose 1,8-diphosphate have been examined by i.r., u.v., and n.m.r. spectroscopy. It was estimated that the compounds were present as their keto forms to the extent of 96, 84, 84, 55, < 1·7, and 0%, respectively.[183a] It was demonstrated by n.m.r. spectroscopy that D-fructose 1,6-diphosphate existed in the furanose form in solution. Both ^{13}C and ^{31}P spectra were used in an attempt to establish the anomeric configuration, and it was concluded that the β-anomer preponderated.

G.l.c. and mass spectrometry of several trimethylsilylated sugar phosphates have been studied, and the techniques were shown to be useful for characterisation purposes.[184] The enzymic condensation of glyceraldehyde and dihydroxyacetone phosphate has been followed by g.l.c. of TMS derivatives of the sugar phosphate.[185]

Nucleoside Phosphates.—Phosphoryl chloride in trimethyl phosphate with unprotected nucleosides gave high yields of 5′-phosphates. Use of the same reagents on 5′-*O*-acetyl nucleosides, however, gave only low yields of the secondary phosphates.[186]

Treatment of ribonucleosides with trimetaphosphate under conditions of high pH gave selective phosphorylation at the secondary positions, and it was suggested that the reaction may have pre-biotic significance.[187] A new method of preparing nucleoside co-enzymes, which can be used

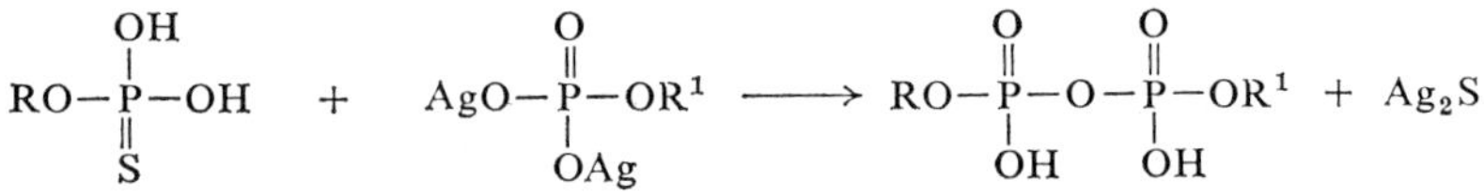

Scheme 22

efficiently on a large scale, involves the reaction shown in Scheme 22; UDPG, for example, has been synthesised in 70% yield by this procedure.[188]

A long paper has been published giving details of the hydrolysis of *cis*-4-hydroxytetrahydrofuran 3-phosphate over a wide range of pH. The cyclic phosphate (89) was the sole initial product at pH > 4, and probably at lower values. The results were used in interpreting certain features of the action of ribonuclease.[189]

[183a] G. R. Gray and R. Barker, *Biochemistry*, 1970, **9**, 2454.
[184] M. Zinbo and W. R. Sherman, *J. Amer. Chem. Soc.*, 1970, **92**, 2105.
[185] C. Bally and F. Leuthardt, *Helv. Chim. Acta*, 1970, **53**, 732.
[186] M. Yoshikawa, T. Kato, and T. Takenishi, *Bull. Chem. Soc. Japan*, 1969, **42**, 3505.
[187] R. Saffhill, *J. Org. Chem.*, 1970, **35**, 2881.
[188] T. Hata and I. Nakagawa, *J. Amer. Chem. Soc.*, 1970, **92**, 5516.
[189] A. Usher, D. I. Richardson, and D. G. Oakenfull, *J. Amer. Chem. Soc.*, 1970, **92**, 4699.

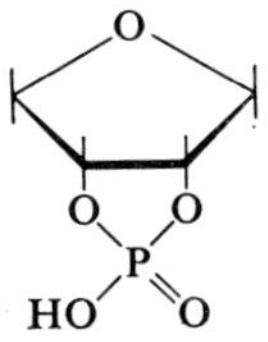

(89)

A general synthesis of nucleoside phosphorothioates (Scheme 23) has been developed to provide a new type of nucleoside analogue.[190] A study has

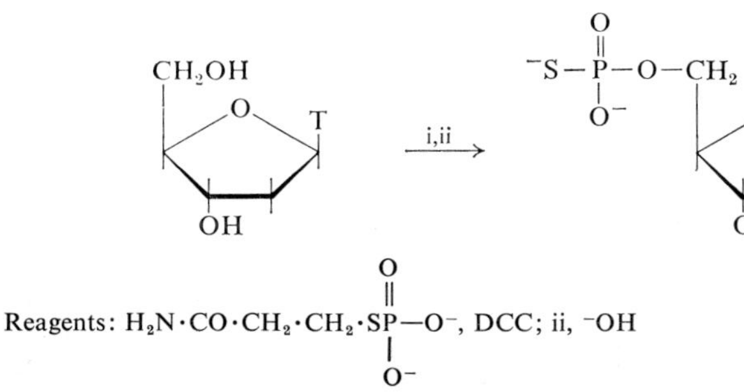

Scheme 23

been made of several nucleoside 5′-phosphorothioates and their properties as enzyme substrates. They were synthesised using tri-imidazolyl-1-phosphine sulphide in pyridine as the esterifying agent with subsequent removal of the imidazole groups with hot aqueous acetic acid. Esterification occurred mainly at the primary positions. Adenosine 5′-phosphorothioates and other 5′-esters were unaffected by alkaline phosphatase, and could be cyclised to give the 3′,5′-cyclic ester by using tri-isopropylbenzenesulphonyl chloride. ^{31}P N.m.r. spectroscopy showed that only one diastereoisomer was formed at the phosphorus atom.[191] 3′,5′-Dithymidine phosphorothioate was also synthesised and was found to be resistant to various esterases. A simple procedure for the synthesis of [^{31}P]3′,5′-cyclic phosphate of adenosine has been described.[192]

Sulphates

Methods for the sulphation of alcohols, some of which are applicable to sugar derivatives, have been reviewed.[193] L-Fucose 4-sulphate has been isolated from fucoidin.[194] The use of sulphuric acid and DCC in DMF on

[190] A. F. Cook, *J. Amer. Chem. Soc.*, 1970, **92**, 190.
[191] F. Eckstein, *J. Amer. Chem. Soc.*, 1970, **92**, 4718.
[192] R. H. Symons, *Biochem. Biophys. Res. Comm.*, 1970, **38**, 807.
[193] E. Gilbert, *Synthesis*, 1969, *No.* 1, 3.
[194] K. Anno, N. Seno, and M. Ota, *Carbohydrate Res.*, 1970, **13**, 167.

a suitably blocked sugar derivative has been recommended as a rapid and convenient means of sugar sulphate formation. The yields were lower than with conventional procedures, but the method appears to be general in scope and suited to the synthesis of ^{35}S-labelled sulphates.[195] A report on the i.r. spectroscopy of the sulphate esters appears on p. 158.

Chlorosulphates, including those of carbohydrate derivatives, have been reviewed.[196] Chlorosulphates have been used as precursors for chlorodeoxy-sugars (see p. 56).

Various *O*-substituted sulphamoyl derivatives of nucleosides have been described.[197]

Sulphonates

Competitive mono-tosylations of methyl 4,6-*O*-benzylidene-α- and -β-D-glucopyranoside with tosyl chloride in pyridine have shown that the C-2-OH group in the α-anomer was more reactive than either hydroxy-group in the β-anomer.[198] Mono-mesylation of methyl 3,6-dideoxy-α-D-*xylo*-hexopyranoside (71),[156] and of the 3-acetamido analogue in the D-*gulo* series (69),[155] gave the 2-ester and the diester in each case.

Methyl β-maltoside with tosyl chloride in pyridine gave the 6,6′-diester (28%), and the 6- and 6′-monoesters in yields of 1 and 18%, respectively.[198a]

The products from treatment of glycoproteins with alkaline sulphite contain a previously undetected product, which was identified as a hexosamine sulphonic acid, with the sulphonic acid grouping probably located at C-3.[199]

It has been shown that carbohydrate tosylates can be cleaved by sodium in liquid ammonia to give the corresponding hydroxy-compounds in high yields.[200]

Displacement Reactions without Participation.—The rates of nucleophilic substitution of some 1,2:5,6-di-*O*-isopropylidene-3-*O*-*p*-tolylsulphonyl-α-D-glycofuranoses with potassium thiobenzoate have been measured for the *gulo*, *allo*, *gluco*, and *galacto* isomers, the rates decreasing in DMF in the order given. The last-named tosylate underwent decomposition but the products isolated from the first three reactions were identified as the C-3 epimeric thiobenzoates.[201] The sequence of reactivity was that predicted from steric considerations. In related work, the tosyloxy-group in the D-*gluco* isomer was displaced, with inversion, by azide and thioacetate ions using reaction times of several days and DMF as solvent at high

[195] R. O. Mumma, C. P. Hoiberg, and R. Simpson, *Carbohydrate Res.*, 1970, **14**, 119.
[196] E. Buncel, *Chem. Rev.*, 1970, **70**, 323.
[197] D. A. Shuman, M. J. Robins, and R. K. Robins, *J. Amer. Chem. Soc.*, 1970, **92**, 3434.
[198] R. D. Guthrie, A. M. Prior, and S. E. Creasey, *J. Chem. Soc.* (*C*), 1970, 1961.
[198a] R. T. Sleeter and H. B. Sinclair, *J. Org. Chem.*, 1970, **35**, 3804.
[199] P. Weber and R. J. Winzler, *Arch. Biochem. Biophys.*, 1970, **137**, 421.
[200] M. A. Miljković, M. Pešić, A. Jokić, and E. A. Davidson, *Carbohydrate Res.*, 1970, **15**, 162.
[201] J. M. Heap and L. N. Owen, *J. Chem. Soc.* (*C*), 1970, 712.

temperature.[202] In both cases, the 3,4-unsaturated sugar was a minor product. Another paper on related compounds with the L-*talo* configuration has shown that the 3-tosyloxy-group can be replaced readily to give the 3-fluoro- and 3-azido-compounds having the L-*ido* configuration. These reactions proceeded so readily that acetonitrile could be used as solvent.[203, 204]

Detailed studies have been made on the $LiAlH_4$ reduction of methyl 4,6-*O*-benzylidene-2,3-di-*O*-*p*-tolylsulphonyl-α-D-glucopyranoside (90): the major product isolated was (91) together with (92) and (93) as minor

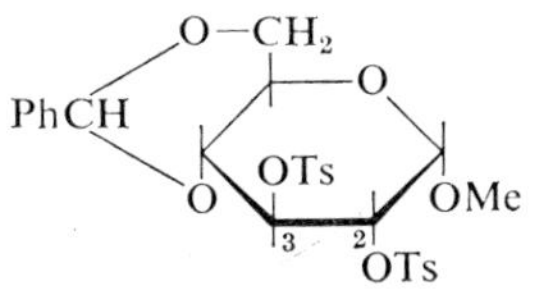

(90)

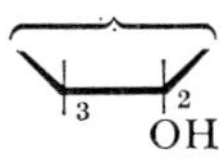

(91)

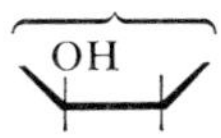

(92)

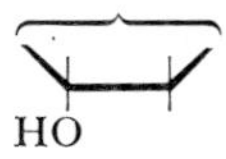

(93)

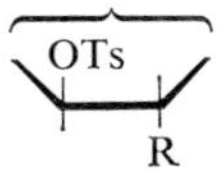

(94) R = H
(95) R = OMe

products. Compounds (94) and (95) underwent no deoxy formation but only detosylation.[205] Similar results have been reported by Czech workers.[206]

Compound (96), prepared in 25% overall yield from D-glucuronolactone, should prove a useful intermediate for the synthesis of 5-substituted derivatives. Its precursor (97) on treatment with lithium bromide in DMF gave (100), and on prolonged reaction (98), which was believed to arise by displacement of the bromine atom in (100) by bromide ion. Compounds (97), (98), and (100) were all readily converted into their methyl glycosides, for example (101). Treatment of (97) with either thioacetic acid, or thiobenzoic acid, or triethyl phosphite all gave the 5-deoxy product (99).[207]

Azide ion displacement of the tosyloxy-group in (102) gave the expected product (103). However, reaction of (102) with sodium benzoate in DMF gave the enol tosylate (104).[208] Unsaturated products were also formed on azidolysis of (105); the ratio of (106) : (107) : (108) was 5 : 1 : 4.[209]

202 U. G. Nayak and R. L. Whistler, *J. Org. Chem.*, 1969, **34**, 3819.
203 J. S. Brimacombe, P. A. Gent, and J. H. Westwood, *Carbohydrate Res.*, 1970, **12**, 475.
204 J. S. Brimacombe, P. A. Gent, and J. H. Westwood, *J. Chem. Soc.* (*C*), 1970, 1632.
205 S. Umezawa, T. Tsuchiya, and H. Hineno, *Bull. Chem. Soc. Japan*, 1970, **43**, 1213.
206 A. Zobáčová, V. Heřmánková, and J. Jarý, *Coll. Czech. Chem. Comm.*, 1970, **35**, 327.
207 T. Irimajiri, H. Yoshida, T. Ogata, and S. Inokawa, *Bull. Chem. Soc. Japan*, 1970, **43**, 3242.
208 J. Hildesheim, A. Gaudemer, and S. D. Géro, *Carbohydrate Res.*, 1970, **14**, 315.
209 A. K. Al-Radhi, J. S. Brimacombe, and L. C. N. Tucker, *Chem. Comm.*, 1970, 363.

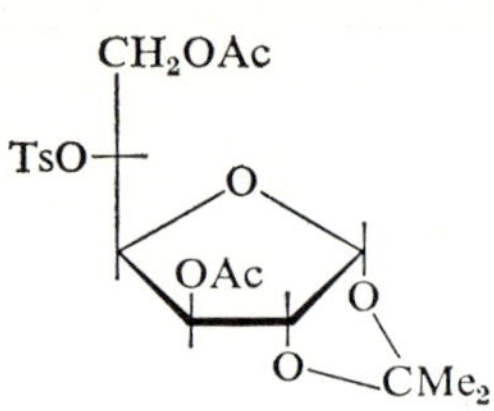

(96)

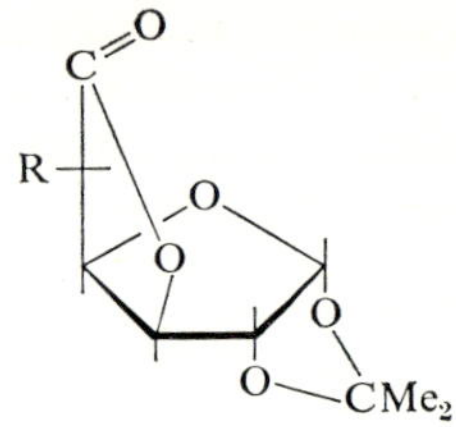

(97) R = OTs
(98) R = Br
(99) R = H

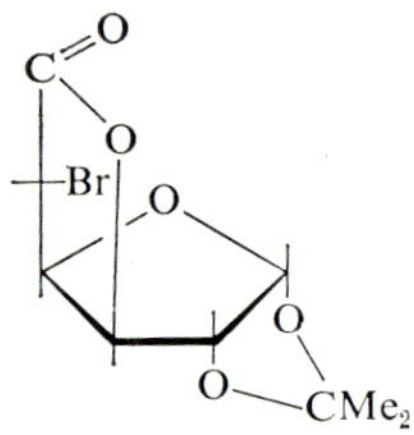

(100)

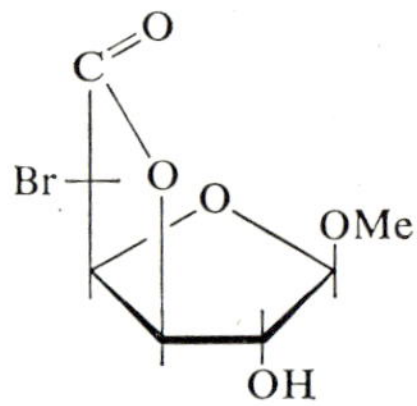

(101)

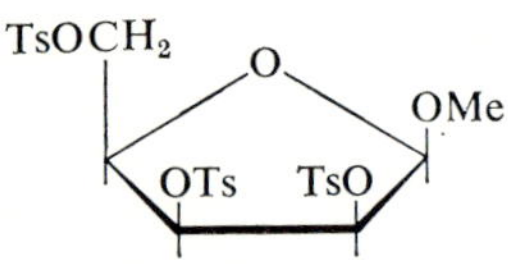

(102)

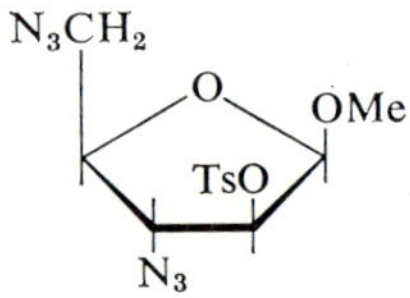

(103)

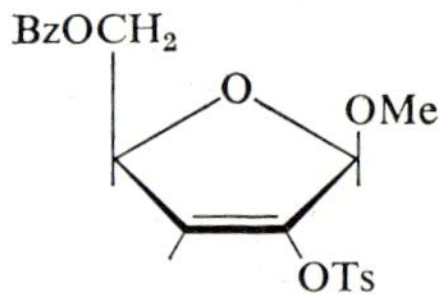

(104)

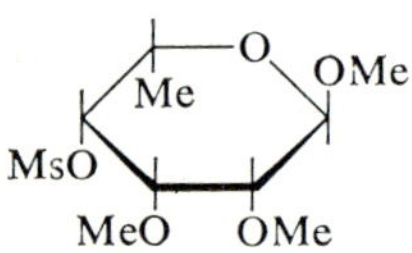

(105)

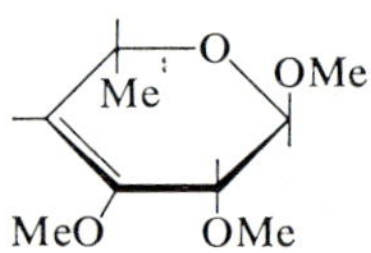

(106)

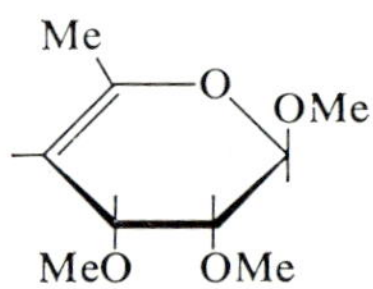

(107)

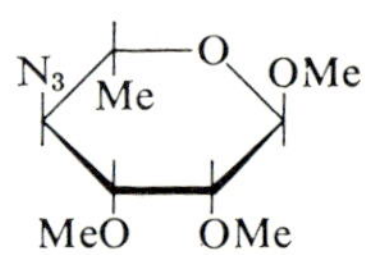

(108)

Reaction of 2-octyl tosylate and certain charge delocalised organolithium reagents, such as allyl-lithium, proceeded in high yield (Scheme 24) and with inversion of configuration.[210] This type of reaction might find use in the synthesis of branched-chain sugars.

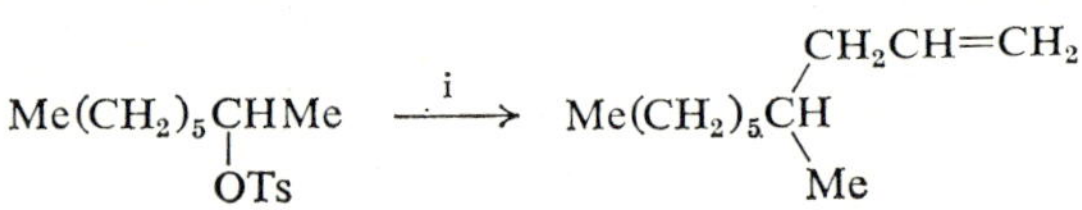

Reagent: i, $CH_2{=}CH{\cdot}CH_2Li$

Scheme 24

[210] W. D. Korte, L. Kinner, and W. C. Kaska, *Tetrahedron Letters*, 1970, 603.

Displacements with Participation of Oxygen Functions.—Brief treatment of 2,3-*O*-isopropylidene-5,6-di-*O*-methanesulphonyl-α-D-mannofuranose with methanolic sodium methoxide gave methyl 5,6-anhydro-2,3-*O*-isopropylidene-β-L-allofuranoside (109) as shown in Scheme 25.[211]

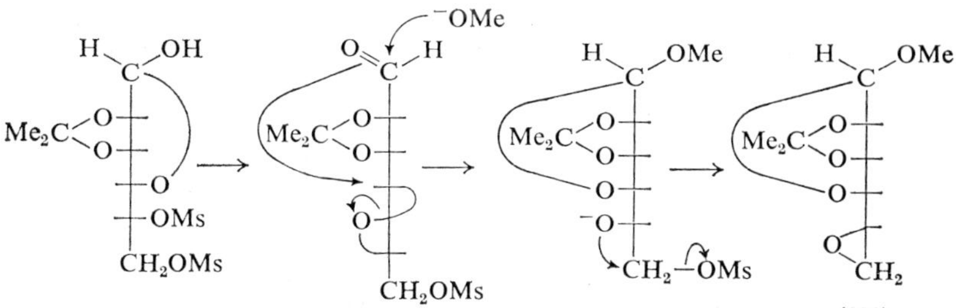

(109)

Scheme 25

Reaction of the 5-*O*-tosyloxy-group in 6-*O*-benzoyl-1,2-*O*-isopropylidene-3,5-di-*O*-*p*-tolylsulphonyl-D-glucofuranose (110) with either acetate or chloride ion in boiling acetic anhydride has been shown to involve

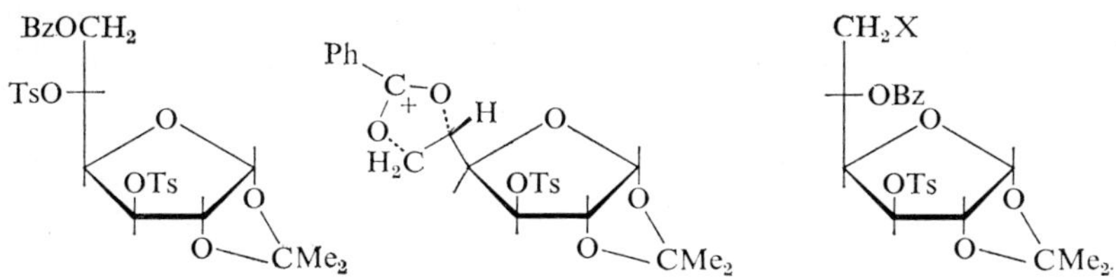

(110) (111) (112)

X = Cl or OAc

participation of the 6-*O*-benzoyl group, *via* the cyclic ion (111), to give 6-substituted-5-*O*-benzoyl products (112).[212] Similar reactions occurred with the 3-*O*-acetyl analogue of (110). The reaction with acetate ion was also reported by another group.[212a]

Reaction of (113) with sodium benzoate in DMF gave, after hydrolysis of the intermediate benzoxonium ion, the D-lyxoside (114) with the D-xyloside (115) as a minor product.[213]

Two examples of participation by a neighbouring benzyloxy-group have been described: one in a furanose system [214] and the other in a pyranose derivative.[215]

211 J. S. Brimacombe, F. Hunedy, and A. Husain, *J. Chem. Soc.* (*C*), 1970, 1273.
212 R. C. Chalk, D. H. Ball, M. A. Lintner, and L. Long, jun., *Chem. Comm.*, 1970, 245.
212a M. A. Miljković and E. A. Davidson, *Carbohydrate Res.*, 1970, **13**, 444.
213 J. Hildesheim, J. Cléophax, and S. D. Géro, *Carbohydrate Res.*, 1970, **14**, 305.
214 O. A. Ching-Puente, *Bol. Soc. quím. Peru*, 1970, **36**, 13 (*Chem. Abs.*, 1970, **73**, 110017*a*).
215 O. A. Ching-Puente, *Bol. Soc. quím. Peru*, 1969, **35**, 121 (*Chem. Abs.*, 1970, **73**, 35692*m*).

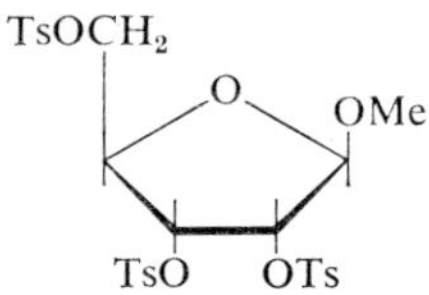

(113)

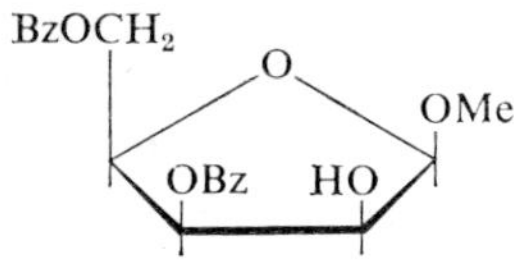

(114)

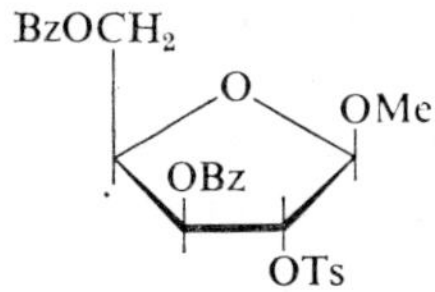

(115)

Treatment of the nucleoside 5′-tosylate (116) with sodium azide in DMF gave the 2′,5′-anhydro-sugar (117) instead of the expected 5′-azide.[216]

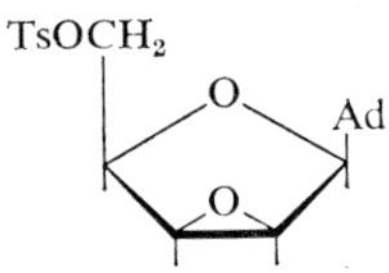

(116)

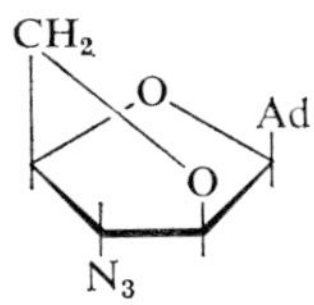

(117)

Displacement Accompanied by Rearrangement.—Reaction of methyl 2,3-*O*-isopropylidene-4-*O*-methanesulphonyl-α-L-rhamnopyranoside (118) with sodium azide afforded preponderantly the 5-azido-α-L-talofuranoside (119) with some of the corresponding β-D-allofuranoside (120) as a minor product. However, (118) reacted with hydrazine to give, after reduction,

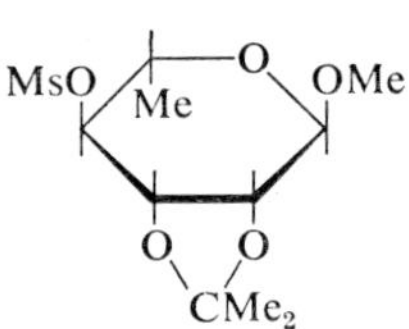

(118)

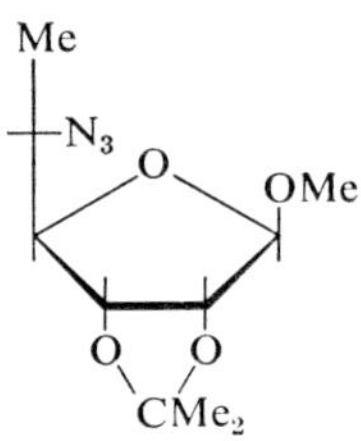

(119)

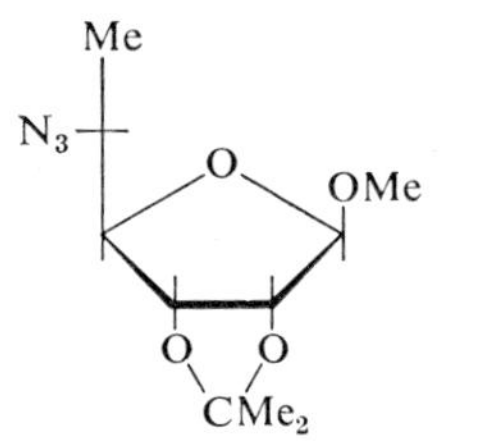

(120)

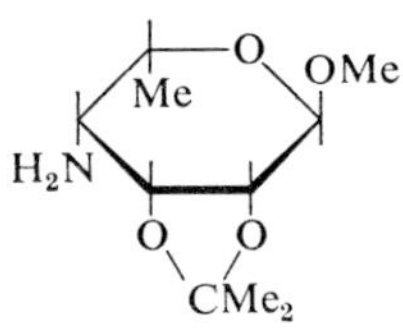

(121)

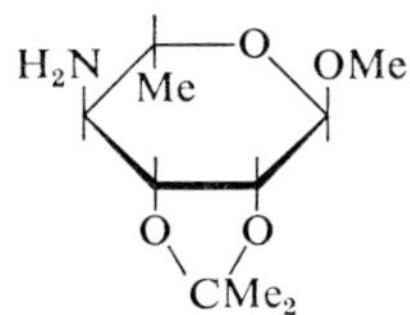

(122)

the 4-amino-4-deoxy-α-L-talopyranoside (121) as the principal product with the corresponding 4-amino-4-deoxy-α-L-mannopyranoside (122) and the

[216] M. Hubert-Habert and L. Goodman, *Canad. J. Chem.*, 1970, **48**, 1335.

amino analogues of (119) and (120) as minor products.[217] In a similar report, it is shown that various 4-*O*-sulphonyl esters of methyl 6-deoxy-2,3-*O*-isopropylidene-α-D-mannopyranoside with either azide ion or phthalimide give 5-substituted α-D-talofuranosides.[218]

[217] J. Jarý, P. Novák, and Z. Samek, *Annalen*, 1970, **740**, 98.
[218] C. L. Stevens, R. P. Glinski, K. G. Taylor, and F. Sirokman, *J. Org. Chem.*, 1970, **35**, 592.

7
Halogenated Sugars

Considerable attention has again been devoted to this class of substituted sugar derivative and, as last year, appreciable developments in the area of fluorinated carbohydrates have been reported.

Glycosyl Halides

The reaction of aldose peracetates with zinc chloride–thionyl chloride has been reported as a general and efficient procedure for the synthesis of acetylated aldosyl chlorides.[219] Alternatively, treatment of peracetylated alkyl glycosides with dichloromethyl methyl ether and zinc chloride gives these compounds, while dibromomethyl methyl ether can be utilised in the synthesis of acetylated glycosyl bromides. Aryl glycoside esters afford lower yields of the glycosyl halides.[220]

3,4,6-Tri-*O*-acetyl-2-*O*-benzyl-α-D-glucopyranosyl bromide has been synthesised from 2-*O*-benzyl-D-glucose; on methanolysis it gave the β-glycoside[221] (*cf.* Vol. 3, p. 60, where methanolysis of such compounds to give α-glycosides is reported). 2,3,4,6-Tetra-*O*-benzoyl-β-D-glucopyranosyl chloride has been prepared from the β-pentabenzoate using titanium tetrachloride in chloroform.[222] Similarly, peresters of both 3-deoxy-D-*xylo*-hexopyranosyl bromide and chloride have been described.[223]

Treatment of the four possible 2-chloro-2-deoxy-glycosyl chlorides (123) with silver tetrafluoroborate in ether under very mild conditions gives mixtures of the corresponding anomeric glycosyl fluoride derivatives (124)

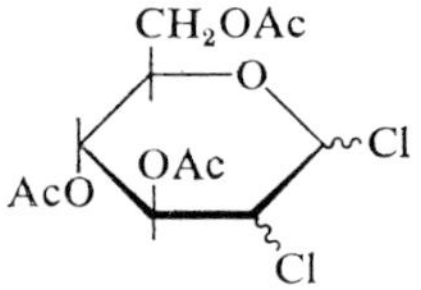

(123)

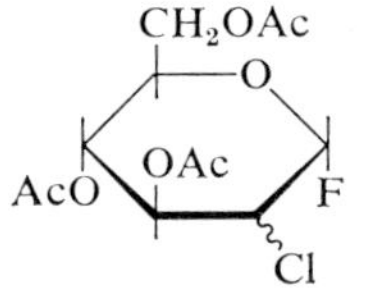

(124)

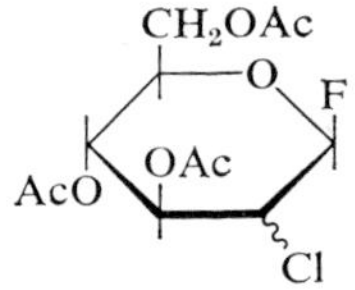

(125)

[219] L. P. Egan, T. G. Squires, and J. R. Vercellotti, *Carbohydrate Res.*, 1970, **14**, 263.
[220] I. Farkas Szabo, I. Farkas, R. Bognar, and H. Gross, *Acta Chim. Acad. Sci. Hung.*, 1970, **64**, 67.
[221] S. Brennan and P. A. Finan, *J. Chem. Soc.* (*C*), 1970, 1742.
[222] Z. Csüros, G. Deák, and L. Fenichel, *Acta Chim. Acad. Sci. Hung.*, 1969, **62**, 121.
[223] H. Zinner and G. Wulf, *J. prakt. Chem.*, 1970, **312**, 314.

and (125). 2,3,4,6-Tetra-*O*-acetyl-α-D-glucopyranosyl chloride did not yield the expected 1,2-*O*-acetoxonium-3,4,6-tri-*O*-acetyl-α-D-glucopyranose tetrafluoroborate on reaction with silver tetrafluoroborate, but instead gave the glycosyl fluorides. Reasons for this were discussed.[224]

Some reactions of glycosyl chlorides with silver perchlorate have been reported.[225] The anomeric 2-deoxy-2-chloro-glucopyranosyl chloride derivatives (126) afforded the perchlorate (127) on treatment with silver perchlorate in toluene, ether, or liquid sulphur dioxide. In contrast,

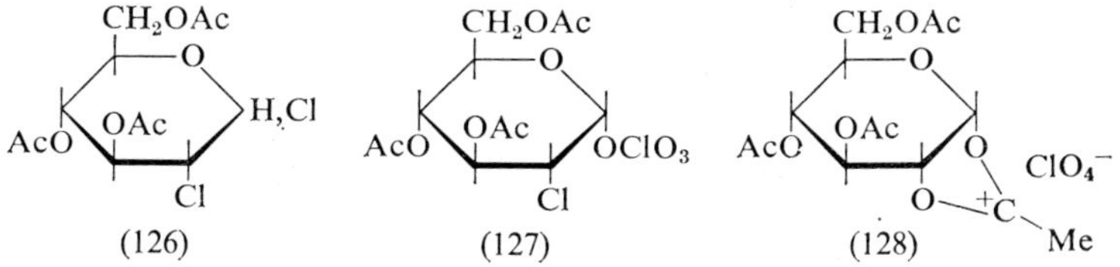

(126) (127) (128)

1,2-*O*-acetoxonium-3,4,6-tri-*O*-acetyl-α-D-glucopyranosyl perchlorate (128) was obtained by reaction of silver perchlorate with 2,3,4,6-tetra-*O*-acetyl-α-D-glucopyranosyl chloride. Methanolysis of both (127) and (128) gave mixtures of α- and β-glycosides, the proportions of which varied with the reaction conditions. However, only β-D-glucopyranosyl chlorides resulted when these compounds were treated with tetraethylammonium chloride.

Treatment of 3,4,6-tri-*O*-acetyl-D-glucal with trifluoro(fluoroxy)methane (see Vol. 3, p. 59) at − 78 °C gave trifluoromethyl 3,4,6-tri-*O*-acetyl-2-deoxy-2-fluoro-α-D-glucopyranoside (129) (26%) and the corresponding vicinal difluoride (130) (34%), together with smaller proportions of the related β-D-mannose derivatives (131) (6%) and (132) (8%).[226] Further studies have shown that treatment of tri-*O*-acetyl-D-galactal with this reagent afforded predominantly trifluoromethyl 3,4,6-tri-*O*-acetyl-2-deoxy-2-fluoro-α-D-galactopyranoside (133) and the corresponding glycosyl fluoride (134) together with traces of the corresponding D-talose derivatives. Hydrolysis of either of the major products liberated 2-deoxy-2-fluoro-D-galactose.[227] Related results have been reported by Kent and his colleagues,[228] who obtained the two *cis*-difluorides and the corresponding glycosides from di-*O*-acetyl-D-xylal. Other additions to glycals affording halogenated products are mentioned in Chapter 14.

The reaction of α-acetobromo-sugars with DMSO has been studied with D-*gluco*- and D-*xylo*-compounds. The main products from tri-*O*-acetyl-α-D-xylopyranosyl bromide were 2,3,4-tri-*O*-acetyl-α- and -β-D-xylose and

[224] K. Igarashi, T. Honma, and J. Irisawa, *Carbohydrate Res.*, 1970, **13**, 49.
[225] K. Igarashi, T. Honma, and J. Irisawa, *Carbohydrate Res.*, 1970, **15**, 329.
[226] J. Adamson, A. B. Foster, L. D. Hall, R. N. Johnson, and R. H. Hesse, *Carbohydrate Res.*, 1970, **15**, 351.
[227] J. Adamson and D. M. Marcus, *Carbohydrate Res.*, 1970, **13**, 314.
[228] R. A. Dwek, P. W. Kent, P. T. Kirby, and A. S. Harrison, *Tetrahedron Letters*, 1970, 2987.

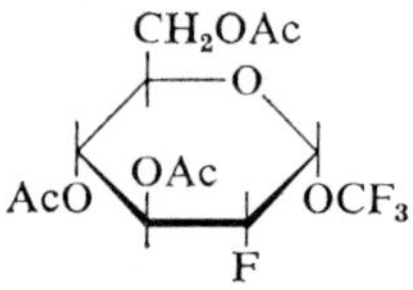

(129)

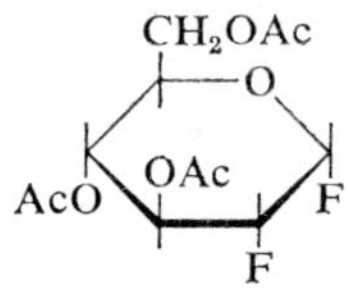

(130)

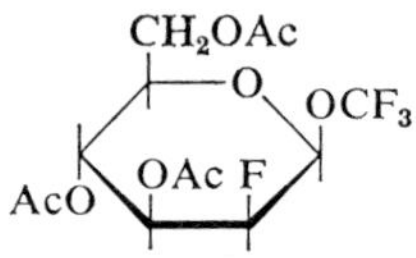

(131)

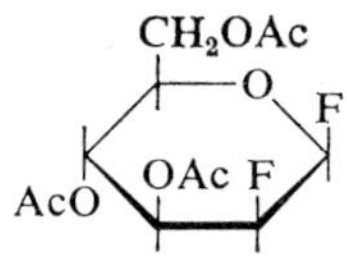

(132)

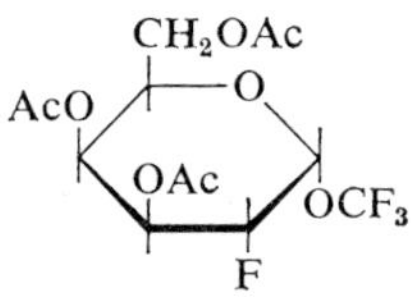

(133)

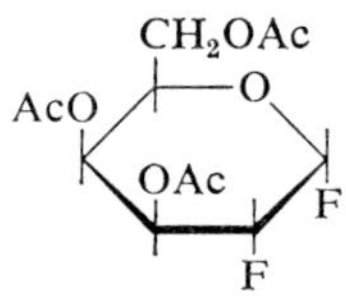

(134)

1,3,4-tri-*O*-acetyl-α-D-xylose. The reaction pathway outlined in Scheme 26 was suggested to account for these products.[229]

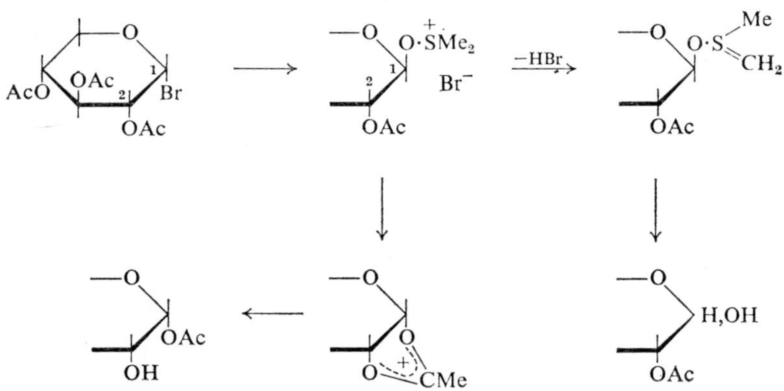

Scheme 26

Polarimetric features of acylated glycosyl halides are reported in Chapter 25.

[229] H. C. Srivastava and K. V. Ramalingam, *Indian J. Chem.*, 1969, **7**, 1206.

Other Halogenated Derivatives

Trichloroacetonitrile and hydrogen halides (Pinner reaction, Scheme 27) have been used to prepare a variety of methyl 2,3,4-tri-*O*-acetyl-6-deoxy-6-halogeno-glycopyranosides in high yield but the application of the reaction

$$ROH + Cl_3CCN + HHal \rightarrow [Cl_3C(=NH.HHal)OR] \rightarrow RHal + Cl_3CCONH_2$$

Scheme 27

to carbohydrate secondary alcohols was not described.[230] A second method for obtaining these primary halogeno-compounds involves treating corresponding sulphonates with halide ions of toluene-solubilised halides (solubilised with hexamethylphosphoric triamide). Preliminary results of applications to secondary sulphonates were also described.[231]

An improved synthesis of 4-deoxy-4-fluoro-D-glucose has been developed (Scheme 28),[232] and yet another procedure utilised direct displace-

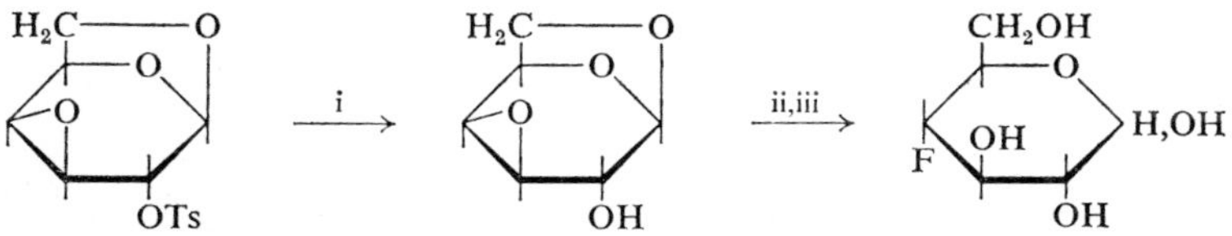

Reagents: i, NaOMe, *hν*; ii, KHF_2; iii, H_3O^+

Scheme 28

ments of the methanesulphonyloxy-group of 4-*O*-methanesulphonyl-D-galactose derivatives. For example, methyl 4-*O*-methanesulphonyl-2,3-di-*O*-methyl-α-D-galactopyranoside on treatment with caesium fluoride in boiling ethylene glycol gave the corresponding 4-deoxy-4-fluoro-α-D-glucopyranoside in moderate yield; a 3—4 Hz coupling was observed between H-1 and the fluorine atom at C-4 in the latter compound. Mass spectra of several 4-fluoro-compounds indicated that the fragmentation patterns were similar to those of acetylated and methylated hexopyranose derivatives.[233] An extension of this type of work has led to the synthesis of 2,4-dideoxy-2,4-difluoro-D-glucose (Scheme 29).[234] A further fluoroglucose derivative

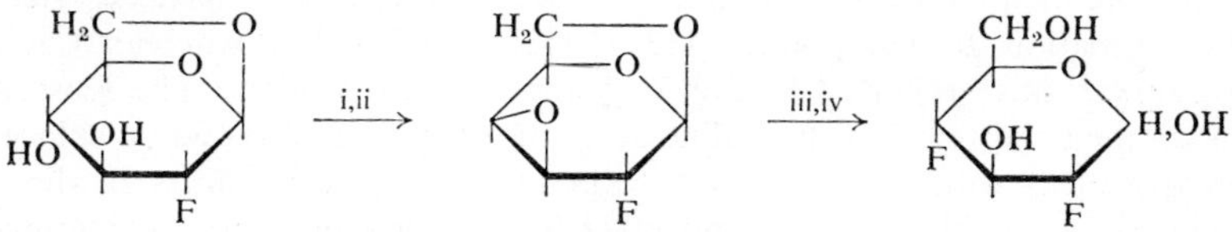

Reagents: i, TsCl, py; ii, MeO^-; iii, KHF_2; iv, H_3O^+

Scheme 29

230 M. L. Shulman, V. N. Yoldikov, and A. Ya. Khorlin, *Tetrahedron Letters*, 1970, 2517.
231 H. B. Sinclair, *Carbohydrate Res.*, 1970, **15**, 147.
232 A. D. Barford, A. B. Foster, and J. H. Westwood, *Carbohydrate Res.*, 1970, **13**, 189.
233 A. B. Foster, R. Hems, and J. H. Westwood, *Carbohydrate Res.*, 1970, **15**, 41.
234 J. Pacák, J. Podešva, and M. Černý, *Chem. and Ind.*, 1970, 929.

to have been prepared is 6-deoxy-6-fluoromuramic acid (135); in this case the fluorine atom was introduced by displacing a tosyloxy-group with

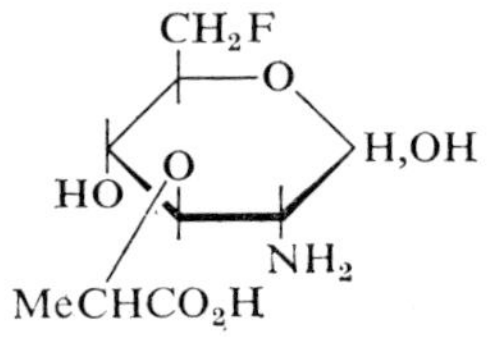

(135)

tetrabutylammonium fluoride.[235] This reagent was also used to displace the tosyloxy-group of 1,2:5,6-di-*O*-isopropylidene-3-*O*-*p*-tolylsulphonyl-*β*-L-talofuranose, so providing a convenient route to 3-deoxy-3-fluoro-L-idose derivatives.[203] In the pentose series, derivatives of 3-deoxy-3-fluoro-D-arabinose and 2-deoxy-2-fluoro-D-xylose were prepared by the route illustrated in Scheme 30.[236] Aromatisation of 1,3,5-tri-*O*-acetyl-2-chloro-2-deoxy-D-arabinofuranose to give 2-acetoxymethyl-4-chlorofuran occurs when the compound is distilled.[237]

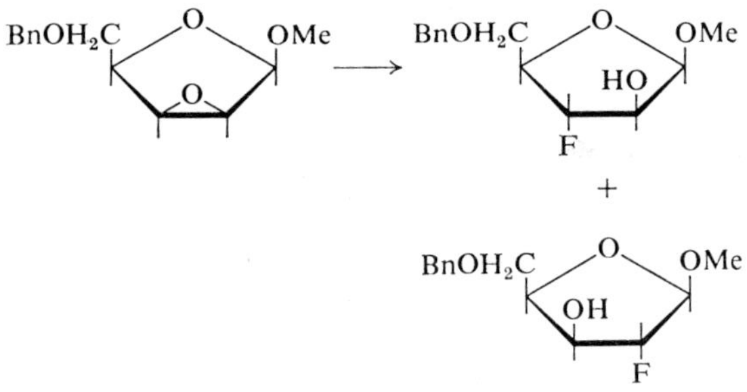

Scheme 30

Two groups have reported on the nature of the reaction undergone by methyl 2,3-*O*-isopropylidene-α-L-rhamnopyranoside (136) with triphenyl phosphite methiodide (*cf.* Vol. 3, p. 66). Russian workers [238] have described the formation of compounds (137), (138), and (139), whereas Czech workers [239] have reported finding only the first two iodides. The Russian group then showed that the iodine atom of compound (138) was displaced on treatment with sodium benzoate in DMF to give products of direct displacement (with retention and inversion of configuration), unsaturated products (amongst which was the expected 5,6-ene), and the manno-

[235] G. D. Diana, *J. Org. Chem.*, 1970, **35**, 1910.
[236] J. A. Wright and J. J. Fox, *Carbohydrate Res.*, 1970, **13**, 297.
[237] J. Kuszmann and P. Sohař, *Carbohydrate Res.*, 1970, **14**, 415.
[238] K. S. Adamyants, A. I. Usov, and N. K. Kochetkov, *Bull. Acad. Sci. U.S.S.R.*, 1970, 650; *Izvest. Akad. Nauk S.S.S.R., Ser. khim.*, 1970, 696.
[239] K. Kefurt, J. Jarý, and Z. Samek, *Coll. Czech. Chem. Comm.*, 1970, **35**, 2613.

pyranoside derivative (137), which was presumably produced by a reaction operating in the reverse way to the ring-contraction by which (138) was originally formed.[240]

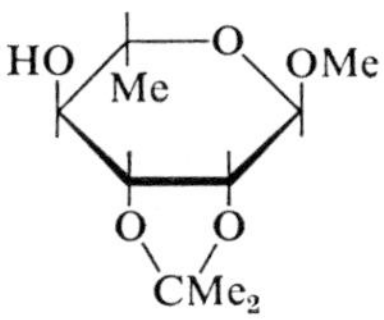

(136)

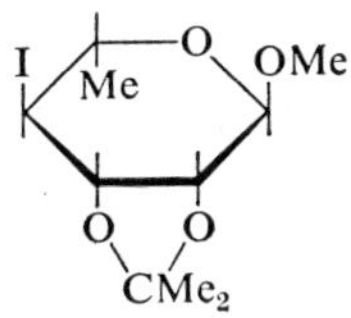

(137)

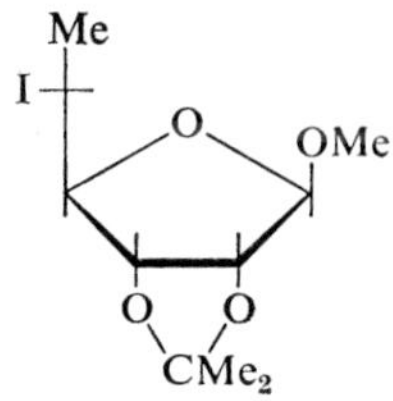

(138)

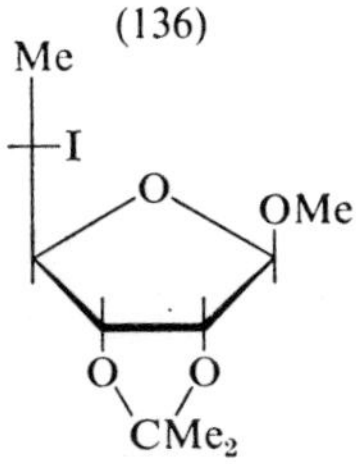

(139)

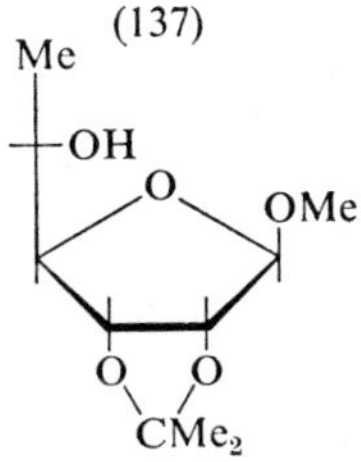

(140)

The iodide (138) was also obtained by treating methyl 6-deoxy-2,3-*O*-isopropylidene-α-L-talofuranoside (140) with triphenyl phosphite methiodide.[240] The same reagent in DMF offers a means of obtaining high yields of 5′-deoxy-5′-iodo-derivatives from suitably protected pyrimidine nucleosides. In some instances, selective iodination can be effected at primary sites. The reagent converted adenosine and 2′,3′-*O*-isopropylidene derivatives of purine nucleosides into N^3,5′-cyclonucleosides, in good yields. Other aspects of the reaction of the reagent with nucleosides were discussed.[241] Continuation of this work by the Syntex group has involved the iodination of various nucleoside derivatives which contain secondary hydroxy-groups. Thymidine derivatives containing 5′-substituents give 3′-deoxy-3′-iodo products with retention of configuration by way of O^2,3′-cyclonucleosides. 5′-Substituted uridines, on the other hand, do not give iodinated products but afford 2′(3′)-*O*-methylphosphonates. 2′,5′-Di-*O*-trityluridine gave the expected 3′-deoxy-3′-iodo-*xylo*-nucleoside and also suffered selective loss of the ether groups. Various related reactions were described in full.[242]

Halogenated derivatives of methyl β-maltoside have been reported following partial tosylation of the glycoside at the primary positions. Thus, the di- (141) and the mono-iodide (142) were prepared, and the latter was

[240] N. K. Kochetkov, A. I. Usov, and K. S. Adamyants, *Izvest. Akad. Nauk S.S.S.R., Ser. khim.*, 1970, 885.
[241] J. P. H. Verheyden and J. G. Moffatt, *J. Org. Chem.*, 1970, **35**, 2319.
[242] J. P. H. Verheyden and J. G. Moffatt, *J. Org. Chem.*, 1970, **35**, 2868.

converted into the anhydro-deoxy-compound (143) following catalytic hydrogenation and deacetylation with alkali.[198a]

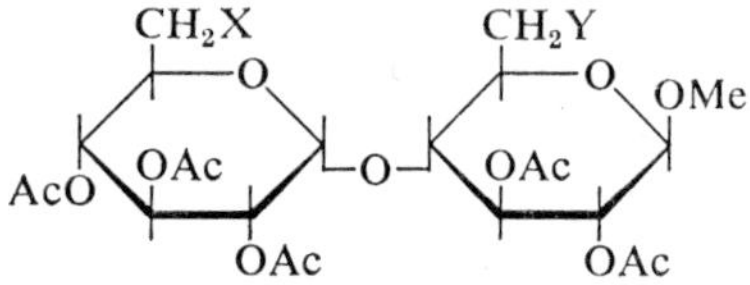

(141) X = Y = I
(142) X = OTs, Y = I

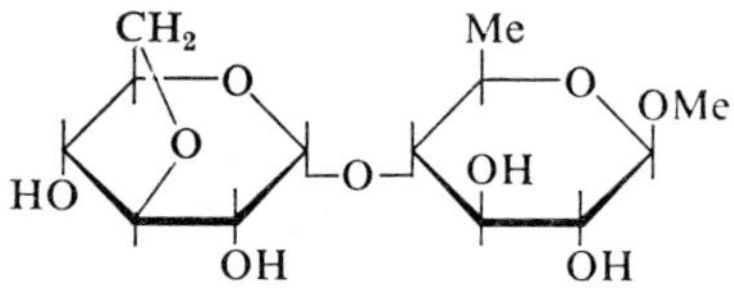

(143)

A useful procedure for the synthesis of 2-chloro-2-deoxy-sugars, which has the virtue of providing a stereospecific means of introducing chlorine, involves treatment of glycosyl halide chlorosulphates with aluminium trichloride. Compound (144) gives (145) with this reagent by a mechanism

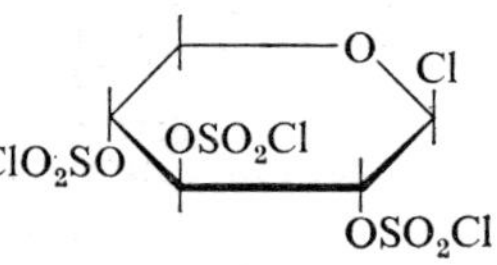

(144)

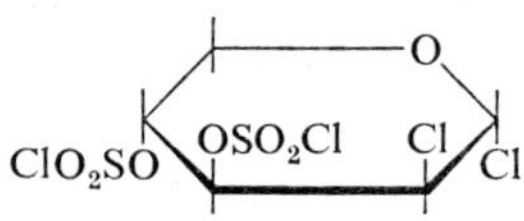

(145)

presumed to involve migration of the anomeric chlorine atom to C-2. Treatment of the product with sodium iodide yielded 2-chloro-2-deoxy-D-lyxose; 2-chloro-2-deoxy-D-xylose can be obtained in similar fashion from the α-D-*lyxo*-isomer of (144).[243]

An unusual bromination has been noted when β-D-glucopyranose penta-acetate is treated with hydrogen bromide and aluminium tribromide. As well as occurring at the anomeric position, substitution also took place at the primary carbon atom to give 2,3,4-tri-*O*-acetyl-6-bromo-6-deoxy-α-D-glucopyranosyl bromide, which was then used to obtain 6-deoxy-compounds (see Scheme 12).[92] The uses of chlorodeoxy-sugars in syntheses of deoxy- and aminodeoxy-sugars have been discussed. Methyl 6-chloro-4,6-dideoxy-α-D-*xylo*-hexopyranoside has been prepared by selective hydrogenation of methyl 4,6-dichloro-4,6-dideoxy-α-D-galacto-pyranoside over Raney nickel in the presence of triethylamine.[244]

[243] H. J. Jennings, *Canad. J. Chem.*, 1970, **48**, 1834.
[244] B. T. Lawton, W. A. Szarek, and J. K. N. Jones, *Carbohydrate Res.*, 1970, **15**, 397.

8

Amino-sugars

Natural Products

4-Amino-4-deoxy-L-arabinose has been identified as a component of some bacterial polysaccharides.[245] Methanolysis of gentamicin A afforded the methyl gentosaminides which were identified as derivatives of 3-methylamino-3-deoxy-D-xylose.[246] Perosamine, obtained on hydrolysis of perimycin, has been confirmed to be 4-amino-4,6-dideoxy-D-mannose by synthesis[246a] (*cf.* Vol. 3, p. 70). Full details of the synthesis of D-halosamine (4-amino-2,4,6-trideoxy-3-*O*-methyl-D-*ribo*-hexose) have been given.[247]

Synthesis

A convenient preparation of 2-amino-2-deoxy-D-galactose from the D-*gluco*-isomer has been described;[248] the key steps are shown in Scheme 31.

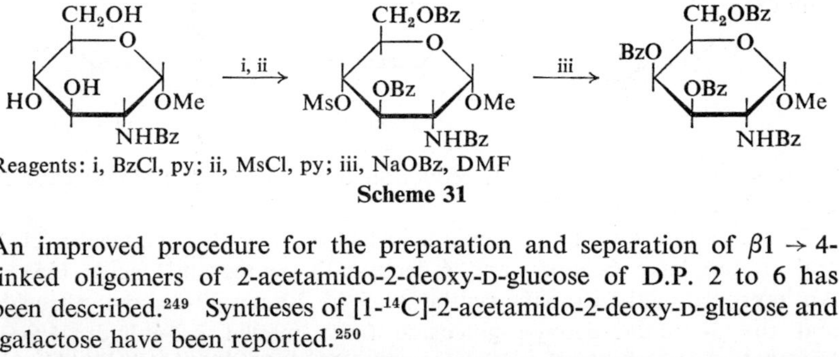

Reagents: i, BzCl, py; ii, MsCl, py; iii, NaOBz, DMF

Scheme 31

An improved procedure for the preparation and separation of $\beta 1 \rightarrow 4$-linked oligomers of 2-acetamido-2-deoxy-D-glucose of D.P. 2 to 6 has been described.[249] Syntheses of [1-^{14}C]-2-acetamido-2-deoxy-D-glucose and -galactose have been reported.[250]

The chromous-chloride-promoted addition of *N*-chlorocarbonates to enol ethers has been used in the synthesis of *N*-alkoxycarbonyl derivatives of 2-amino-2-deoxy-sugars as illustrated in Scheme 32.[251] The α-chloride

245 W. A. Volk, C. Galanos, and O. Luderitz, *F.E.B.S. Letters*, 1970, **8**, 161.
246 H. Maehr and C. P. Schaffner, *J. Amer. Chem. Soc.*, 1970, **92**, 1697.
246a C. L. Stevens, R. P. Glinski, K. G. Taylor, P. Blumbergs, and S. K. Gupta, *J. Amer. Chem. Soc.*, 1970, **92**, 3160.
247 M. M. Janot, Q. Khuong-Huu, C. Monneret, I. Kaboré, J. Hildesheim, S. D. Géro, and R. Goutarel, *Tetrahedron*, 1970, **26**, 1695.
248 M. W. Horner, L. Hough, and A. C. Richardson, *J. Chem. Soc.* (*C*), 1970, 1336.
249 B. Capon and R. L. Foster, *J. Chem. Soc.* (*C*), 1970, 1654.
250 J. F. Kennedy, *J. Labelled Compounds*, 1970, **6**, 201.
251 J. Lessard, H. Driguez, and J. P. Vermes, *Tetrahedron Letters*, 1970, 4887.

(146) was the major product obtained from tri-*O*-acetyl-D-glucal, although the β-chloride and adducts having the D-*manno* configuration were also formed. Compound (146) was converted into β-glycosides by Koenigs–Knorr methods.

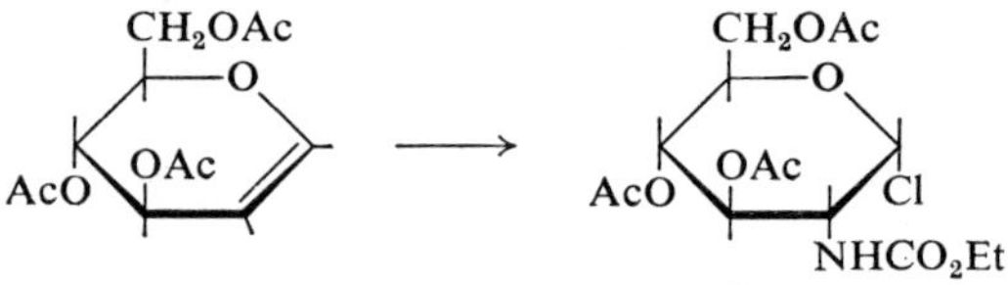

(146)

Reagent: $EtO_2C\cdot NHCl$, $CrCl_2$

Scheme 32

2-Acetamido-2-deoxy-β-D-glucosylamine has been conveniently prepared from 2-acetamido-3,4,6-tri-*O*-acetyl-2-deoxy-β-D-glucosyl azide by de-acetylation followed by catalytic hydrogenation.[252] D-*threo*-Pentulose was found to react with α-amino-acids to give mixtures of derivatives of 2-amino-2-deoxy-D-xylose and -lyxose and of 1-deoxy-1-amino-D-*threo*-pentulose.[253] Compound (147) was readily converted into (148), which was

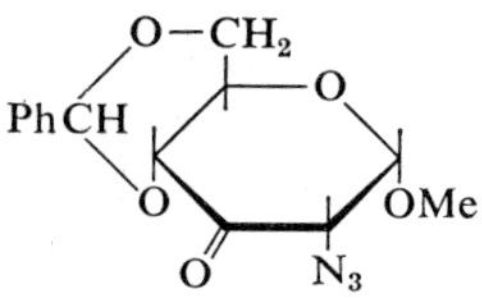

(147)

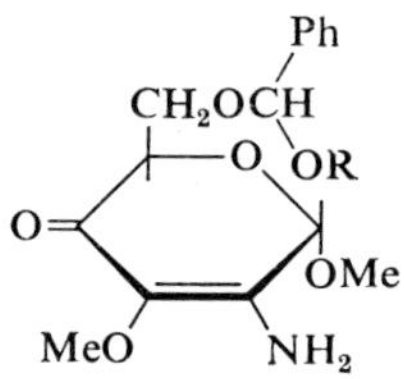

(148)

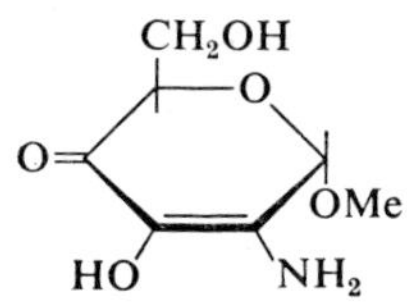

(149)

hydrolysed with acid to (149); reduction of the latter compound gave 2-amino-2-deoxy-D-allose.[254]

2-Amino-2-deoxy-D-mannuronic acid has been synthesised (see p. 118), and so too have 6-deoxy- and 6-deoxy-6-fluoro-muramic acid (135).[235] An improved method for the synthesis of the 2-amino-2-deoxy-D-altroside and the 3-amino-3-deoxy-D-glucoside from methyl 2,3-anhydro-4,6-*O*-benzylidene-α-D-alloside has been reported.[255]

A synthesis of 3-acetamido-2,3-dideoxy-DL-tetrose has been accomplished from DL-aspartic acid and ethyl malonate.[256] 3-Acetamido-3,5-dideoxy-D-ribose has been synthesised from both compounds (150) and (151) by the routes shown in Scheme 33.[257]

[252] M. Kiyozumi, K. Kato, T. Komori, A. Yamamoto, T. Kawasaki, and H. Tsukamoto, *Carbohydrate Res.*, 1970, **14**, 355.
[253] K. Heyns and W. Beilfuss, *Chem. Ber.*, 1970, **103**, 2873.
[254] W. Meyer Zu Reckendorf, *Chem. Ber.*, 1970, **103**, 2418.
[255] K. Čapek, Z. Kefurtová, and J. Jarý, *Coll. Czech. Chem. Comm.*, 1970, **35**, 1930.
[256] S. David and A. Veyrières, *Carbohydrate Res.*, 1970, **13**, 203.
[257] H. Ando and J. Yoshimura, *Bull. Chem. Soc. Japan*, 1970, **43**, 2966.

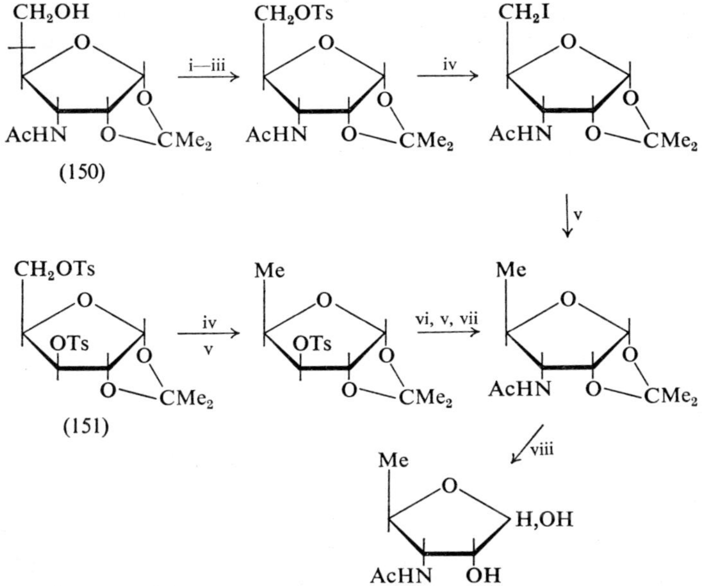

Reagents: i, $NaIO_4$; ii, $NaBH_4$; iii, TsCl, py; iv, NaI, MeCN; v, H_2, Ni; vi, NH_2NH_2; vii, Ac_2O, py; viii, H_3O^+

Scheme 33

A new synthesis of methyl 3-amino-3-deoxy-β-D-ribofuranoside has been devised (Scheme 34).[258] The 5-*O*-trityl ether of 1,2-*O*-isopropylidene-α-D-xylofuranose has been used as the starting material for a synthesis of

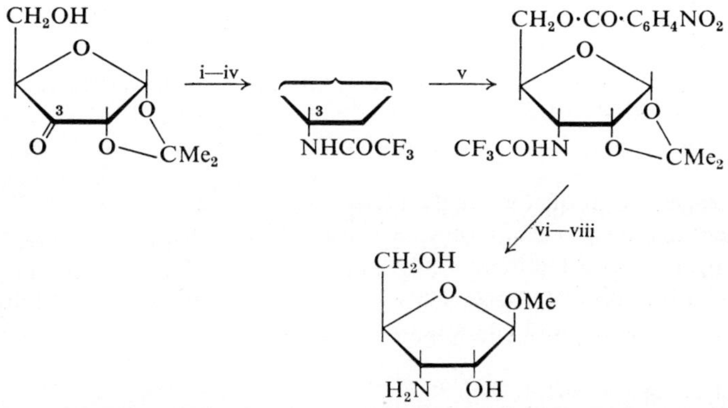

Reagents: i, $NH_2OH \cdot HCl$, py, EtOH; ii, LAH, THF; iii, $(CF_3CO)_2O$, py; iv, MeOH; v, *p*-$NO_2C_6H_4COCl$, py; vi, TFA, H_2O; vii, MeOH, HCl; viii, $BuNH_2$, MeOH

Scheme 34

[258] A. N. Fujiwara, E. M. Acton, and L. Goodman, *J. Heterocyclic Chem.*, 1970, **7**, 891.

methyl 3-amino-3-deoxy-α-D-xylopyranoside and its β-anomer.[174] Reduction of the oxime (152) with LAH gave the D-*lyxo* derivative (153) with high

(152) (153)

stereospecificity.[259] Opening of the epoxide ring in 9-(2,3-anhydro-β-D-lyxofuranosyl)adenine with azide ion and subsequent reduction gave the 3′-amino-3′deoxy-D-*arabino* and 2′-amino-2′-deoxy-D-*xylo* analogues of adenosine.[260] The ratio of ring-opening at C-2′ and C-3′ was about 15 : 1.

Methyl 4-amino-2,3,4-trideoxy-α-D-*erythro*-hex-2-enopyranosiduronic acid, the carbohydrate component of blasticidin S, has been synthesised as shown in Scheme 35.[261, 261a] The formation of 4- and 5-amino-sugars from pyranoside 4-sulphonates has been mentioned earlier.[217, 218]

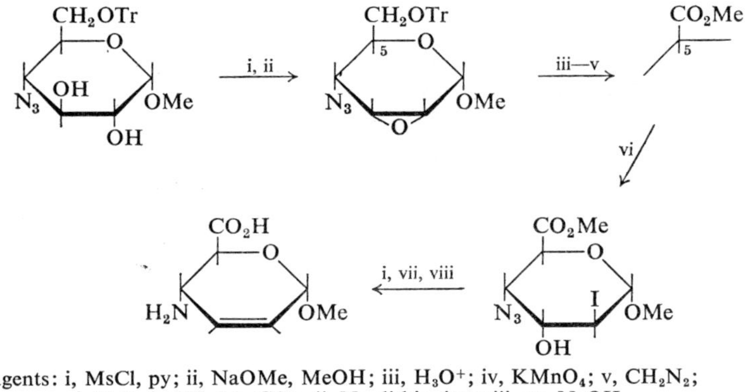

Reagents: i, MsCl, py; ii, NaOMe, MeOH; iii, H_3O^+; iv, $KMnO_4$; v, CH_2N_2; vi, NaI, AcOH, Me_2CO; vii, Na dithionite; viii, aq. NaOH

Scheme 35

Hydrogenolysis of the azide (154) with Raney nickel in hydrazine afforded the methanesulphonic acid salt of 5-amino-5,6-dideoxy-1,2-*O*-isopropylidene-3-*O*-methanesulphonyl-β-L-idofuranose (155) presumably by way of the epimine intermediate (156).[262] Partial hydrogenation of (157) gave (158), which could be dimerised to the heptacyclic compound (159).[263]

[259] J. M. J. Tronchet and R. Graf, *Helv. Chim. Acta*, 1970, **53**, 851.
[260] A. P. Martinez, D. F. Calkins, E. J. Reist, W. W. Lee, and L. Goodman, *J. Heterocyclic Chem.*, 1970, **7**, 713.
[261] R. S. Goody, K. A. Watanabe, and J. J. Fox, *Tetrahedron Letters*, 1970, 293.
[261a] K. A. Watanabe, R. S. Goody, and J. J. Fox, *Tetrahedron*, 1970, **26**, 3883.
[262] C. F. Gibbs and L. Hough, *Carbohydrate Res.*, 1970, **15**, 29.
[263] H. Paulsen and F. Kownatzki, *Chem. Ber.*, 1970, **103**, 1631.

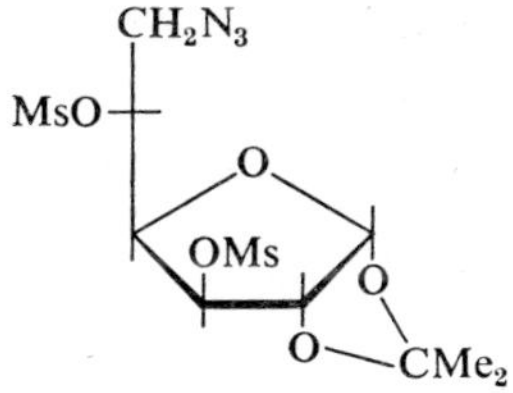

(154)

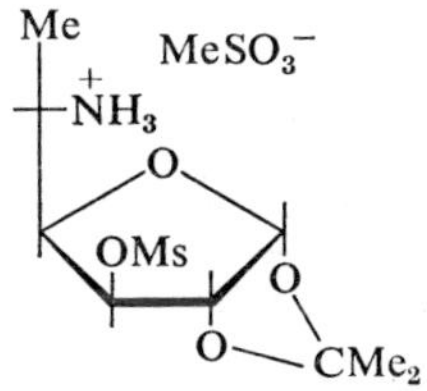

(155)

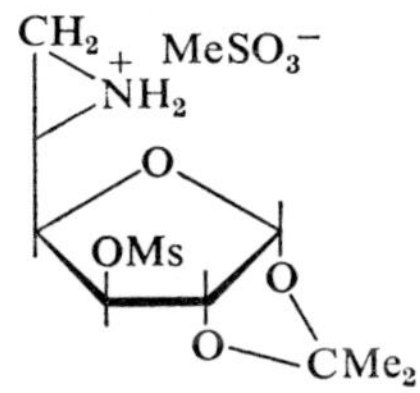

(156)

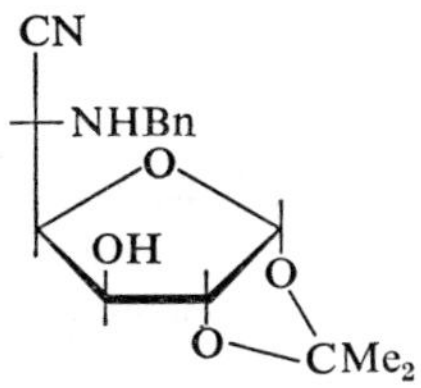

(157)

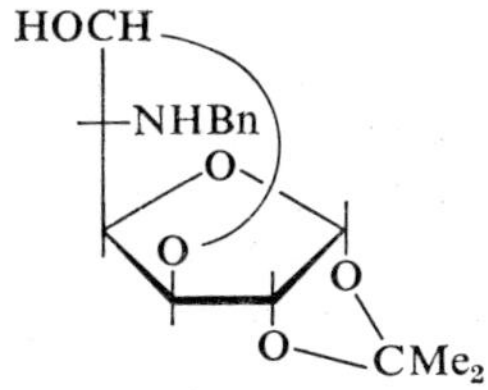

(158)

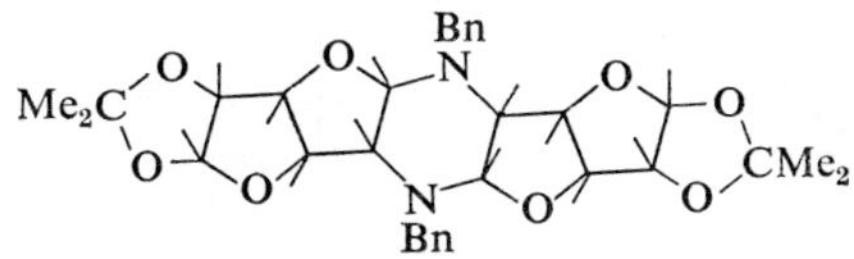

(159)

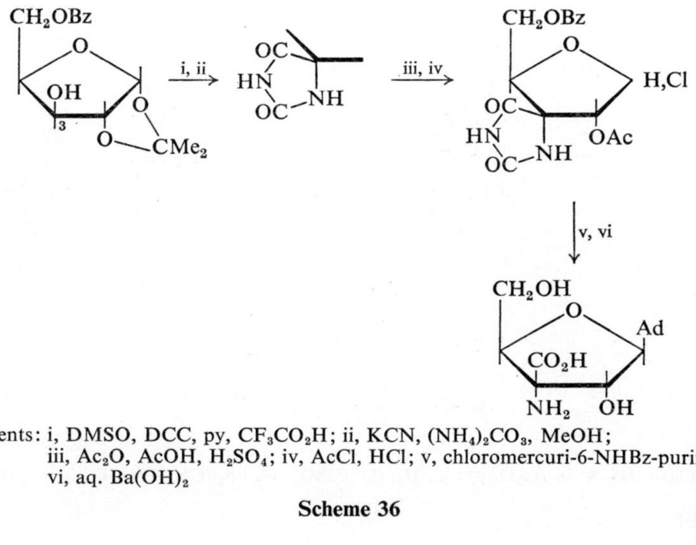

Reagents: i, DMSO, DCC, py, CF_3CO_2H; ii, KCN, $(NH_4)_2CO_3$, MeOH; iii, Ac_2O, AcOH, H_2SO_4; iv, AcCl, HCl; v, chloromercuri-6-NHBz-purine; vi, aq. $Ba(OH)_2$

Scheme 36

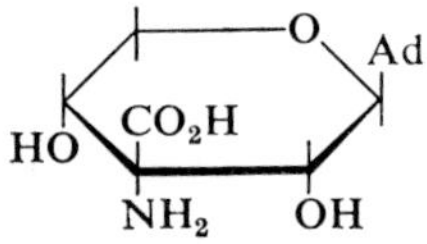

(160)

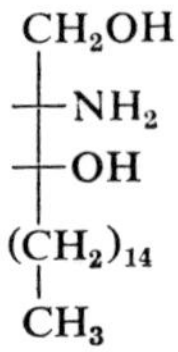

(161)

A general synthesis of 6-amino-6-deoxy-α-D-glucopyranosides has been described based on the addition of nitrosyl chloride to 3,4-di-*O*-acetyl-6-*O*-*p*-tolylsulphonyl-D-glucal and the reaction of the adduct with alcohols. The 2-oximino-group was then converted into a keto-group and reduced, after which the amino-group was introduced by displacement of the 6-

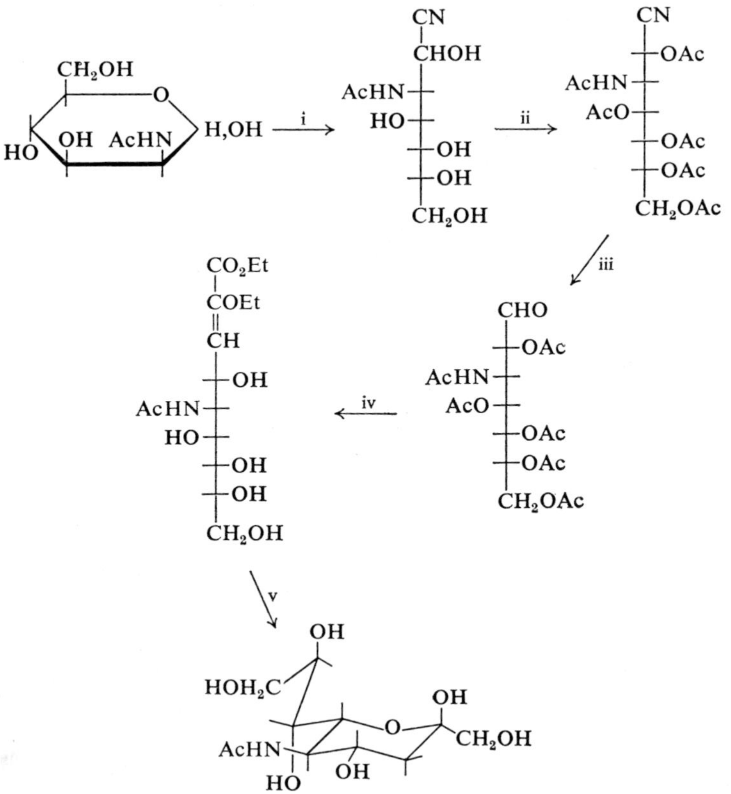

Reagents: i, HCN; ii, Ac_2O, py; iii, H_2, Pd–$BaSO_4$; iv, $Ph_3\overset{+}{P}\overset{-}{C}H(OEt)CO_2Et$; v, pH 9

Scheme 37

tosyloxy-group with azide ion and reduction.[264] Other references to the glycosides of amino-sugars can be found in Chapter 3.

Several papers on branched-chain amino-sugars have appeared and a new class of nucleoside derivative has been prepared by the route shown in Scheme 36. The pyranose analogue (160) was prepared by similar methods.[265] Other examples having an amino-group substituted onto the carbon atom forming the branch are described in Chapter 15.

Dihydrosphingosine (161) has been synthesised by the approach pioneered by Gigg and his group for such compounds.[266] Two new syntheses of *N*-acetylneuraminic acid from 2-acetamido-2-deoxy-D-mannose have been reported;[267] one of the routes used is illustrated in Scheme 37.

Reactions

Treatment of 2-amino-2-deoxy-D-glucose with *p*-nitrophenyl esters in DMSO containing triethylamine provides a method for selective *N*-acylation; the *N*-acetyl, *N*-benzoyl, and *N*-stearoyl derivatives were obtained in high yield.[267a] 2-Deoxy-2-iodoacetamido-D-glucose has been prepared from the tetra-acetate and iodoacetic acid using DCC, followed by de-*O*-acetylation.[268] 2-Amino-2-deoxyaldoses and isothiocyanates have been reacted together to give as the first product the thiourea derivative (162); this cyclised to form (163) which then lost a molecule of water giving (164).[269]

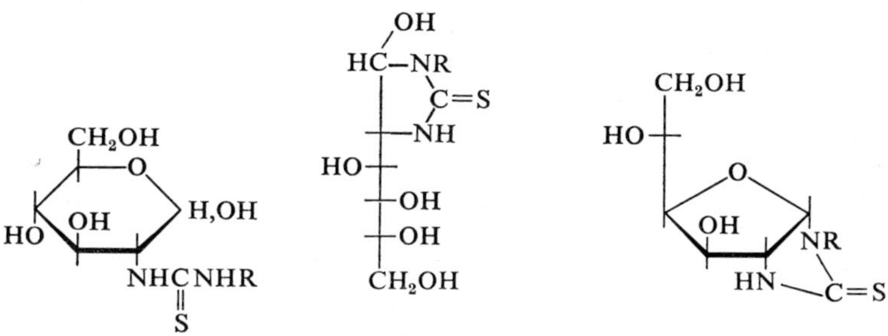

(162) (163) (164)

Several 4,6-cyclic acetals of 2-acetamido-2-deoxy-D-glucose have been synthesised by condensations in acetonitrile using hydrochloric acid as catalyst (with aliphatic aldehydes), and either without solvent or in dioxan

[264] T. L. Nagabhushan, *Canad. J. Chem.*, 1970, **48**, 257.
[265] H. Yanagisawa, M. Kinoshita, S. Nakada, and S. Umezawa, *Bull. Chem. Soc., Japan*, 1970, **43**, 246.
[266] E. J. Reist and P. H. Christie, *J. Org. Chem.*, 1970, **35**, 3521.
[267] M. N. Mirzayanova, L. P. Davydova, and G. I. Samokhvalov, *J. Gen. Chem. (U.S.S.R.)*, 1970, **40**, 693.
[267a] H. Mukerjee and P. R. Pal, *J. Org. Chem.*, 1970, **35**, 2042.
[268] P. W. Kent, J. P. Ackers, and R. J. White, *Biochem. J.*, 1970, **118**, 73.
[269] J. E. Scott, *Carbohydrate Res.*, 1970, **14**, 389.

using zinc chloride as catalyst (with aromatic aldehydes). Methylation and hydrolysis of the resulting acetals yielded 2-amino-2-deoxy-3-*O*-methyl-D-glucose hydrochloride.[270]

A new blocking group has been used for *eq,eq*-hydroxyamino-groups; methyl 2-amino-2-deoxy-α-D-glucopyranoside, for example, gave the 2,3-carbonate with *p*-nitrophenoxycarbonyl chloride[271] (see also p. 129). The deamination of 1-amino-1-deoxypentitols has been described earlier.[130] Deamination of methyl 4-amino-4-deoxy-2,3-*O*-isopropylidene-α-L-talopyranoside (165) in 90% acetic acid gave mainly the L-mannose derivatives (166) and (167) but no corresponding derivatives of L-talose were detected.

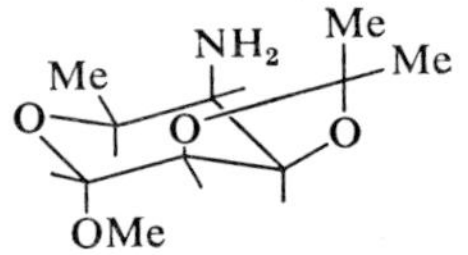

(165)

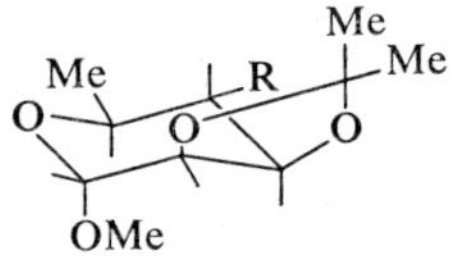

(166) R = OH
(167) R = OAc

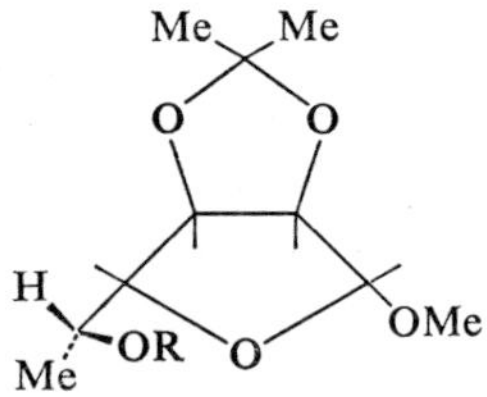

(168) R = H
(169) R = Ac

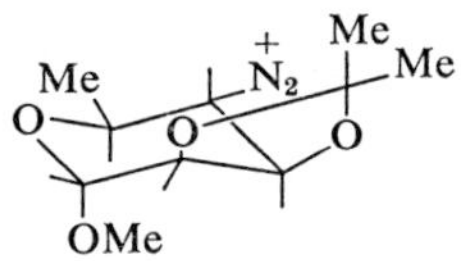

(170)

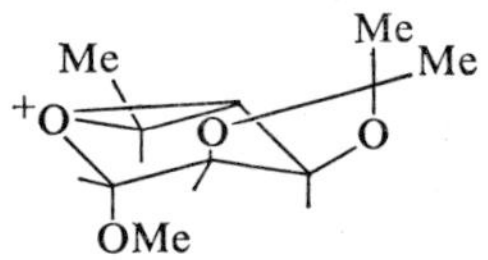

(171)

Deamination of the epimeric L-*manno* amine gave (166), (167), and the rearranged products (168) and (169), which were presumed to be formed from the oxonium ion (171) arising as a result of participation by the ring-oxygen atom in the decomposition of the diazonium ion (170).[272]

The oxidation of amino-sugars with periodate has been studied (see p. 144) and so has the Elson–Morgan reaction between 2-amino-2-deoxy-D-galactose and pentane-2,4-dione.[273] Seven of the eight possible 2-amino-2-deoxyhexoses have been separated by chromatography on a cationic resin using a borate buffer; acidic and neutral amino-acids were eluted before the hexosamines.[274]

Di- and Poly-amino-sugars

Two routes to 2,3-diamino-2,3-dideoxy-D-galactose have been described; one of the routes used is outlined in Scheme 38[275] and the other route is

[270] L. Holmquist, *Acta Chem. Scand.*, 1970, **24**, 173.
[271] S. Umezawa, T. Tsuchiya, and Y. Takagi, *Bull. Chem. Soc. Japan*, 1970, **43**, 1602.
[272] A. K. Al-Radhi, J. S. Brimacombe, and L. C. N. Tucker, *Chem. Comm.*, 1970, 1250.
[273] F. Serafini-Cessi and C. Cessi, *Biochem. J.*, 1970, **120**, 873.
[274] M. Yaguchi and M. G. Perry, *Canad. J. Biochem.*, 1970, **48**, 386.
[275] W. Meyer Zu Reckendorf and N. Wassiliadou-Micheli, *Chem. Ber.*, 1970, **103**, 37.

similar.[276] Oxidation of (172) with iodine gave a triulose derivative which was isolated as (173) after treatment with phenylhydrazine. The latter

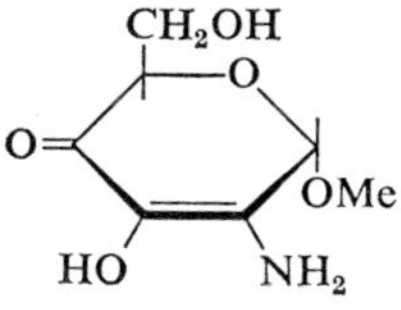

(172)

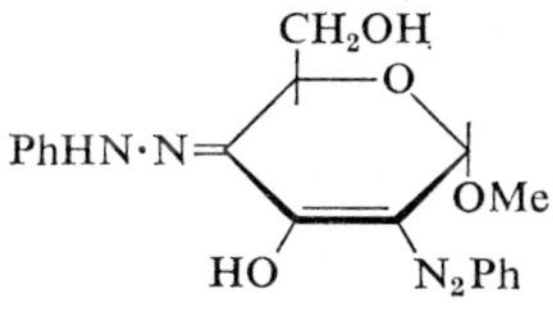

(173)

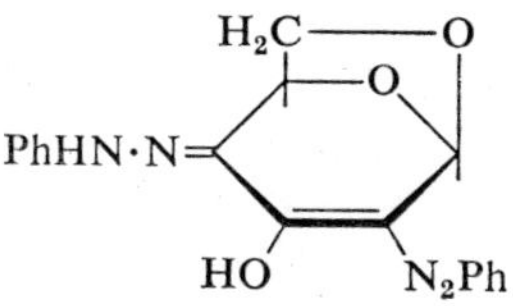

(174)

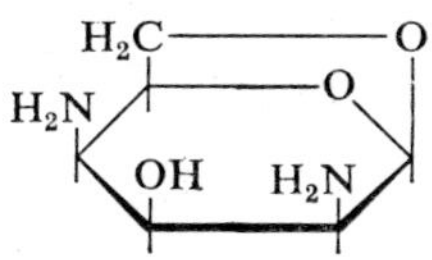

(175)

compound gave (174) spontaneously on acid hydrolysis, and both compounds afforded 2,4-diamino-1,6-anhydro-2,4-dideoxy-D-talose (175) on hydrogenation.[277]

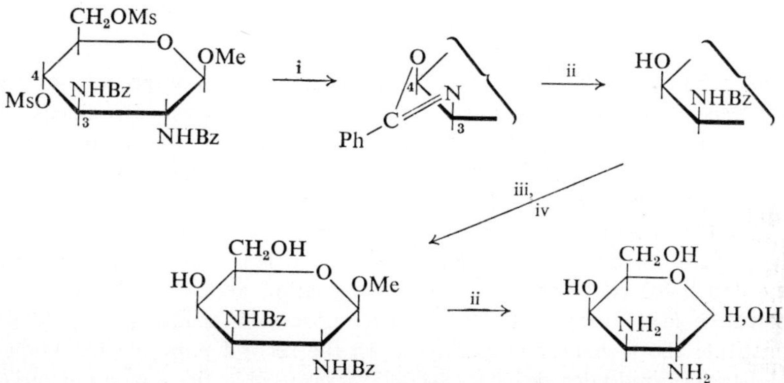

Reagents: i, NaOAc, EtOH; ii, HCl; iii, NaOAc, DMF; iv, NaOMe, MeOH

Scheme 38

Methyl 2,4-diacetamido-3,6-di-*O*-acetyl-2,4-dideoxy-α-D-idopyranoside (176) has been prepared by hydrazinolysis, reduction, and acetylation of either compound (177) or methyl 2,6-di-*O*-benzoyl-3,4-di-*O*-methanesulphonyl-α-D-glucopyranoside.[278] The product must presumably arise

[276] W. Meyer Zu Reckendorf, *Chem. Ber.*, 1970, **103**, 995.
[277] W. Meyer Zu Reckendorf, *Chem. Ber.*, 1970, **103**, 2424.
[278] T. Suami and T. Shoji, *Bull. Chem. Soc. Japan*, 1970, **43**, 2948.

from participation by the ester groups. Reaction of (178) with hydrazine gave, after reduction and acetylation, the 4,6-diacetamido derivative (179). Hydrazinolysis proceeded by way of the cyclic compound (180), which was also isolated and characterised.[278]

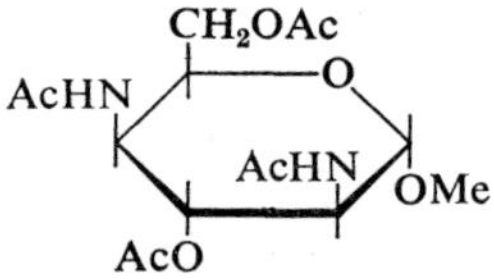

(176)

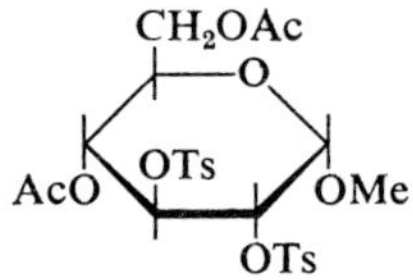

(177)

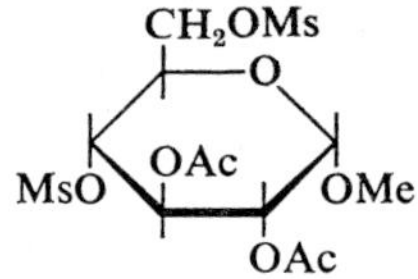

(178)

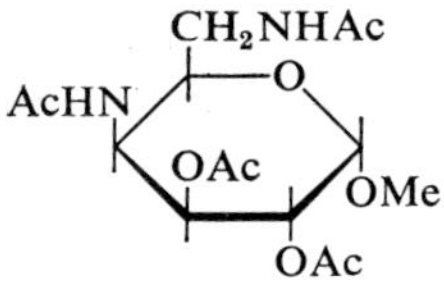

(179)

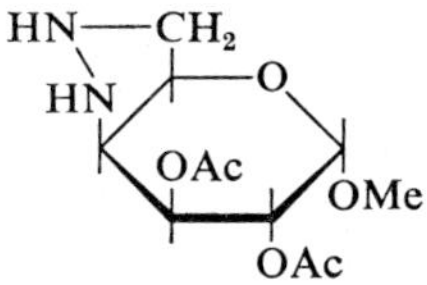

(180)

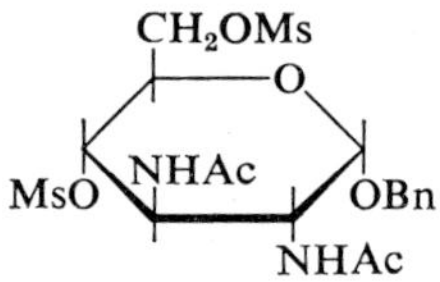

(181)

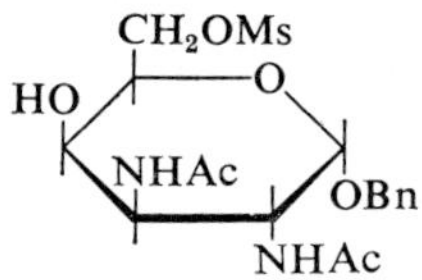

(182)

Baer has extended the use of unsaturated nitro-compounds to the synthesis of 2,3-diamino- and 2,3,4-triamino derivatives of D-glucose.[279] Nitro-compounds were also used as precursors in syntheses of methyl 2,3,4-triacetamido-2,3,4,6-tetradeoxy-α-L-glucopyranoside and the corresponding L-mannopyranoside[280] (*cf.* Vol. 3, p. 78).

2,3,4,6-Tetra-amino-2,3,4,6-tetradeoxy-D-glucose[281] and its methyl α-glycoside[282] have been described (*cf.* Vol. 3, p. 79). In the first paper, the key step involved inversion of the configuration at C-4 in the D-glucose derivative (181) to give (182), so that subsequent azidolysis would reconstitute the D-*gluco* configuration. In the second paper,[282] the known 2,3,4-triacetamido derivative (see ref. 279) was used as the starting material. 2,3,4,6-Tetra-amino-2,3,4,6-tetradeoxy-D-galactose has been prepared from benzyl 2-acetamido-2-deoxy-α-D-allopyranoside by standard reactions.[283]

[279] H. H. Baer and F. Rajabalee, *Carbohydrate Res.*, 1970, **12**, 241.
[280] F. W. Lichtenthaler and W. Fischer, *Chem. Comm.*, 1970, 1081.
[281] W. Meyer Zu Reckendorf, *Tetrahedron Letters*, 1970, 287.
[282] H. H. Baer and M. Bayer, *Carbohydrate Res.*, 1970, **14**, 114.
[283] W. Meyer Zu Reckendorf, *Chimia* (*Switz.*), 1970, **24**, 16.

9
Hydrazones, Osazones, and Formazans

Two papers[284, 285] have discussed critically the proposed cyclic structure for sugar osazones (see vol. 3, p. 82). A careful analysis of *all* the available information suggests that the usually accepted acyclic structure (*e.g.* 183) originally proposed by Fieser and Fieser is still to be favoured.

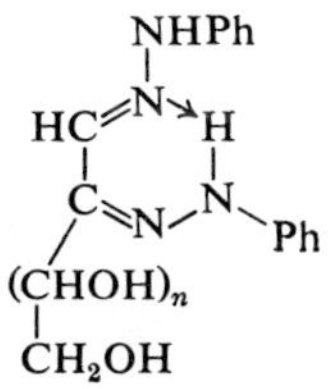
(183)

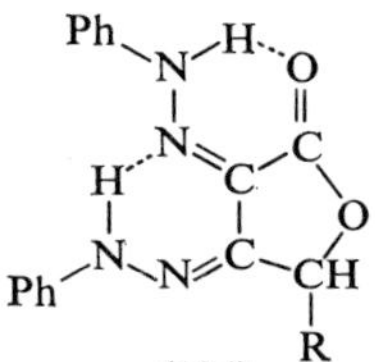
(184)

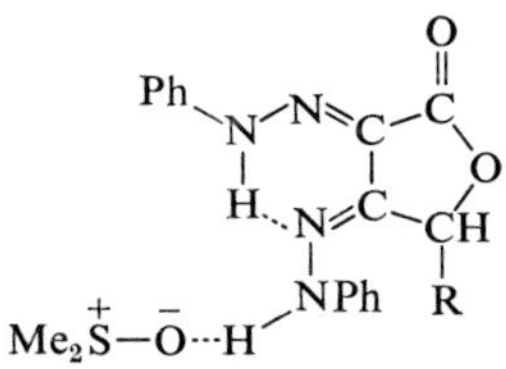
(185)

A study of L-ascorbic acid osazones by spectroscopic methods showed that they are γ-lactones and not δ-lactones as was claimed previously[286] (see vol. 2, p. 95). In dimethyl sulphoxide solution, a 'mutarotation' occurred in which the chelated form (184) was converted into the chelated form (185). The fact that acetylation and benzoylation of dehydro-L-ascorbic acid phenylhydrazone (186) afforded the unsaturated derivative (187) was also compatible with a γ-lactone structure for (186).[287] The derivative (189) was obtained by treating (188) with sodium iodide in acetone, whereas (190) gave (191) under similar conditions.

The ascorbic acid homologue (192) reacted with phenylhydrazine to give (193) and (194), and the analogue (195) afforded (196) with phenylhydrazine.[288]

Oxidation of sugar monobenzoylhydrazone acetates with iodine and mercuric oxide has been shown to yield saccharide oxadiazoles.[289] Thus,

[284] J. Buckingham, *Tetrahedron Letters*, 1970, 951.
[285] L. Mester, H. El Khadem, and D. Horton, *J. Chem. Soc.* (*C*), 1970, 2567.
[286] J. M. Rao and P. M. Nair, *Tetrahedron*, 1970, **26**, 3833.
[287] H. El Khadem and S. H. El Ashry, *Carbohydrate Res.*, 1970, **13**, 57.
[288] H. El Khadem, and I. El Kholy, Z. M. El-Shafei, and M. El Sekeki, *Carbohydrate Res.*, 1970, **15**, 179.
[289] H. El Khadem, M. A. E. Shaban, and M. A. M. Nassr, *Carbohydrate Res.*, 1970, **13**, 470.

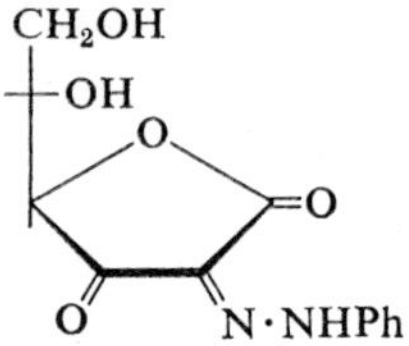

(186)

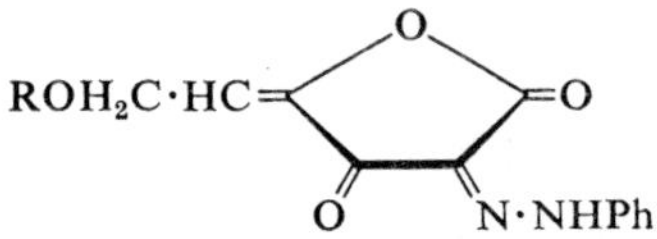

(187) R = Ac or Bz

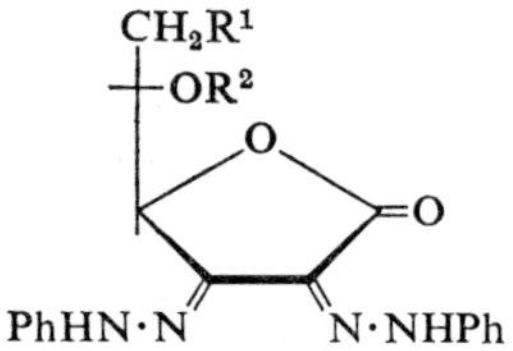

(188) R^1 = OTs, R^2 = H
(190) R^1 = OTs, R^2 = Ts
(191) R^1 = I, R^2 = Ts

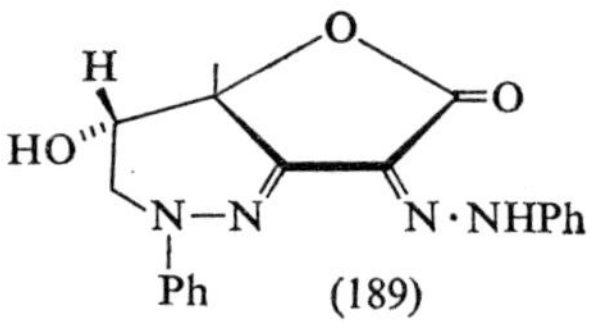

(189)

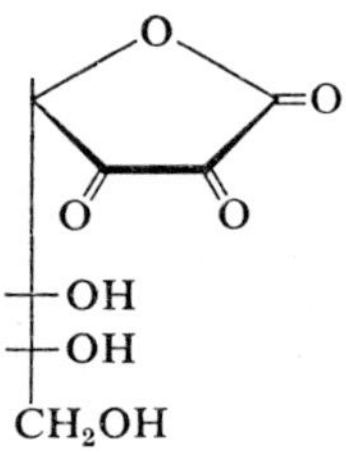

(192)

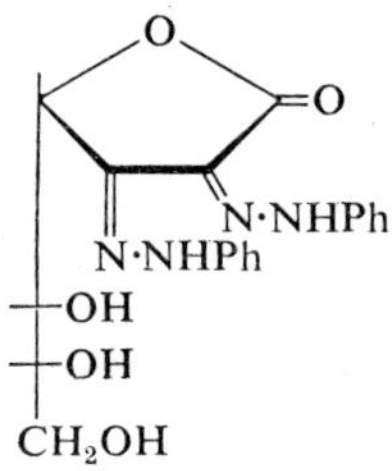

(193)

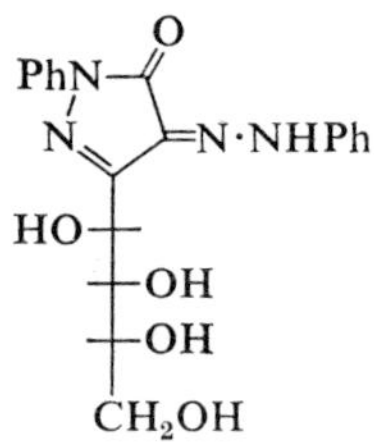

(194)

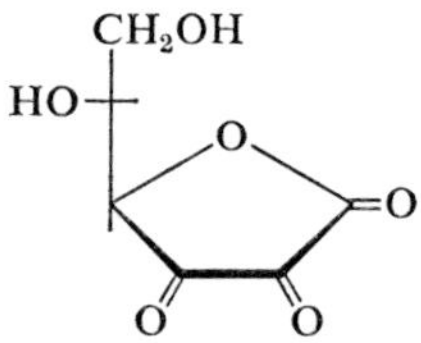

(195)

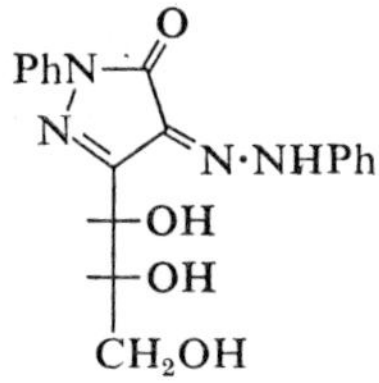

(196)

penta-*O*-acetyl-*aldehydo*-D-galactose benzoylhydrazone (197) was converted into (198), and an oxadiazole was obtained from tetra-*O*-acetyl-*aldehydo*-D-arabinose benzoylhydrazone in similar fashion. Deacetylation of (198) afforded the iminolactone (199).

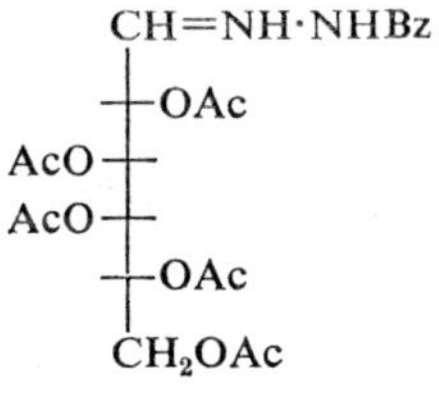

(197)

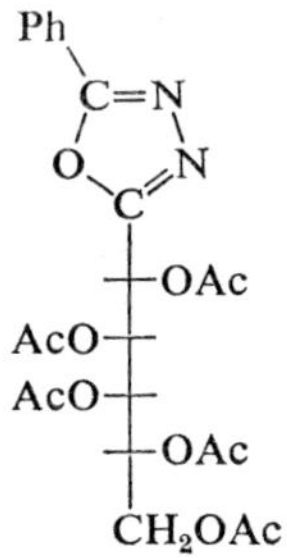

(198)

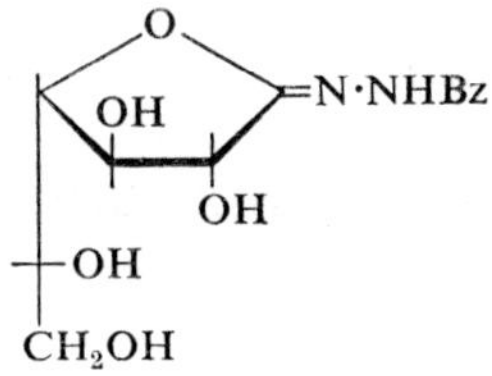

(199)

The highly crystalline bis(benzoylhydrazone) of 3-deoxy-D-*erythro*-hexos-2-ulose has proved a useful derivative in the purification of the 2-ulose.[290] Separation has been achieved of the diastereoisomeric 3-methyl-2-(polyhydroxyalkyl)-benzothiazolines (*e.g.* 200) resulting from the reaction of aldoses such as D-arabinose with *N*-methylbenzothiazoline.[291]

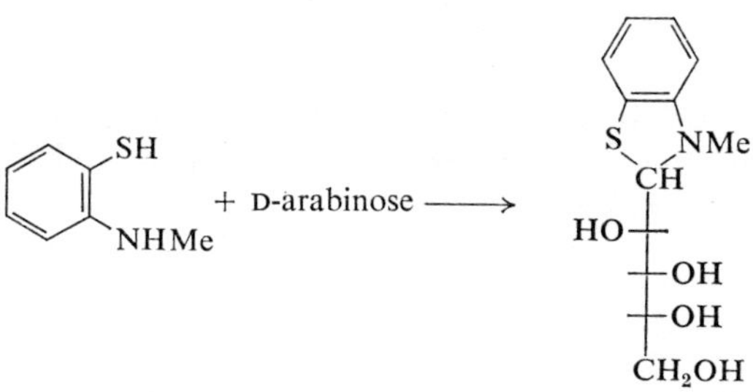

(±)-(200)

The preparation of D-apiose phenylosotriazole, a crystalline derivative useful for identifying radioactive D-apiose in biosynthetic experiments, has been described.[292] The acetonation of D-glucose phenylosotriazole has been investigated.[144]

[290] H. El Khadem, D. Horton, M. H. Meshreki, and M. A. Nashed, *Carbohydrate Res.*, 1970, **13**, 317.
[291] L. Szilágyi and R. Bognár, *Carbohydrate Res.*, 1970, **15**, 371.
[292] P. K. Kindel, *Carbohydrate Res.*, 1970, **12**, 466.

10
Miscellaneous Nitrogen-containing Compounds

Glycosylamines and Related Compounds

A simple synthesis of 2,3-*O*-isopropylidene-β-D-ribofuranosylamine (201) has been achieved by reaction of D-ribopyranosylamine with acetone, 2,2-dimethoxypropane, and *p*-tolylsulphonic acid as a catalyst;[293] (201) is a useful intermediate in nucleoside synthesis. *N*-Acetylglycosylamines, containing both furanose and pyranose rings, have been isolated from the ammonolysis of β-D-mannopyranose penta-acetate and α-L-rhamnopyranose tetra-acetate.[294] The synthesis of 2-acetamido-2-deoxy-β-D-glucosylamine has been described.[252]

Derivatives of *N*-(L-aspart-4-oyl)-4-*O*-(β-D-galactopyranosyl)-β-D-glucopyranosylamine, its 2-acetamido-2-deoxy-analogue, and the 6-*O*-linked

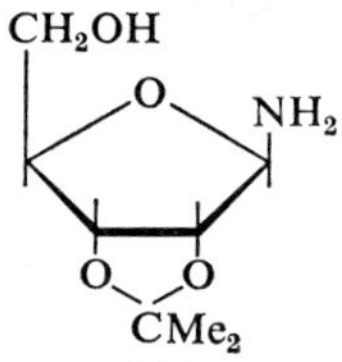

(201)

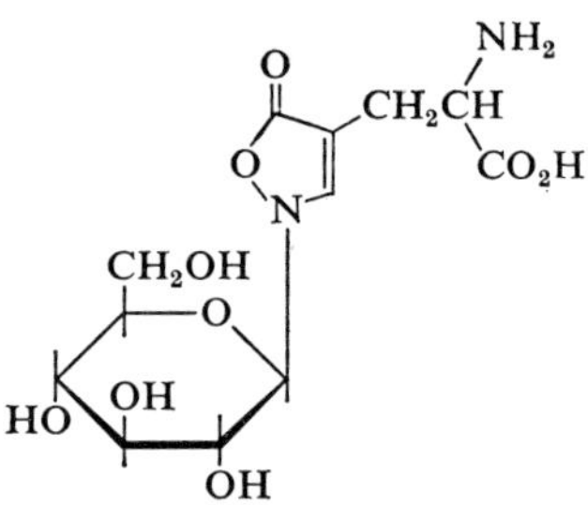

(202)

isomer of the latter compound have been prepared.[295] *N*-Glycosides of *p*-aminopropiophenone have been reported.[296]

Many papers have appeared on *N*-glycosides of heterocyclic ring systems. Wagner and his group continue to be very active in this field.[297–301] Papers

[293] N. J. Cusack and G. Shaw, *Chem. Comm.*, 1970, 1114.
[294] A. B. Zanlungo, J. O. Deferrari, and R. A. Cadenas, *Carbohydrate Res.*, 1970, **14**, 245.
[295] M. Spinola and R. W. Jeanloz, *Carbohydrate Res.*, 1970, **15**, 361.
[296] J. Sykulski, *Roczniki Chem.*, 1970, **44**, 1967.
[297] P. Nuhn and G. Wagner, *J. prakt. Chem.*, 1970, **312**, 97.
[298] D. Heller and G. Wagner, *Z. Chem.*, 1970, **10**, 111, 114.
[299] P. Nuhn and G. Wagner, *J. prakt. Chem.*, 1970, **312**, 90.
[300] G. Wagner and D. Goebel, *Z. Chem.*, 1970, **10**, 265.
[301] G. Wagner and H. Florestedt, *Pharmazie*, 1970, **25**, 144.

have appeared on derivatives of benzotriazole,[302] benzimidazolone,[303] benzimidazolethione,[304] and benzimidazole and its 5(6)-nitro- and 2-methyl-derivatives.[305] Compound (202) has been isolated from pea seedlings.[306] The pK_b values of *N*-glucosides derived from pyrazoles and imidazoles have been determined.[307]

3,5-Dihydroxy-2-methyl-5,6-dihydropyran-4-one was formed in good yield by treatment of D-glucose with piperidine salts. It was suggested that the reaction involved formation of the D-glucosylamine, its Amadori rearrangement, followed by a 1,4-elimination and cyclisation.[308] Comparisons have been made of the reaction of D-glucose and analogues having H, $CONH_2$, and CO_2K at C-5 with *p*-toluidine, and the Amadori rearrangement of the products; electron-attracting groups at C-5 facilitated the latter process.[309]

The kinetics of the formation of *N*-glycosylamines have been studied, but no firm conclusions could be drawn.[310, 311] The i.r. spectra of the glycosylamines of D-xylose, L-arabinose, and their triacetates have been investigated. It was concluded that the compounds contained a pyranose ring and that α- and β-anomers could be distinguished.[312] The o.r.d. of *N*-*p*-anisyl-D-glucopyranosylamine and of some of its derivatives have been recorded.[313] The kinetics of the mutarotation of several *N*-glycosylamines catalysed by variously substituted benzoic acids have been used to determine the pK_a's of the latter.[314] Autoxidative decomposition of *N*-aryl-D-glucosylamines has been shown to give isatin derivatives.[314a]

Azides

Syntheses of methyl 3-azido-3-deoxy-α- and -β-D-xylopyranosides have been described.[174] 2,3-Di-*O*-benzoyl-4,6-di-*O*-methanesulphonyl-α-D-glucopyranosyl 2,3-di-*O*-benzoyl-4,6-di-*O*-methanesulphonyl-α-D-glucopyranoside (an α,α-trehalose derivative) has been converted into the corresponding galactosyl galactoside tetra-azide with sodium azide in HMPT.[315] Treat-

[302] G. García-Muñoz, J. Iglesias, R. Madroñero, and M. C. Saldaña, *Anales de Quim.*, 1970, **66**, 383.
[303] H. Zinner and K. Peseke, *J. prakt. Chem.*, 1970, **312**, 307.
[304] H. Zinner and K. Peseke, *J. prakt. Chem.*, 1970, **312**, 185.
[305] J. Jasinska and J. Sokolowski, *Zesz. Nauk. Wyzsz. Szk. Pedagog. Gdansku: Met., Fiz., Chem.*, 1969, **9**, 115 (*Chem. Abs.*, 1970, **72**, 90820*f*).
[306] F. Lambein and R. Van Parijs, *Biochem. Biophys. Res. Comm.*, 1970, **40**, 557.
[307] J. Jasinska and J. Sokolowski, *Roczniki Chem.*, 1970, **44**, 1913.
[308] G. A. M. Van Den Ouweland and H. G. Peer, *Rec. Trav. chim.*, 1970, **89**, 750.
[309] K. Heyns, T. Chiemprasert, and W. Baltes, *Chem. Ber.*, 1970, **103**, 2877.
[310] S. Kolka and J. Sokolowski, *Zesz. Nauk. Wyzsz. Szk. Pedagog. Gdansku: Met., Fiz., Chem.*, 1969, **9**, 143 (*Chem. Abs.*, 1970, **72**, 55799*j*).
[311] S. Kolka and J. Sokolowski, *Roczniki Chem.*, 1970, **44**, 85.
[312] Z. Smiatacz and J. Sokolowski, *Roczniki Chem.*, 1970, **44**, 757.
[313] J. Szafranek, Z. Fialkiewicz, and J. Sokolowski, *Roczniki Chem.*, 1970, **44**, 1183.
[314] N. Galicka, K. Smiataczowa, and T. Jasinski, *Roczniki Chem.*, 1970, **44**, 411.
[314a] T. Ozawa and N. Kinae, *Yakugaku Zasshi*, 1970, **90**, 665.
[315] Y. Ali, L. Hough, and A. C. Richardson, *Carbohydrate Res.*, 1970, **14**, 181.

ment of (116) with sodium azide in DMF gave (117), rather than the expected 5-azide.[216]

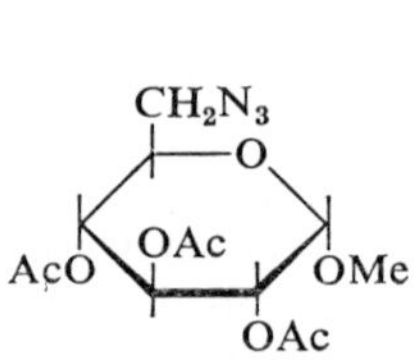

(203)

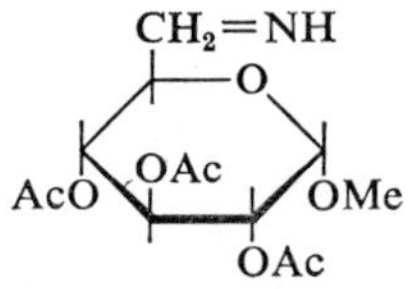

(204)

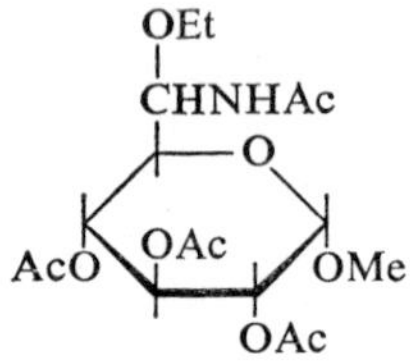

(205)

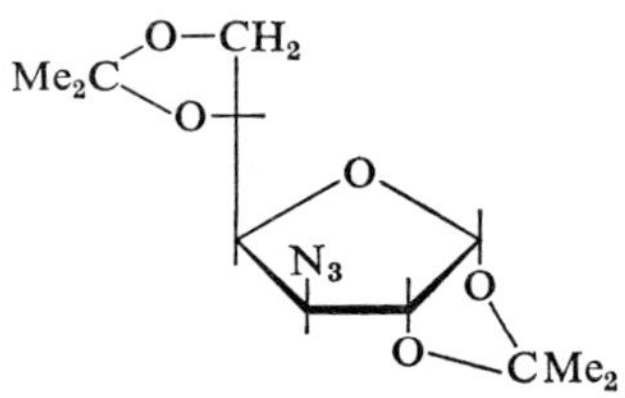

(206)

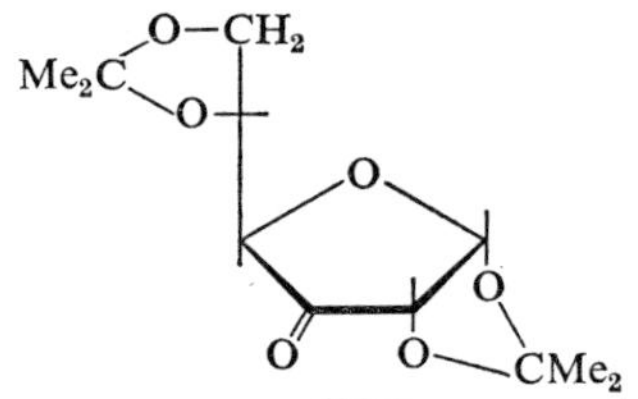

(207)

Photolysis of (203) in either cyclohexene or ethanol afforded principally (204) and, in the latter solvent, a product which on acetylation gave (205).[316] Photolysis of secondary azides gave the corresponding ketones but in low yield, *e.g.* (206) gave (207).[317]

Nitro-compounds

Addition of nitromethane to L-idose gave the expected 1-deoxy-1-nitroheptitols with the L-*glycero*-D-*gulo* isomer preponderant.[318] The stereochemistry of this reaction was discussed generally and it was noted that the major products have C-2 and C-4 in the *threo* configuration. This was accounted for in terms of the preferred conformations of the reactive acyclic sugar modifications, which were assumed to adopt chair-like shapes. A rationalisation could also be developed if a zig-zag conformation was assumed, with the substituents at C-3 reducing the access of the nucleophile to one side of the carbonyl group.[318]

4-*O*-Acetyl-1,2-dideoxy-3,5-*O*-ethylidene-1-nitro-D-*erythro*-pent-1-enitol (208) has been prepared as outlined in Scheme 39.[319] The nitro-olefin was not obtainable from the nitro-acetate by conventional methods.

Treatment of the nitro-olefin (209) with sodium hypochlorite in THF gave the epoxide (210), whereas when the solvent was changed to acetone, (211) was formed. Other branched compounds (212), (213), and (214) were prepared from (209).[320] Addition of anthranilic acid to (209) gave mainly

[316] R. L. Whistler and A. K. M. Anisuzzaman, *J. Org. Chem.*, 1969, **34**, 3823.
[317] D. M. Clode and D. Horton, *Carbohydrate Res.*, 1970, **14**, 405.
[318] J. Kovář and H. H. Baer, *Canad. J. Chem.*, 1970, **48**, 2377.
[319] K. D. Carlson, C. R. Smith jun., and I. A. Wolff, *Carbohydrate Res.*, 1970, **13**, 391.
[320] T. Sakakibara, S. Kumazawa, and T. Nakagawa, *Bull. Chem. Soc. Japan*, 1970, **43**, 2655.

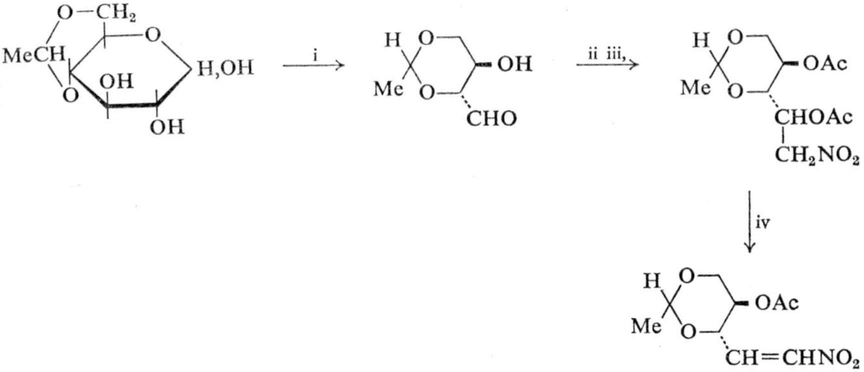

(208)

Reagents: i, $NaIO_4$; ii, $MeNO_2$, MeONa, MeOH; iii, Ac_2O, py; iv, SiO_2

Scheme 39

the D-*manno* adduct (215) instead of the expected D-*gluco* compound. However, in the presence of catalytic amounts of potassium hydroxide the addition followed the normal stereochemical pathway.[321] The addition of *m*- and *p*-aminobenzoic acids to (209) gave only D-*gluco* adducts, and anthranilic acid added similarly to the α-anomer of (209).[322]

3-Deoxy-3-nitro-D-glucose can be converted by standard reactions into its acetobromo-derivative, which can then be used in Koenigs–Knorr syntheses.[323]

Nitryl iodide, prepared *in situ* in ether solution from silver nitrate and iodine, has been added to various unsaturated sugars. For example, (216) gave (217), which was further converted into the nitro-olefin (218). Borohydride reduction of (218) gave (219).[324]

Nitro-compounds have been used as precursors in the synthesis of polyamino sugars.[279, 280]

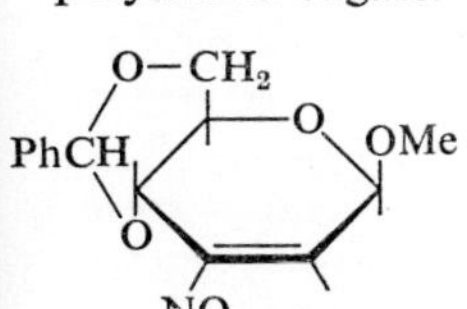

(209)

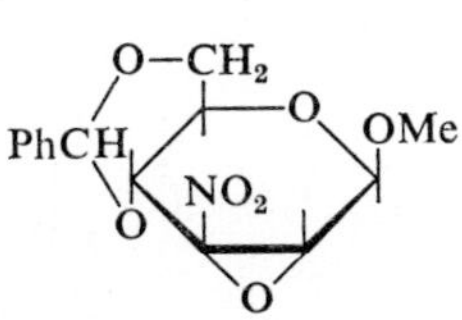

(210)

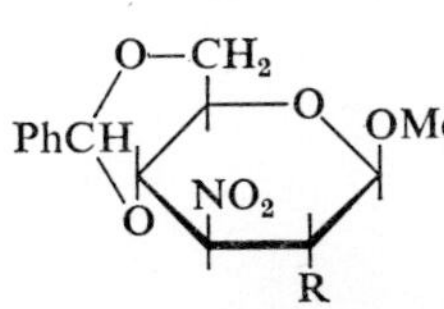

(211) R = $\cdot CH_2COMe$
(212) R = $\cdot CH(CN)_2$
(213) R = $\cdot CH(CO_2Et)_2$

(214) R =

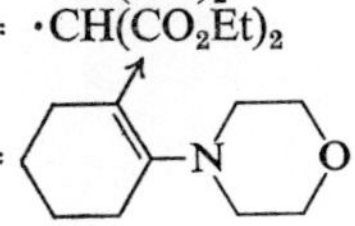

[321] H. H. Baer and F. Kienzle, *J. Org. Chem.*, 1969, **34**, 3848.
[322] H. H. Baer, E. Rajabalee, and F. Kienzle, *J. Org. Chem.*, 1969, **34**, 4204.
[323] H. H. Baer, W. Rank, and F. Kienzle, *Canad. J. Chem.*, 1970, **48**, 1302.
[324] W. A. Szarek, D. G. Lance, and R. L. Beach, *Carbohydrate Res.*, 1970, **13**, 75.

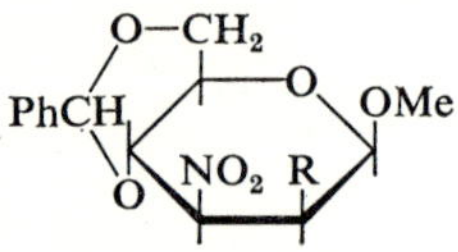

(215) R = $\cdot C_6H_4CO_2H$

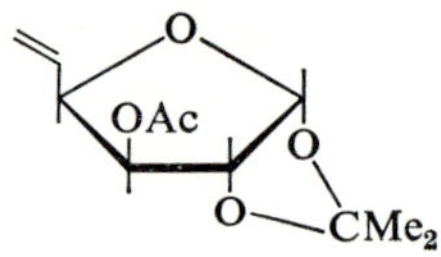

(216)

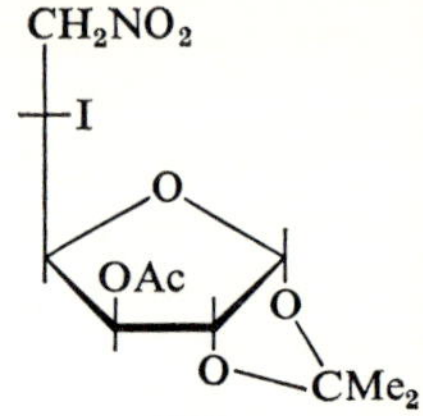

(217)

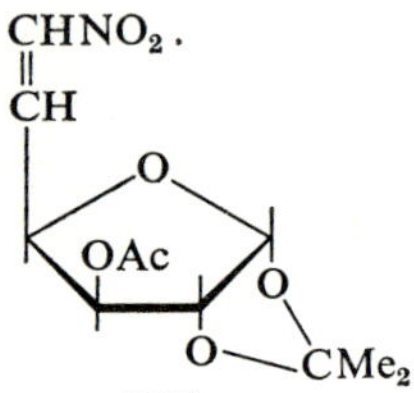

(218)

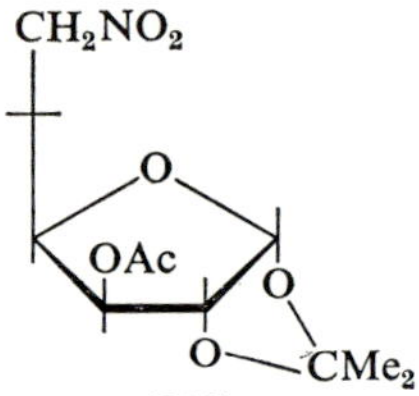

(219)

Epimines

5,6-Dideoxy-5,6-epimino-L-iditol derivatives have been prepared from the corresponding derivatives of 5,6-di-*O*-*p*-tolylsulphonyl-D-glucitol by sequential displacement of the 6-tosyloxy-group with azide ion, and $LiAlH_4$ reduction.[325] Reaction of the epimine with chloride ion gave 5-amino-6-chloro-derivatives. 3,4-Dideoxy-3,4-epimino-1,2:5,6-di-*O*-isopropylidene-D-iditol has been obtained by a similar approach,[326] and a preparation of benzyl 5,6-acetylepimino-5,6-dideoxy-2,3-*O*-isopropylidene-β-L-gulofuranoside has also been described.[327] Several reactions of the epimine (220) have been reported, namely hydrogenolysis (to the α-amino-acid derivative), acetolysis, and hydrolysis.[328] Several *N*-substituted methyl 2,3-dideoxy-2,3-epimino-β-D-lyxofuranosides have been prepared, *via* azido-tosylate

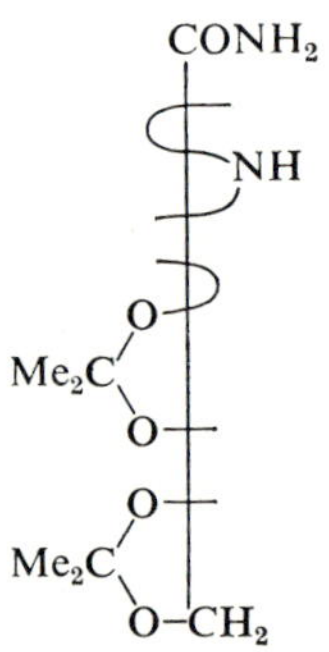

(220)

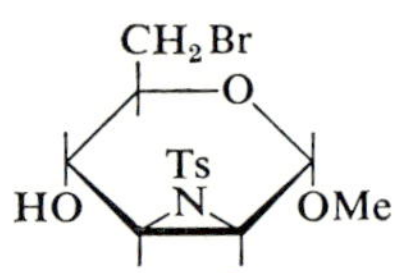

(221)

[325] A. D. Barford and A. C. Richardson, *Carbohydrate Res.*, 1970, **14**, 217.
[326] A. D. Barford and A. C. Richardson, *Carbohydrate Res.*, 1970, **14**, 231

precursors.[329] Further examples of *N*-amino-epimines [330] and a synthesis of the *N*-tosylepimine (221) [331] have been described.

Reaction of methyl 2,3-*N*-aroylepimino-4,6-*O*-benzylidene-2,3-dideoxy-α-D-mannoside in DMF with or without sodium iodide gave the oxazoline (222), although the yield was higher in the former case. With potassium thiocyanate in dioxan, the same epimines did not give oxazolines but 2-arylamido-2,3-dideoxy-3-thiocyanates, which were desulphurised to give the 2-arylamido-2,3-dideoxy-compounds.[332] Compound (224) has been isolated as a by-product in the acetolysis of (223).[276]

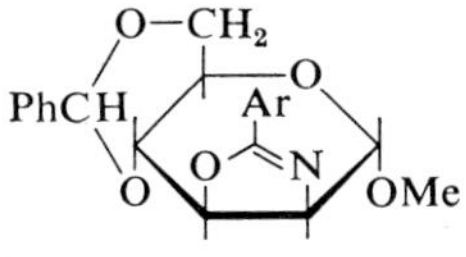

(222)

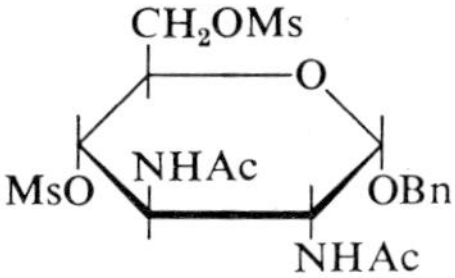

(223)

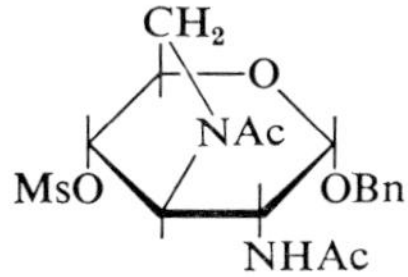

(224)

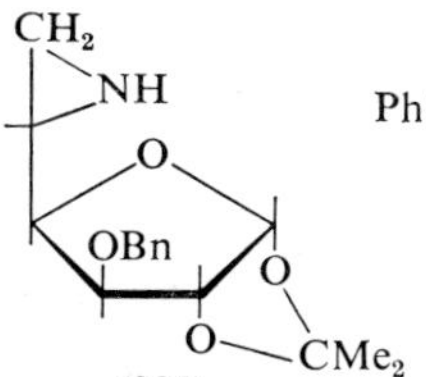

(225)

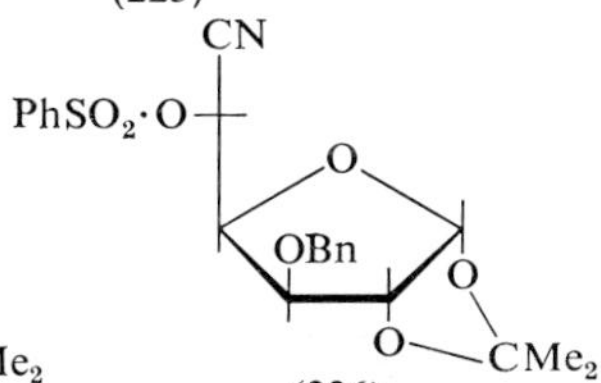

(226)

The epimine (225) has been prepared by $LiAlH_4$ reduction of (226) obtained from the pentodialdose derivative.[333]

Other Heterocyclic Derivatives

1,3-Dipolar cycloadditions of alkenes and alkynes to sugar nitrile oxides have led to the formation of 3-glycosyl-isoxazoline and -isoxazole derivatives, respectively, as shown in Scheme 40. Alternatively, aryl nitrile oxides could be added to unsaturated sugars (Scheme 41) to form 5-glycosyl-isoxazolines.[334] 1,3-Dipolar cycloadditions of alkenes and alkynes to sugar nitrilimines (Scheme 42) gave 3-glycosyl-pyrazolines and -pyrazoles, respectively.

Several papers on oxazolines have appeared. Methyl β-D-lyxofuranosides with an oxazoline ring bridging C-2 and C-3 have been prepared either by

[327] J. S. Brimacombe, F. Hunedy, and M. Stacey, *Carbohydrate Res.*, 1970, **13**, 447.
[328] B. A. Dmitriev, N. E. Bairamova, and N. K. Kochetkov, *Bull. Acad. Sci. U.S.S.R.*, 1970, 597; *Izvest. Akad. Nauk S.S.S.R.*, 1970, 650.
[329] J. Cleophax, S. D. Géro, J. Hildesheim, A. M. Sepulchre, R. D. Guthrie, and C. W. Smith, *J. Chem. Soc.* (*C*), 1970, 1385.
[330] H. Paulsen and M. Budzis, *Chem. Ber.*, 1970, **103**, 3794.
[331] T. L. Hullar and S. B. Siskin, *J. Org. Chem.*, 1970, **35**, 225.
[332] Z. M. El Shafei and R. D. Guthrie, *J. Chem. Soc.* (*C*), 1970, 843.
[333] K. Ichimura, *Bull. Chem. Soc. Japan*, 1970, **43**, 2501.
[334] J. M. J. Tronchet, A. Jotterand, N. LeHong, F. Perret, S. Thorndahl-Jaccard, J. Tronchet, J. M. Chalet, L. Faivre, C. Hausser, and S. Sebastian, *Helv. Chim. Acta*, 1970, **53**, 1484.

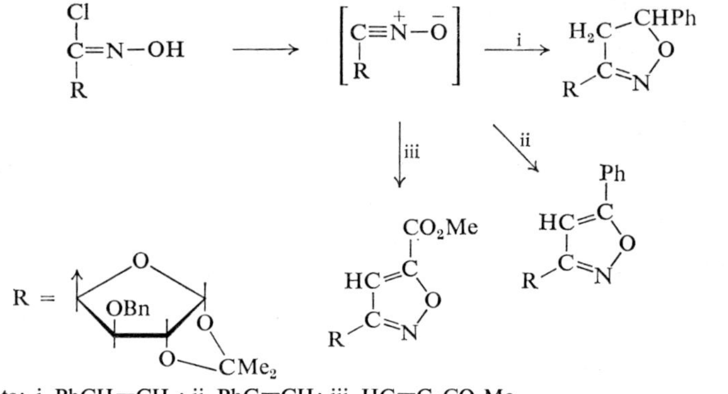

Reagents: i, $PhCH{=}CH_2$; ii, $PhC{\equiv}CH$; iii, $HC{\equiv}C{\cdot}CO_2Me$

Scheme 40

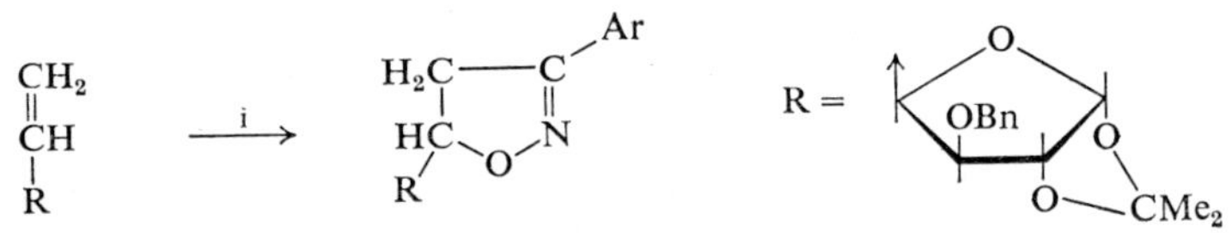

Reagent: i, $Ar{-}C{\equiv}N^{+}{-}O^{-}$

Scheme 41

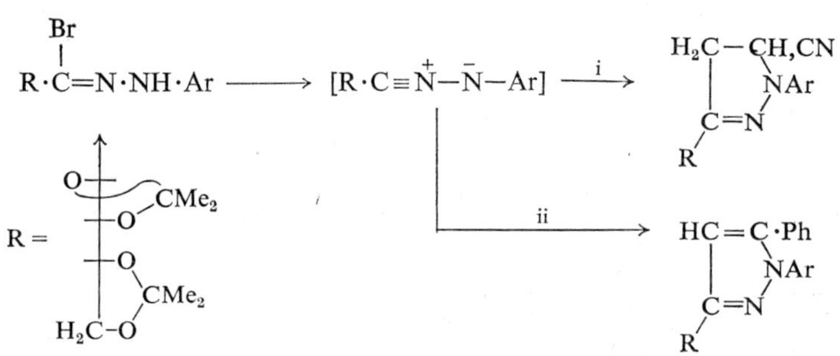

$Ar = {\cdot}C_6H_4{\cdot}NO_2$

Reagents: i, $CH_2{=}CHCN$; ii, $PhC{\equiv}CH$

Scheme 42

rearrangement of an *N*-benzoylepimine using sodium iodide in acetonitrile, or from a *trans*-3-benzamido-2-sulphonate precursor. The former method gave preponderantly the compound with the C-2–N bond.[335] The epimine (225) has also been converted into the oxazoline (227) by formation and

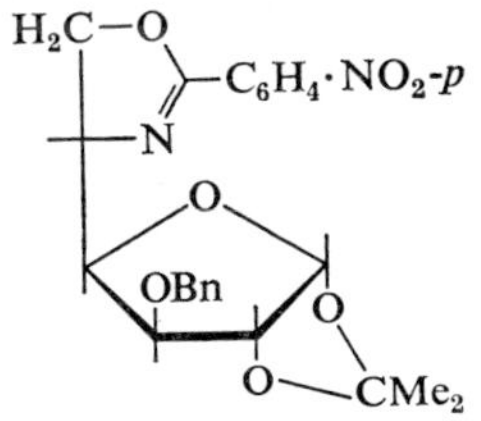

(227)

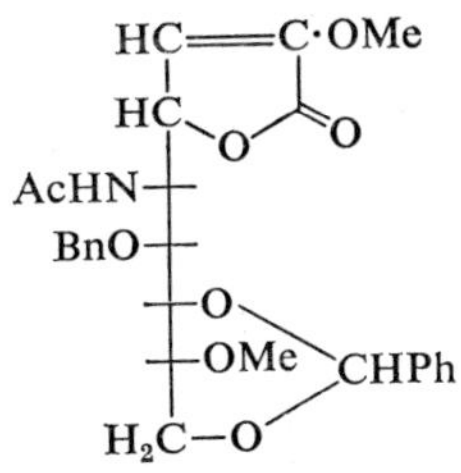

(228)

rearrangement of the *N*-aroyl derivative.[333] The oxazoline (228) has been prepared as a precursor for a synthesis of methylated derivatives of *N*-acetylneuraminic acid.[336]

Other Nitrogen-containing Compounds

The phenylazo-ene (229) and its isomer (230) have been prepared by elimination of benzoic acid from the appropriate 2[or 3]-*O*-benzoyl-3[or 2]-ulose phenylhydrazone. Both (229) and (230) underwent addition reactions with such reagents as acetonitrile. Moreover, (229) underwent 1,4-addition

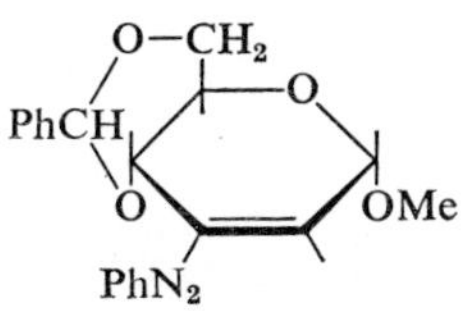

(229)

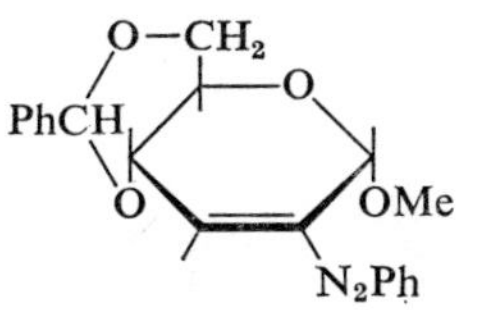

(230)

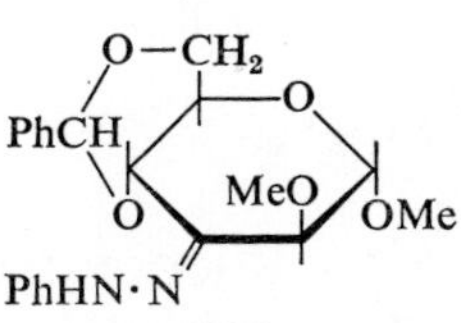

(231)

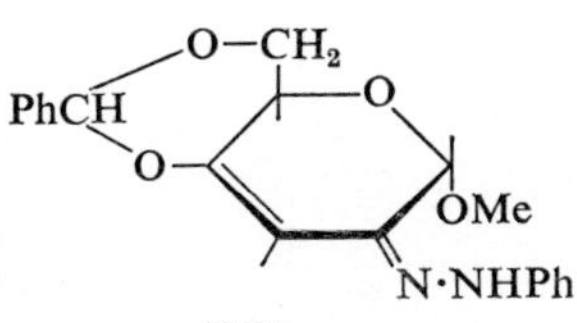

(232)

reactions with a variety of nucleophiles; for example, sodium methoxide gave (231) whereas (230) yielded the rearranged product (232).[337]

335 S. D. Géro, J. Hildesheim, E. Walczak, R. D. Guthrie, and C. W. Smith, *J. Chem. Soc.* (*C*), 1970, 1402.
336 A. Ya. Khorlin and I. M. Privalova, *Carbohydrate Res.*, 1970, **13**, 373.
337 P. M. Collins, D. Gardiner, S. Kumar, and W. G. Overend, *Chem. Comm.*, 1970, 1433.

D-Gluconamide has been shown to react with benzoyl chloride in pyridine to give mixtures of products depending on the conditions used (Scheme 43).[338]

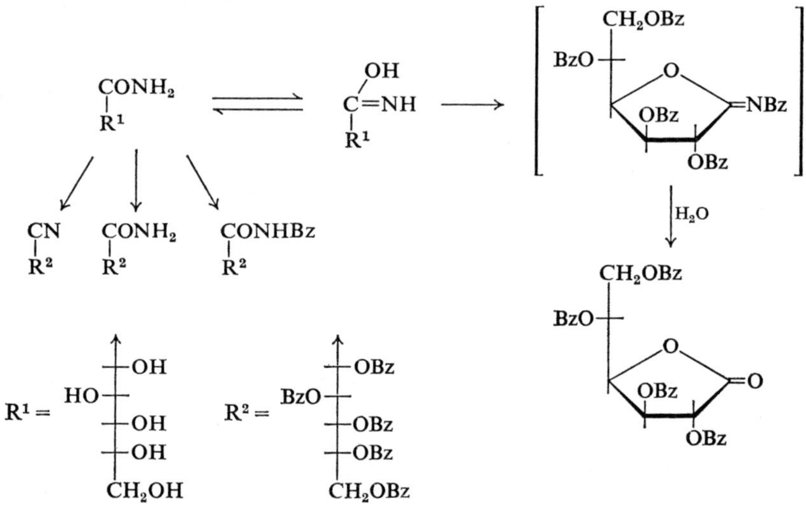

Scheme 43

5′-Phosphoribosyl-*N*-formylglycinamide, an important intermediate in the biosynthesis of purine ribonucleotides, has been synthesised by the improved procedure shown in Scheme 44.[339]

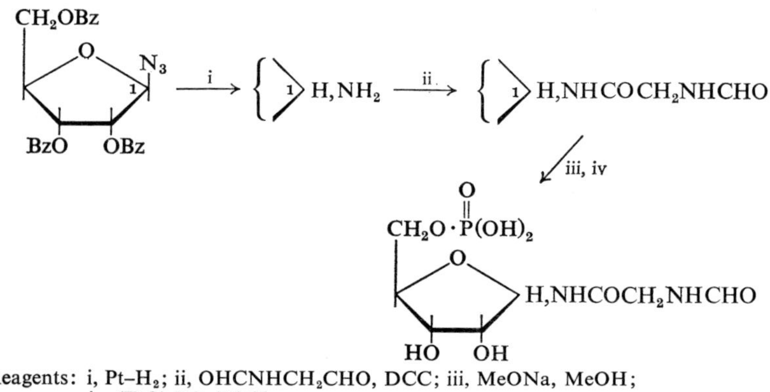

Reagents: i, Pt–H_2; ii, $OHCNHCH_2CHO$, DCC; iii, MeONa, MeOH; iv, $(EtO)_3PO$, $POCl_3$

Scheme 44

[338] J. O. Deferrari, R. M. de Lederkremer, B. Matsuhiro, and M. I. Litter, *Carbohydrate Res.*, 1970, **14**, 103.

[339] S. Y. Chu and J. F. Henderson, *Canad. J. Chem.*, 1970, **48**, 2306.

The reactions of hydrazines with 1,2-*O*-isopropylidene-α-D-*xylo*-hexofuranurono-3,6-lactone-5-ulose have been described.[340] Photochemical irradiation of D-galactose azine did not afford the expected D-galactononitrile, but instead gave D-galactose and D-lyxose.[21a] Wolff rearrangement of the diazo-D-*gluco*-heptulose derivative (233) gave (234) but not (235).[341]

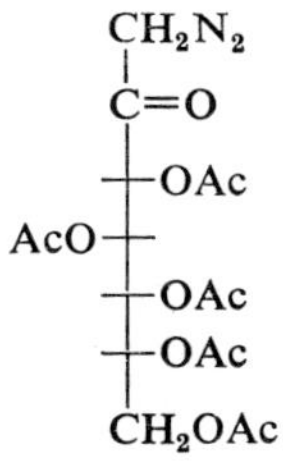

(233)

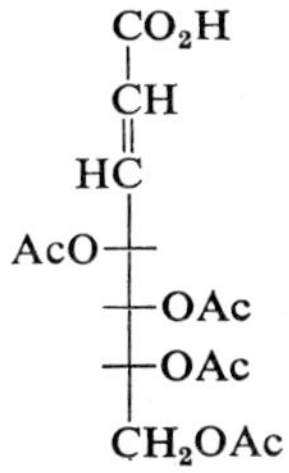

(234)

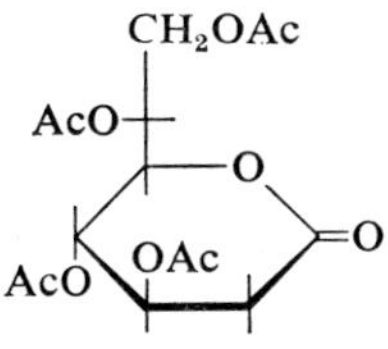

(235)

Improved methods for the synthesis of oximes of D-glucose and D-fructose have been investigated.[342] Direct irradiation of D-galactose oxime has given D-lyxose as the major product; the reaction is effectively a 'photochemical Wohl degradation'. The imino-lactone (236) was isolated in good yield from the reaction mixture, and the sequence shown in Scheme 45 was proposed to account for the products.[22]

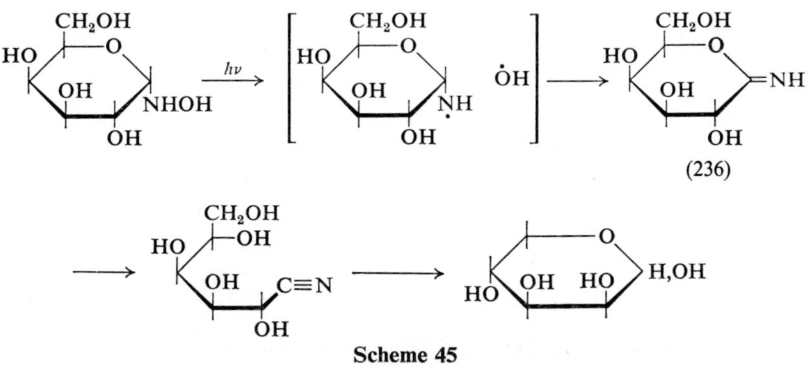

Scheme 45

Acyl nitrites added to tri-*O*-acetyl-D-glucal in a similar fashion to nitrosyl chloride to give dimeric products having the D-*gluco* configuration. Partial hydrogenation of the adducts gave the corresponding azoxy-derivatives.[343]

[340] H. Paulsen and H. Kuhne, *Carbohydrate Res.*, 1970, **13**, 284.
[341] I. Dijong and W. van der Heydt, *Annalen*, 1970, **735**, 138.
[342] M. Tutsumi, M. Iio, and H. Omura, *Eiyo To Skokuryo*, 1969, **22**, 462.
[343] J. H. Jordaan, *Carbohydrate Res.*, 1970, **12**, 69.

11
Thio-sugars

Diphenyl dithioacetals of D-ribose, D-xylose, and D- and L-arabinose have been described and their conformations deduced by n.m.r. studies. Acetonation of the D-arabinose derivative gave the 2,3:4,5-diacetal, which on treatment with the sodium methylsulphinyl carbanion gave the unsaturated sugar (237).[344]

The product resulting from treatment of 3,4,5,6-tetra-*O*-benzoyl-D-glucose diethyl dithioacetal (238) with ethanethiol and hydrochloric acid

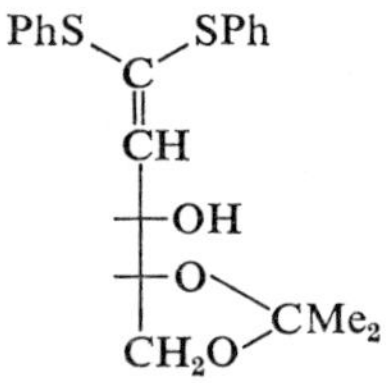

(237)

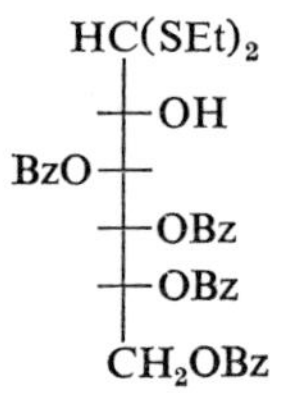

(238)

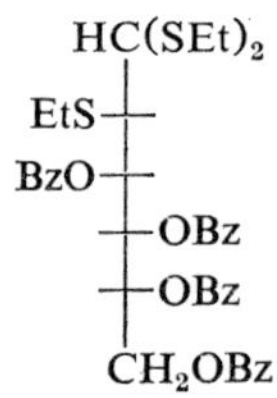

(239)

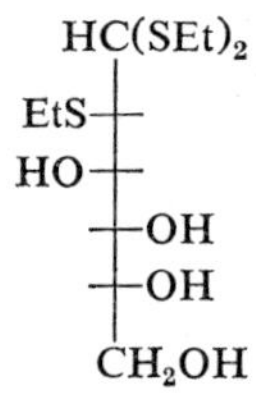

(240)

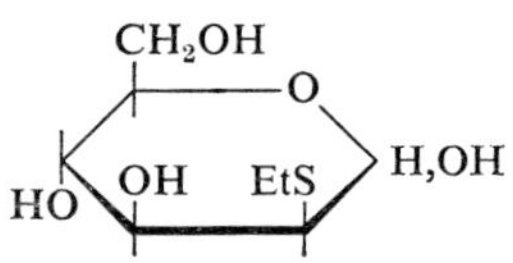

(241)

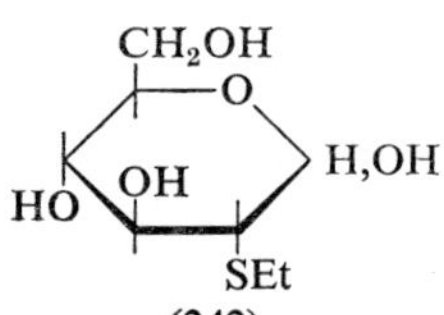

(242)

has the D-*manno* configuration (239). Extended treatment of the de-esterified product (240) with hot, aqueous mercuric chloride in the presence of barium carbonate gave not only (241) but also some of the epimeric 2-*S*-ethyl-2-thio-D-glucose (242). The formation of (239) from (238) was presumed to involve a 1,2-episulphonium ion intermediate.[345]

A detailed report on the reaction of glycosyl halides with potassium methyl and benzyl xanthates has been published. In alcohol at room temperature, the major products were the corresponding glycosyl methyl

[344] D. Horton and J. D. Wander, *Carbohydrate Res.*, 1970, **13**, 33.
[345] B. Berrang and D. Horton, *Chem. Comm.*, 1970, 1038.

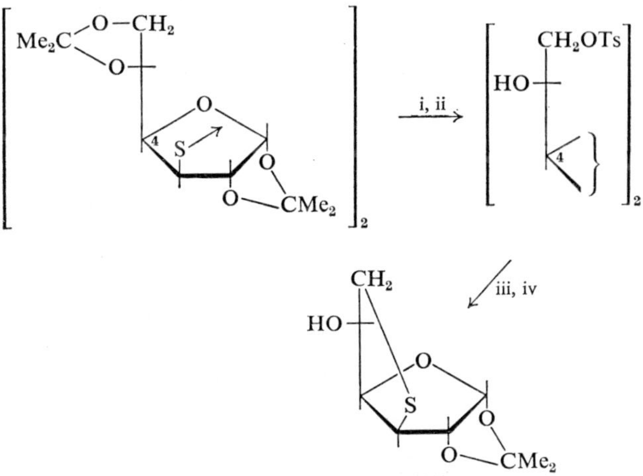

(243)

Reagents: i, aq Pr^{i}OH, H_2SO_4; ii, TsCl, py; iii, $LiAlH_4$; iv, NaOH

Scheme 46

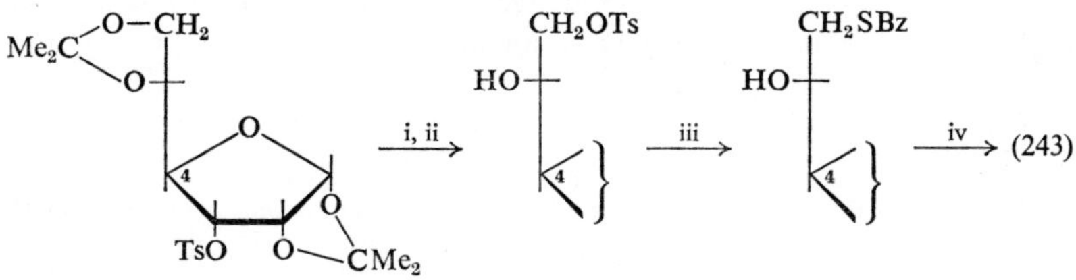

Reagents: i, aq AcOH; ii, TsCl, py; iii, KSBz, EtOH; iv, NaOMe, MeOH

Scheme 47

and benzyl xanthates, whereas in acetone the products were 1-thioglycosides and diglycosylsulphides.[346]

3,6-Dideoxy-3,6-epithio-1,2-*O*-isopropylidene-α-D-glucofuranose (243) has been synthesised by the routes shown in Schemes 46 and 47.[347] Unlike the 3,6-anhydro-derivative, the pyranose 3,6-epithio-derivative (244),

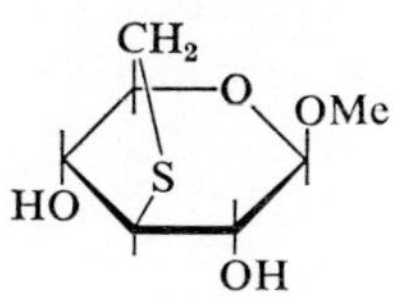

(244)

prepared by a similar route, did not rearrange to the isomeric furanoside when treated with methanolic hydrogen chloride. The difference in behaviour was attributed to the greater length of the C—S bonds.

[346] M. Sakata, M. Haga, and S. Tejima, *Carbohydrate Res.*, 1970, **13**, 379.
[347] J. M. Heap and L. N. Owen, *J. Chem. Soc.* (*C*), 1970, 707.

3-*O*-(*S*-Iododithiocarbonyl)-1,2:5,6-di-*O*-isopropylidene-α-D-glucofuranose has been identified as a by-product in the oxidation of 1,2:5,6-di-*O*-isopropylidene-3-*O*-(*S*-sodium dithiocarbonyl)-α-D-glucofuranose with iodine.[348]

Treatment of 1,6-anhydro-2-*O*-*p*-tolylsulphonyl-β-D-glucopyranose with sodium benzyl sulphide gave (245) and (246) in yields of 22% and 52%, respectively, by way of the 2,3-epoxide; contrary to expectation, the

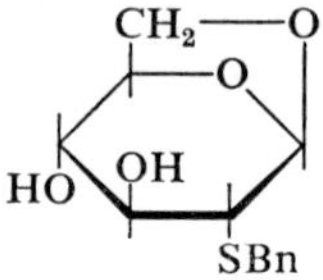

(245)

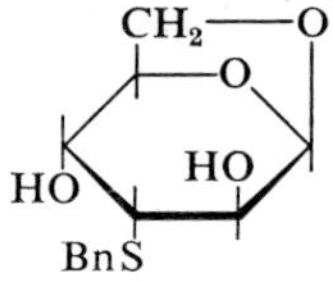

(246)

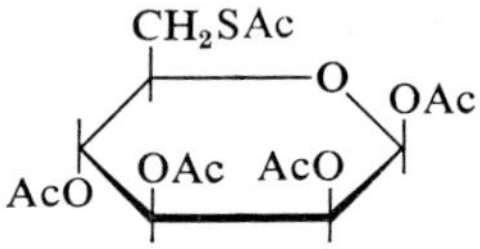

(247)

diequatorial D-*altro*-isomer was the main product of ring-opening of the epoxide. Several derivatives (including disulphides) of these products were described.[349] The synthesis of the 6-thio-D-mannose derivative (247) from the 6-methanesulphonate has been reported.[350]

Thio-sugars derived from unsaturated sugars are described in Chapter 14, and several 3-thiobenzoyl derivatives of 1,2:5,6-di-*O*-isopropylidene-α-D-glycofuranoses have already been noted.[201]

[348] B. S. Shasha, W. M. Doane, C. R. Russell, and C. E. Rist, *Carbohydrate Res.*, 1970, **13**, 457.
[349] E. Hardegger and W. Schüep, *Helv. Chim. Acta*, 1970, **53**, 951.
[350] J. Fernández-Bolaños and R. Guzman Fernández-Bolaños, *Anales de Quim.*, 1969, **65**, 1193.

12

Derivatives with Sulphur or Nitrogen in the Sugar Ring

A review (in Japanese) of sugars containing sulphur, nitrogen, and phosphorus atoms in the ring has been published.[351]

Nitrogen Derivatives

Most reports of the chemistry of sugars containing nitrogen in the ring continue to come from Paulsen and his co-workers. The seven-membered ring compound (248) has been synthesised as shown in Scheme 48, and its

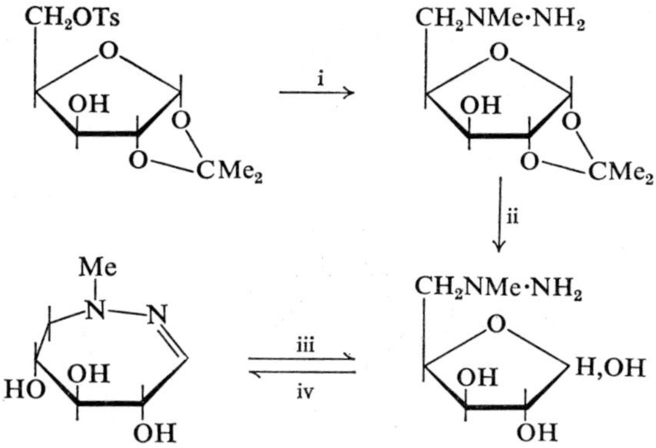

Reagents: i, $MeNHNH_2$; ii, HCl; iii, ^-OH; iv, H_3O^+

Scheme 48

reactions have been discussed.[352] A related intramolecular hydrazone (250) has been prepared by reaction of the tosylate (249) with hydrazine followed by treatment with alkali.[352a] Hydrogenation of (250) afforded a mixture of the diamine (251) and the cyclic derivative (252).

Hydrolysis of the dimethyl acetal (253) with acid gave (254) and (255), whereas the corresponding *N*-methyl derivative (256) gave (257).[353] On the

[351] S. Inokawa, *Kagaku* (*Kyoto*), 1969, **24**, 901.
[352] H. Paulsen and G. Steinert, *Chem. Ber.*, 1970, **103**, 475.
[352a] H. Paulsen and G. Steinert, *Chem. Ber.*, 1970, **103**, 1834.
[353] H. Paulsen, K. Steinert, and G. Steinert, *Chem. Ber.*, 1970, **103**, 1846.

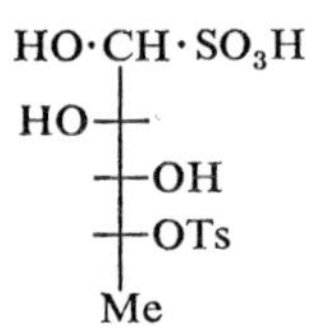

(249)

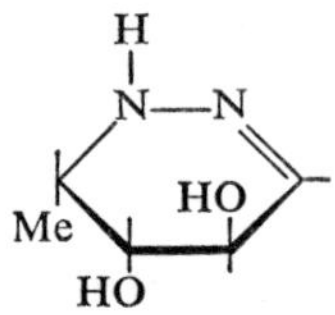

(250)

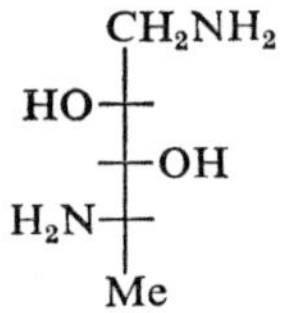

(251)

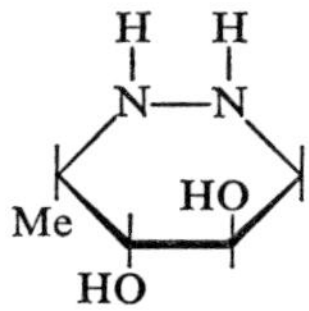

(252)

other hand, compound (258) was converted into (259) on hydrolysis with acid.

Displacement of the tosyloxy-group in (260) with azide ion did not provide an efficient conversion into 4-amino-derivatives since an elimination

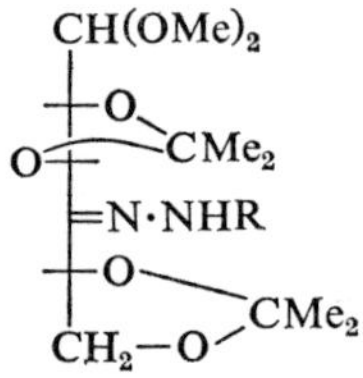

(253) R = H
(256) R = Me

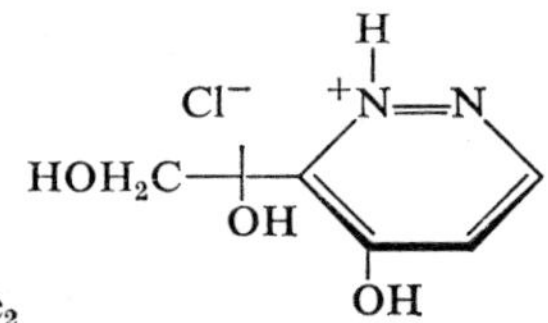

(254)

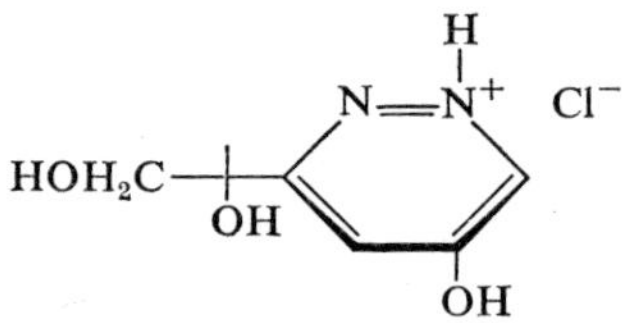

(255)

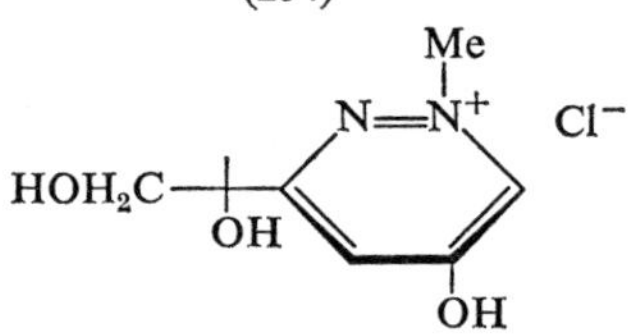

(257)

reaction preponderated over the displacement reaction.[354] However, reduction of the oxime derived from the corresponding 4-ulose yielded equal

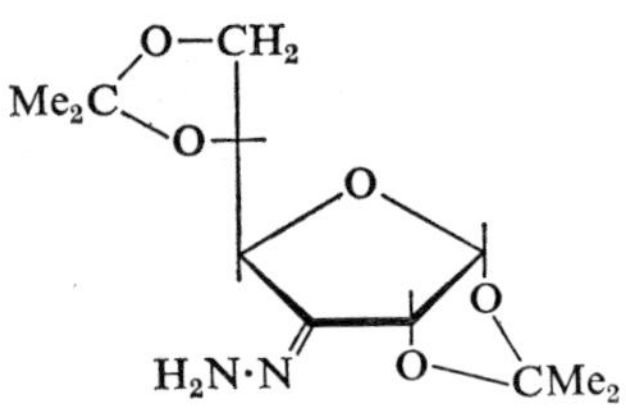

(258)

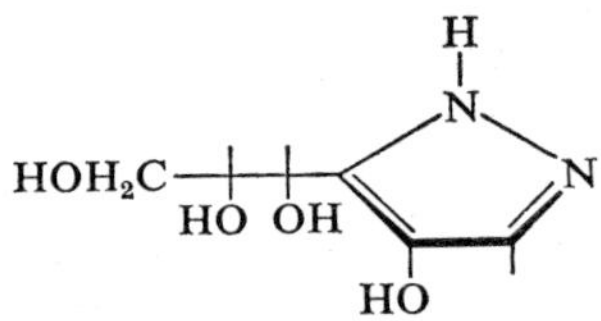

(259)

[54] H. Paulsen, K. Steinert, and K. Heyns, *Chem. Ber.*, 1970, **103**, 1599.

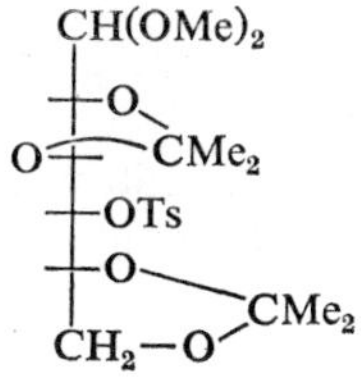

(260)

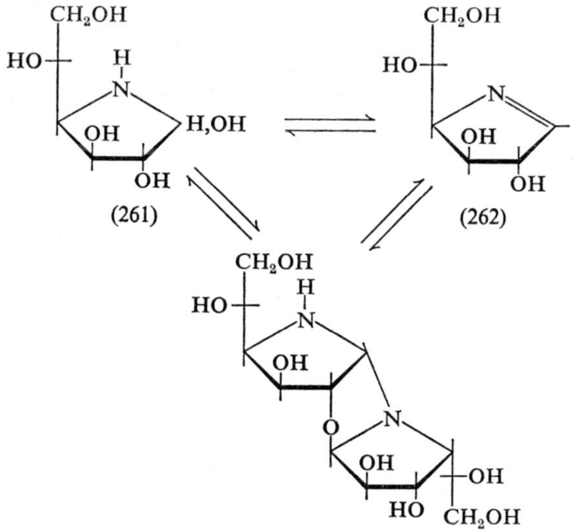

(261) (262) (263)

proportions of the 4-amino-4-deoxy-D-*gluco*- and -D-*galacto*-derivatives. In alkaline solution, 4-amino-4-deoxy-D-glucose existed as an equilibrium mixture of the species (261), (262), and (263), whereas in acid solution it existed as the hydrochloride of the pyranose form. The D-galactose epimer behaved analogously.

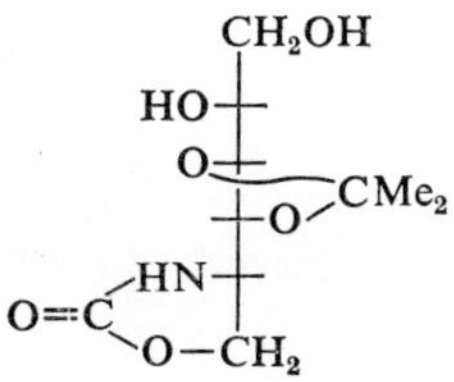

(264)

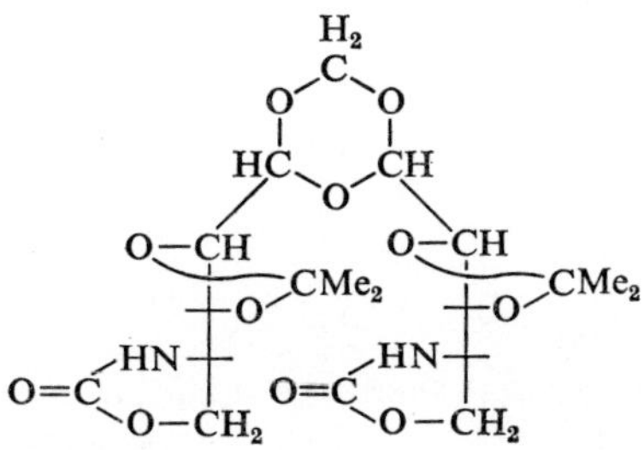

(265)

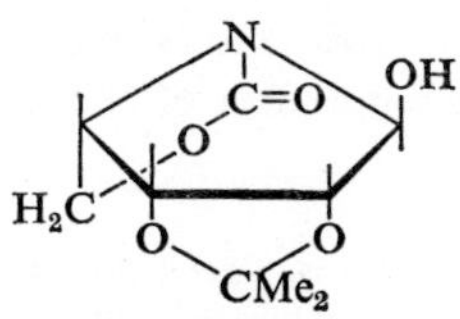

(266)

The dimer (265), formed by oxidation of (264) with periodate, underwent thermal rearrangement to (266), which was then converted into 4-amino-4-deoxy-L-lyxose.[355] The free amino-sugar existed in solution as the usual mixture of pyrrolidine, bis-pyrrolidine, and pyrroline forms.

The formation of a 2,5-anhydro-L-arabinitol derivative having a nitrogen atom in the ring (Scheme 49) has been reported,[356] and 2,6-epimino-derivatives have also been synthesised (Scheme 50).[331]

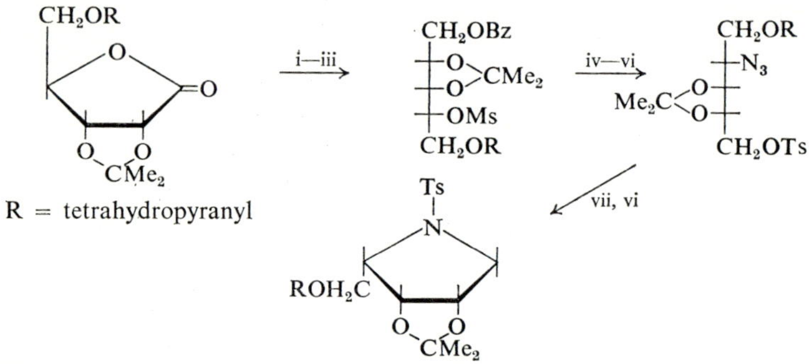

Reagents: i, $NaBH_4$, i-C_3H_7OH; ii, BzCl; iii, MsCl, py; iv, NaN_3, DMF; v, catalytic debenzoylation; vi, TsCl, py; vii, H_2–Pt

Scheme 49

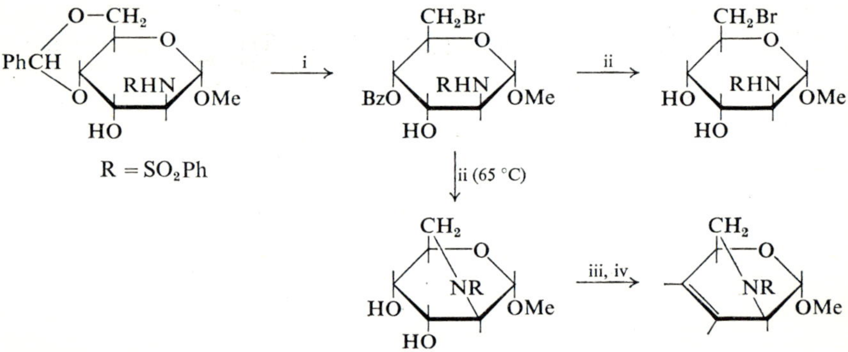

Reagents: i, NBS; ii, MeO^-; iii, TsCl, py; iv, NaI, Zn, DMF

Scheme 50

5-Acetamido-5,6-dideoxy-L-idose has been prepared and it has been found to exist in solution as a mixture of the furanose and six-membered ring forms.[262]

[355] H. Paulsen, J. Brüning, and K. Heyns, *Chem. Ber.*, 1970, **103**, 1621.

[356] A. Gateau, A.-M. Sepulchre, A. Gaudemer, and S. D. Géro, *Carbohydrate Res.*, 1970, **15**, 322.

Sulphur Derivatives

Whistler and his co-workers have been responsible for most of the reports of sugars having a sulphur atom in the ring. The methyl 2-deoxy-4-thio-D-*erythro*-pentofuranosides have been obtained by the route outlined in Scheme 51,[357] and the methyl 4-thio-D-arabinofuranosides by that shown in Scheme 52.[358] The latter compounds were synthesised as possible precursors of nucleosides. A synthesis of 6-amino-6-deoxy-5-thio-D-glucopyranose has been described (Scheme 53).[359] The addition of toluenethiol to the nitro-olefin (267) was stereospecific, yielding only the D-glucose derivative (268). Electronic and stereochemical factors were invoked to explain the observed specificity.

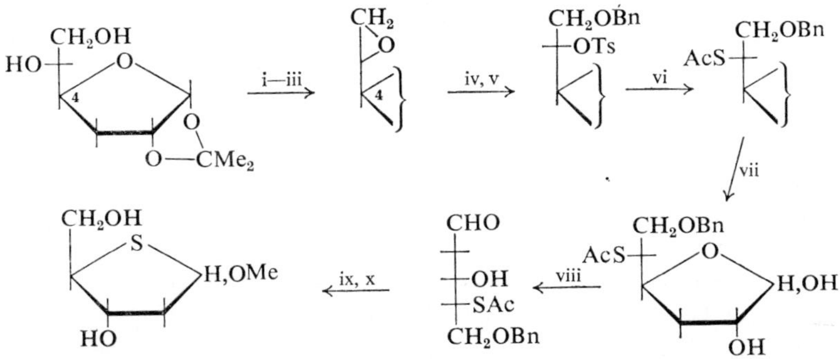

Reagents: 1, BzCl; ii, TsCl; iii, NaOMe; iv, NaOBn; v, TsCl; vi, AcS^-; vii, AcOH; viii, IO_4^-; ix, MeOH, HCl; x, Na, NH_3

Scheme 51

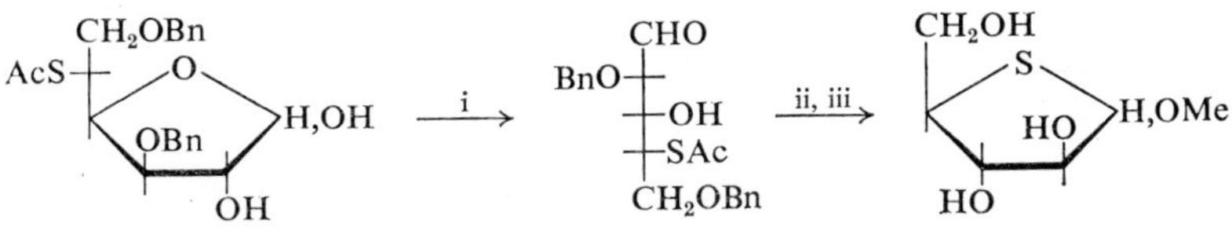

Reagents: i, IO_4^-; ii, MeOH, H^+; iii, Na, NH_3

Scheme 52

Two 5-thio-glucopyranose derivatives have been prepared.[360] 5-Thio-D-glucopyranosyl phosphate was obtained by treating acetylated 5-thio-glucopyranosyl bromide with silver diphenylphosphate, followed by hydrogenolysis of the phenyl groups and deacetylation. Methyl 5-thio-2,3,4,6-tetra-*O*-(trimethylsilyl)-α-D-glucopyranoside was converted into

[357] U. G. Nayak and R. L. Whistler, *Annalen*, 1970, **741**, 131.
[358] R. L. Whistler, U. G. Nayak, and A. W. Perkins, *J. Org. Chem.*, 1970, **35**, 519.
[359] R. L. Whistler and R. E. Pyler, *Carbohydrate Res.*, 1970, **12**, 201.
[360] R. L. Whistler and J. H. Stark, *Carbohydrate Res.*, 1970, **13**, 15.

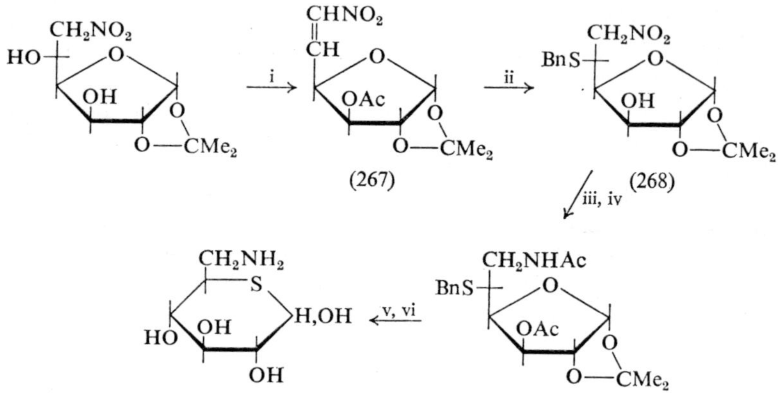

Reagents: i, Ac_2O, NaOAc; ii, BnSH, py; iii, $LiAlH_4$; iv, Ac_2O; v, Na, NH_3; vi, H_3O^+

Scheme 53

5-thio-D-glucopyranose 6-phosphate by preferential removal of the 6-*O*-(trimethylsilyl) group by methanolysis, phosphorylation with diphenyl phosphorochloridate, and subsequent removal of the protecting groups.

Phosphorus Derivatives

There have been no reports this year of sugar derivatives containing a phosphorus atom in the ring.

13
Deoxy-sugars

2-Deoxy-L-*erythro*-pentose (2-deoxy-L-ribose) has been synthesised from 2-deoxy-L-*lyxo*-hexose (Scheme 54), and a similar sequence of reactions applied to 2-deoxy-D-*arabino*-hexose gave 2-deoxy-D-*threo*-pentose.[361] In the hexose series, 3-deoxy-D-*xylo*-hexose (3-deoxy-D-galactose) was prepared from D-glucose (Scheme 55)[362] and on acetylation with acetic

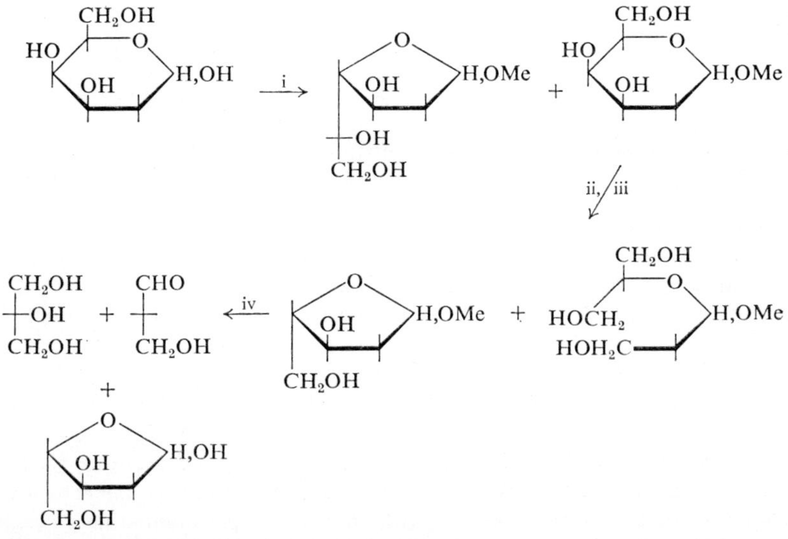

Reagents: i, MeOH, H^+; ii, $NaIO_4$; iii, $NaBH_4$; iv, H^+

Scheme 54

anhydride in the presence of several catalysts gave mixtures of the various cyclic tetra-acetates. Both pyranose tetra-acetates afforded methyl 2,4,6-tri-*O*-acetyl-3-deoxy-β-D-*xylo*-hexopyranoside when treated with hydrogen chloride followed by methanol; the furanoid glycoside perester was prepared similarly.[363] Zinner and Wulf have also reported the pyranosyl

[361] S. D. Schimmel and R. D. Bevill, *Analyt. Biochem.*, 1970, **37**, 385.
[362] H. Zinner and G. Wulf, *J. prakt. Chem.*, 1970, **312**, 385.
[363] H. Zinner and G. Wulf, *J. prakt. Chem.*, 1970, **312**, 192.

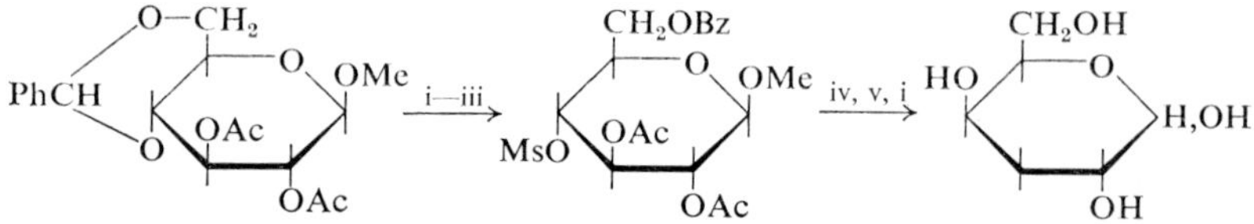

Reagents: i, H^+; ii, BzCl, py; iii, MsCl, py; iv, MeO^-; v, $LiAlH_4$

Scheme 55

bromide and chloride esters of the sugar.[223] A synthesis of 2-deoxy-L-*arabino*-hexose has been developed by standard application of the Wittig reaction to 2,3:4,5-di-*O*-isopropylidene-*aldehydo*-L-arabinose.[364]

DL-Chalcose (270) has been synthesised from 1,6;2,3-dianhydro-4-deoxy-DL-ribopyranose (269), which had previously been obtained from acrolein dimer (Scheme 56), and further comments were made on the

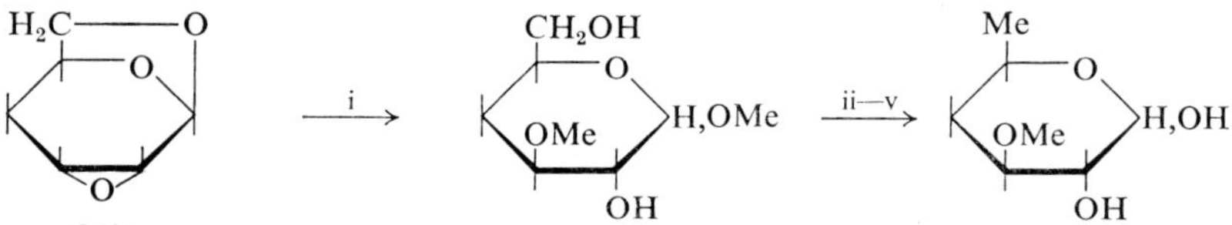

(269) (270)

Reagents: i, MeOH, H^+; ii, TsCl, py; iii, $LiAlH_4$; iv, NaOMe; v, H^+

Scheme 56

reactions leading to (269).[365] Another 4-deoxy-sugar to be described is 4-deoxy-D-*arabino*-hexose, which is a constituent of several lipopolysaccharides.[366] Methylated deoxy-sugars are referred to in Chapter 4, and a general reaction for synthesising 3-deoxyglycal derivatives is reported in Chapter 14.

1-Deoxy-L-*xylo*-hex-2-ulose (272) (as the pyranose form) has been prepared from 1,3:2,4-di-*O*-ethylidene-5,6-di-*O*-methanesulphonyl-D-glucitol by way of the aminoaziridine intermediate (271) (Scheme 57).[330]

The methyl glycosides of abequose have been prepared (Scheme 58) for immunological purposes. Each glycoside was partially acetylated to give a mixture of the 2- and 4-esters, with the former predominating in each case.[367] In related work, a facile synthesis of 4,6-dideoxy-D-*xylo*-hexose was achieved (Scheme 59),[368] and selective hydrogenolysis of the dichloro-derivative (273) over Raney nickel in the presence of triethylamine yielded methyl 6-chloro-4,6-dideoxy-α-D-*xylo*-hexopyranoside.[244] The reaction of methyl 4,6-*O*-benzylidene-2,3-di-*O*-*p*-tolylsulphonyl-α-D-glucopyranoside (90) with $LiAlH_4$ to give 2(3)-deoxy-sugars has been noted in Chapter 6.

[364] M. F. Shostakovskii, N. N. Aseeva, and A. I. Polyakov, *Izvest. Akad. Nauk S.S.S.R., Ser. khim.*, 1970, 892.
[365] R. M. Srivastava and R. K. Brown, *Canad. J. Chem.*, 1970, **48**, 830.
[366] J. Keleti, H. Mayer, I. Fromme, and O. Lüderitz, *European J. Biochem.*, 1970, **16**, 284.
[367] H. F. G. Beving, H. B. Borén, and P. J. Garegg, *Acta Chem. Scand.*, 1970, **24**, 919.
[368] B. T. Lawton, W. A. Szarek, and J. K. N. Jones, *Carbohydrate Res.*, 1970, **14**, 255.

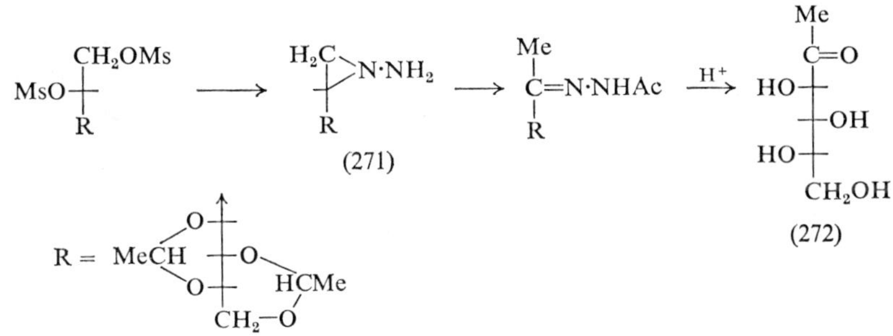

Scheme 57

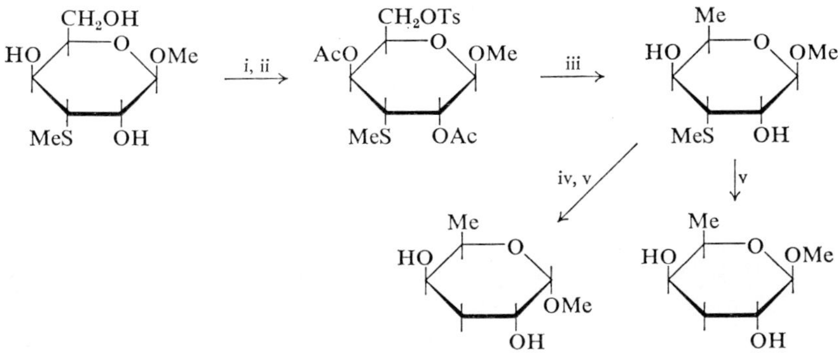

Reagents: i, TsCl, py; ii, Ac_2O, py; iii, $LiAlH_4$; iv, MeOH, HCl; v, Ni, H_2

Scheme 58

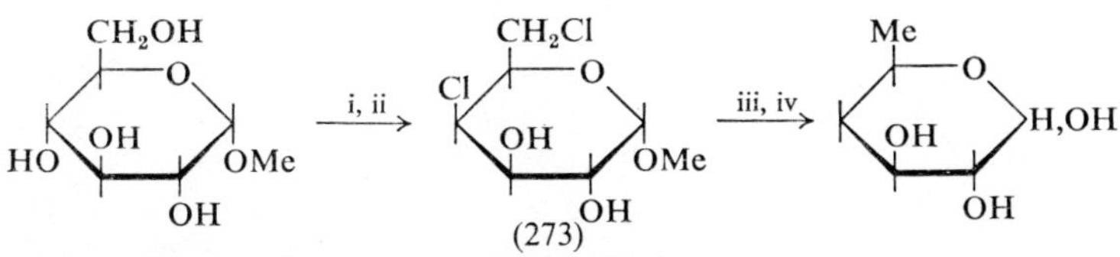

Reagents: i, SO_2Cl_2, py; ii, dechlorosulphonylation; iii, Ni–H_2; iv, H^+

Scheme 59

Hydrogenation of the epoxide (274) afforded the 3,6- and 4,6-dideoxy-derivatives (275) and (276) in the ratio 9 : 91, and the latter compound was converted into the *lyxo*-epoxide (277). The isomeric epoxide (278) was also obtained by treating the dimesylate (279) with alkali.[369]

The Reformatsky reaction has been used in the synthesis of deoxy-sugars;[370] thus, the 5-aldehyde (280) was converted into (281) on treatment with zinc and ethyl bromoacetate.

[369] K. Čapek and J. Jarý, *Coll. Czech. Chem. Comm.*, 1970, **35**, 1727.

[370] Yu. A. Zhdanov, Yu. E. Alekseev, and Kh. A. Kurdanov, *Zhur. obshchei Khim.*, 1970, **40**, 943.

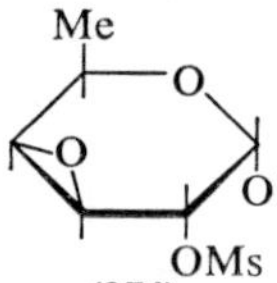

(274)

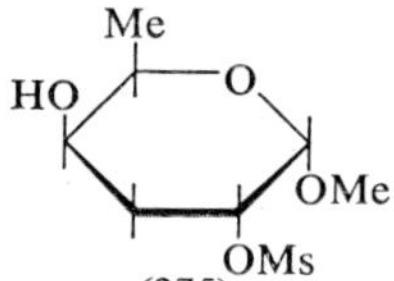

(275)

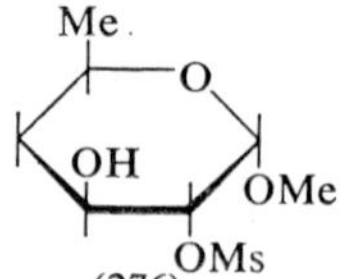

(276)

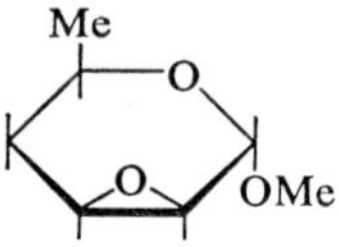

(277)

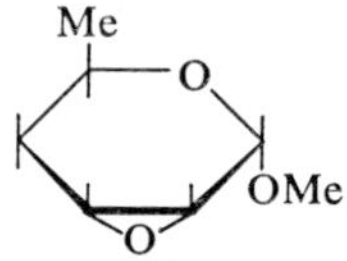

(278)

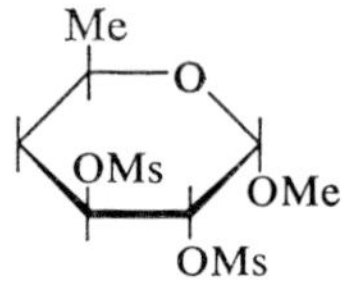

(279)

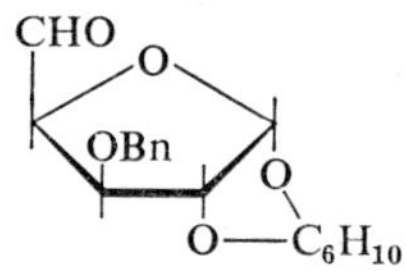

(280)

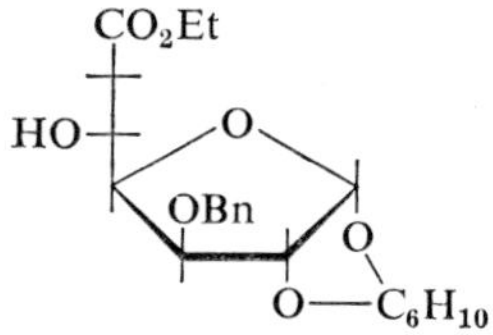

(281)

Various deoxy-sugar derivatives have been prepared from non-carbohydrate precursors. Thus, syntheses of 2,3-dideoxypentose derivatives have been reported (Scheme 60),[371] and the *cis*- and *trans*- forms of alkyl 6-*O*-acetyl-2,3,4-trideoxyhexopyranosides have also been described.[372]

Deoxy-sugars were formed on γ-radiolysis of aqueous solutions of carbohydrates.[43]

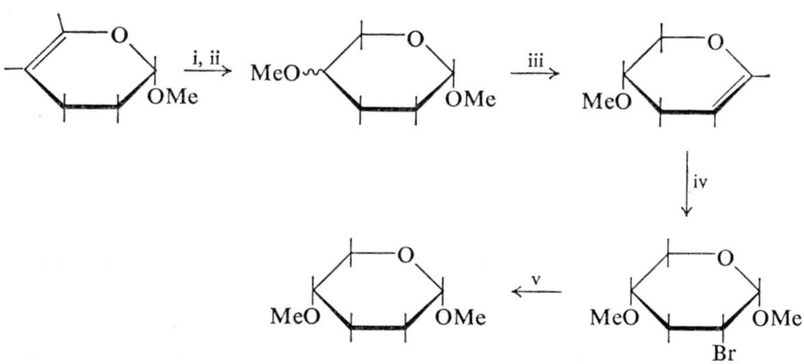

Reagents: i, B_2H_6, H_2O_2; ii, NaH, MeI; iii, P_2O_5, heat; iv, $Me_2CNBr{\cdot}CO{\cdot}NBrCO$, MeOH; v, H_2–Pd

Scheme 60

[371] R. M. Srivastava and R. K. Brown, *Canad. J. Chem.*, 1970, **48**, 2334.
[372] J. Jurczak, A. Konowal, and A. Zamojski, *Roczniki Chem.*, 1970, **44**, 1587.

14
Unsaturated Derivatives

The considerable interest already shown in this class of compounds has been sustained and several new facets of the chemistry of unsaturated sugars have been reported.

Glycals

Addition reactions of glycals continue to be studied and some results have already been reported in Chapter 7. Japanese workers have been active in this field and their results are all summarised here, although halogenated products resulted in some cases.

The ionic chlorination of tri-*O*-acetyl-D-glucal with gaseous chlorine has been examined further and the stereochemistry of addition was shown to be highly solvent dependent. In non-polar media, *cis*-products (*i.e.* α-*gluco*- and β-*manno*-adducts) predominated, whereas in polar solvents β-*gluco*- and α-*manno*-derivatives were mainly formed. The mechanisms of the additions were discussed.[373] The free-radical chlorination of the same glycal with iodobenzene dichloride gave 3,4,6-tri-*O*-acetyl-2-chloro-2-deoxy-α-D-mannopyranosyl chloride in 75% yield together with the α- and β-*gluco*-isomers (11 and 16%, respectively). In the presence of oxygen the radical reaction was replaced by a heterolytic process yielding products in the usual ratios.[374] Radical addition of thioacetic acid to tri-*O*-acetyl-D-glucal using cumene hydroperoxide as initiator gave (282) and (283) in 70 and 30% yield, respectively; these compounds were also formed when oxygen was present and the unsaturated sugars (284), (285), and (286) were then also found among the products.[374a] The addition of *N*-chlorocarbamate to tri-*O*-acetyl-D-glucal has been studied (see Scheme 32).[251]

A report has appeared on various methods which are available for hydroxylating glycals. D-Galactal with hydrogen peroxide in the presence of osmium tetroxide gave D-galactose and D-talose in the ratio 4 : 1. In the presence of selenium dioxide, vanadium pentoxide, or chromium trioxide, the ratio of these sugars was 2 : 1; however, in the reactions where 2,3-*trans*-products were formed preferentially, the overall yields were less than

[373] K. Igarashi, T. Honma, and T. Imagawa, *J. Org. Chem.*, 1970, **35**, 610.
[374] K. Igarashi and T. Honma, *J. Org. Chem.*, 1970, **35**, 617.
[374a] K. Igarashi and T. Honma, *J. Org. Chem.*, 1970, **35**, 606.

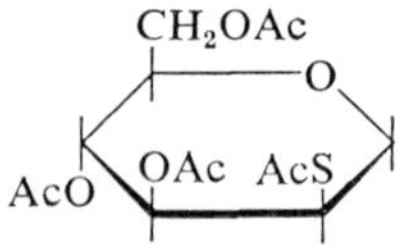

(282)

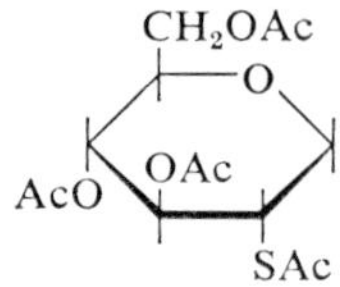

(283)

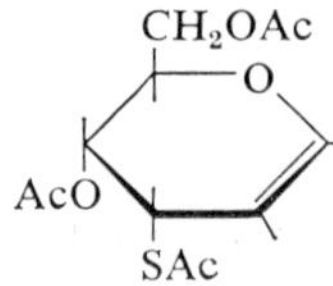

(284)

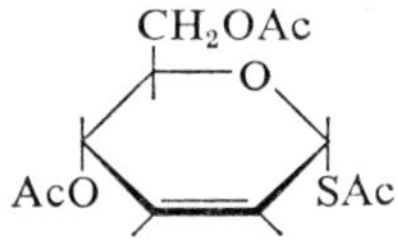

(285)

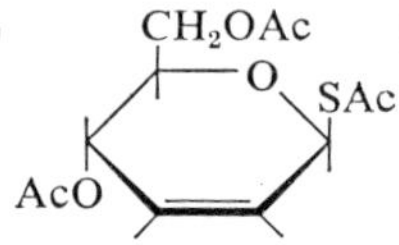

(286)

10%. Hydroxylation in the presence of either molybdenum trioxide or tungsten trioxide, on the other hand, afforded almost exclusively 2,3-*cis*-products (*i.e.* D-talose from D-galactal) in substantially better yields (*ca.* 90%). It was suggested that the high specificity resulted partly from the directing effect of the hydroxy-group at C-3, which complexed with the oxidising agents.[24]

Addition of dioxolane to tetra-*O*-acetyl-2-hydroxy-D-glucal in the presence of u.v. light gave 2′-dioxolanyl 2,3,4,6-tetra-*O*-acetyl-D-glucopyranoside in high yield, but a similar reaction with tri-*O*-acetyl-D-glucal afforded a mixture of products having the dioxolane ring attached at position 1 or 2 of the sugar and through its own acetal or non-acetal carbon atoms.[375]

D-Glucal is reported to be selectively oxidised at the allylic centre to give the enone (287) on treatment with silver carbonate–Celite.[376] The reaction

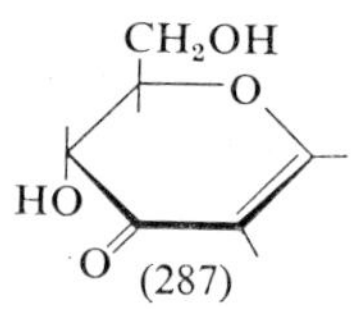

(287)

gave the enone in 60—80% yield so is to be preferred to the platinum-catalysed oxidation previously reported (K. Heyns and H. Gottschalck, *Chem. Ber.*, 1966, **99**, 3718).

Fraser-Reid and Radatus have reported on several novel findings. The reaction of the iodomethylallal derivative (288) in either neutral water or methanol seems to occur by way of the ionic intermediate (289) to give the acyclic sugar derivative (290) in the first case, and equimolar amounts of the glycosides (291) and (292) in the second. Under basic conditions, direct elimination occurred to give the conjugate diene.[377] The C-3 epimer of (288) also hydrolysed extremely rapidly to give (293) presumably by way of

[375] K. Matsuura, S. Maeda, Y. Araki, Y. Ishido, and T. Murai, *Tetrahedron Letters*, 1970, 2869.
[376] J. M. J. Tronchet, J. Tronchet, and A. Birkhäuser, *Helv. Chim. Acta*, 1970, **53**, 1489.
[377] B. Fraser-Reid and B. Radatus, *Canad. J. Chem.*, 1970, **48**, 2146.

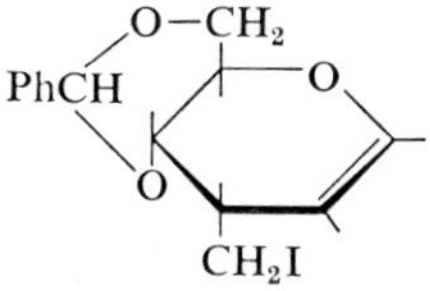

(288)

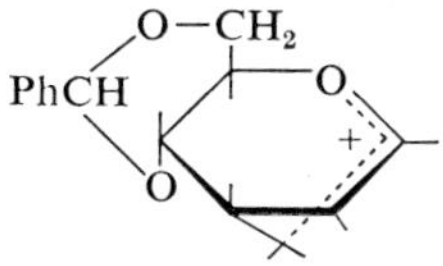

(289)

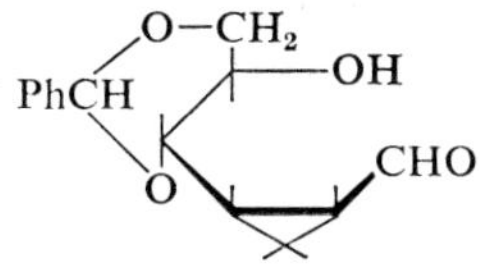

(290)

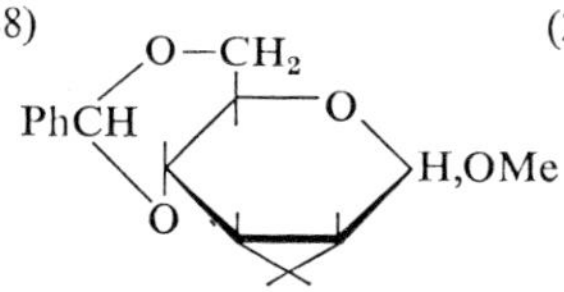

(291) α-anomer
(292) β-anomer

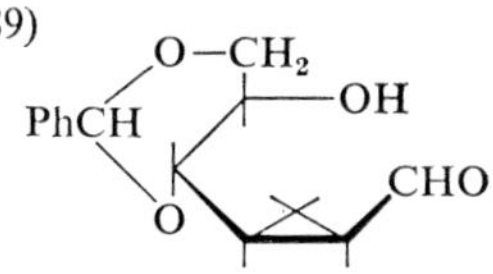

(293)

the ionic analogue of (289).[378] The interesting observation was also made that heating tri-*O*-acetyl-D-glucal or ethyl 4,6-di-*O*-acetyl-2,3-dideoxy-α-D-*erythro*-hex-2-enopyranoside in aqueous dioxan gave 'di-*O*-acetyl-pseudoglucal' (294) as expected, but an acyclic aldehyde, shown to be the *trans*-isomer (295), was also found. Compound (295) was isolated after

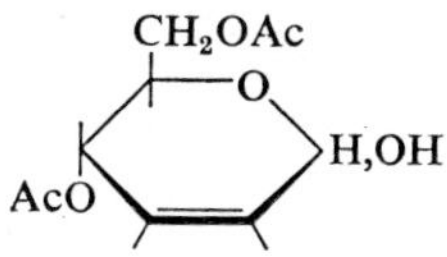

(294)

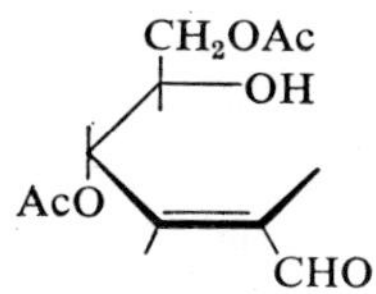

(295)

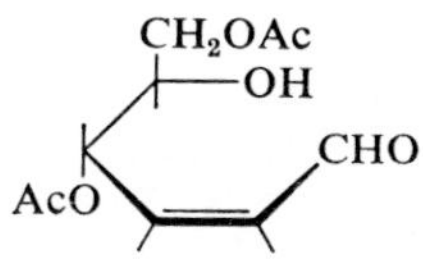

(296)

acetylation of the hydroxy-group and is probably formed from the *cis*-isomer (296) by a photochemical process. When the hydrolyses were carried out in the dark or in the presence of hydroquinone, the yield of compound (295) was markedly reduced. It was suggested that the presence of (295) may have been responsible for the reducing properties of tri-*O*-acetyl-D-glucal, which caused Fischer to give the compound the 'al' suffix. Compound (296) would also have given positive reducing tests.[379]

A new reaction in this area, which seems to be generally applicable, involved the reductive rearrangement of 2,3-unsaturated glycopyranosides to analogous 3-deoxyglycal derivatives (Scheme 61). The reactions were

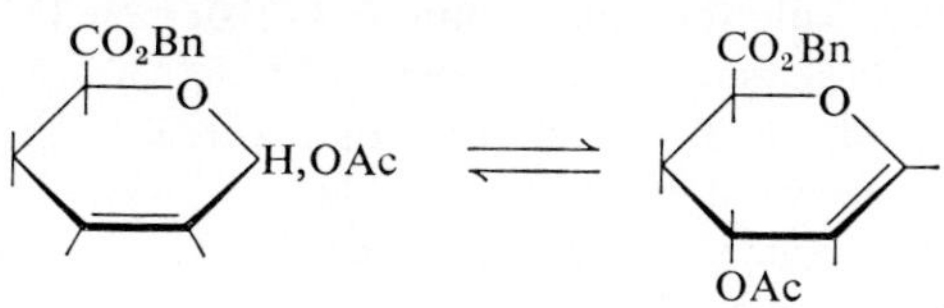

Scheme 61

[378] B. Fraser-Reid and B. Radatus, *Chem. Comm.*, 1970, 779.
[379] B. Fraser-Reid and B. Radatus, *J. Amer. Chem. Soc.*, 1970, **92**, 5228.

carried out under reflux with lithium aluminium hydride in an inert solvent and mainly caused rupture of the glycosidic bond. Compound (297) was thus produced in 95% yield, the other product of reaction being the vinyl

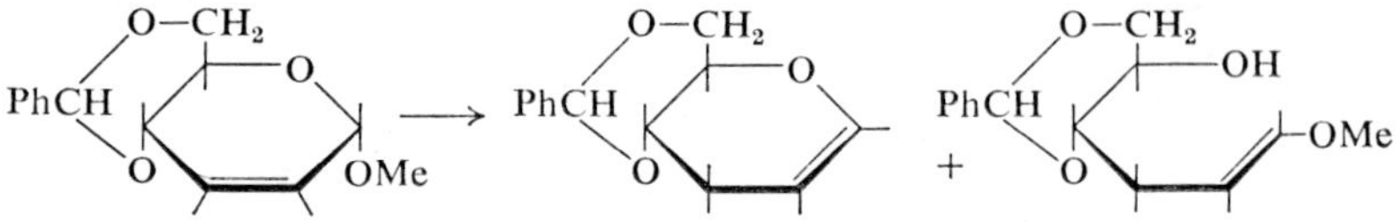

(297) (298)

ether (298). Isotope studies gave results consistent with the occurrence of cyclic transition states such as (299).[380]

The system illustrated in Scheme 62 has been studied as a model for the glycal–pseudoglycal system. After heating in acetic anhydride, the 2,3-unsaturated anomers were present at equilibrium to the extent of 42%,

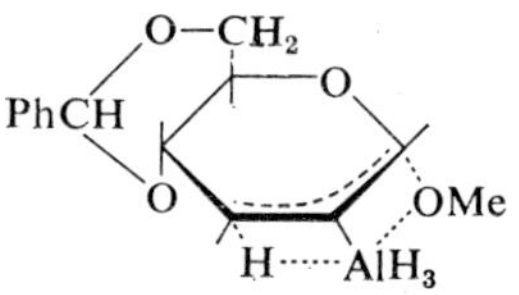

(299)

Scheme 62

whereas the *cis*- and *trans*-glycal analogues comprised 39% and 19%, respectively, of the mixture.[381]

Further studies on the reactions of unsaturated carbohydrates with anhydrous hydrogen fluoride at − 70 °C have shown that compounds (300)—(302) yielded initially the ion (303), which on work up gave (304). After longer reaction times the ion (303) added hydrogen fluoride to give (305), and this when worked up afforded (306). This illustrates the preferential formation of axial esters on hydrolysis of dioxolanium ions fused to six-membered rings.[382]

A continuation of this work had led to a report on the reaction of 2-hydroxyglycal esters with hydrogen chloride and hydrogen bromide. Compounds (307) and (308) gave (309) on treatment with hydrogen chloride in benzene solution, and compound (310) gave (311) when treated similarly. With 1 mol. equiv. of hydrogen bromide, (307) and (308) behaved as in the reaction with hydrogen chloride but, with an excess of the reagent, further reaction occurred to give unstable products. When these were treated with silver benzoate they gave mainly (312) together with (313), (314), and other minor products.[383]

[380] B. Fraser-Reid and B. Radatus, *J. Amer. Chem. Soc.*, 1970, **92**, 6661.
[381] Z. Zwierzchowska and A. Zamojski, *Roczniki Chem.*, 1970, **44**, 1609.
[382] I. Lundt and C. Pedersen, *Acta Chem. Scand.*, 1970, **24**, 240.
[383] K. Bock and C. Pedersen, *Acta Chem. Scand.*, 1970, **24**, 2465.

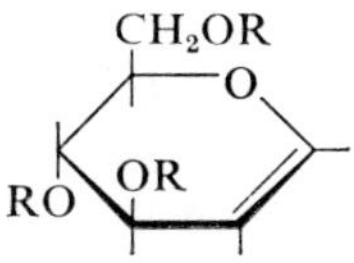

(300)

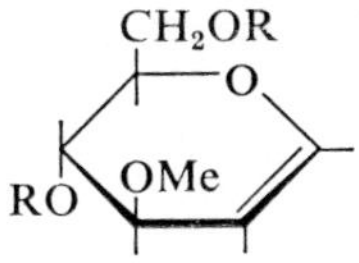

(301)

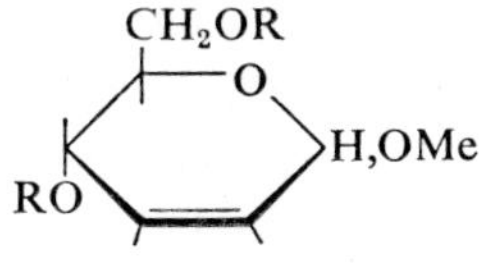

(302)

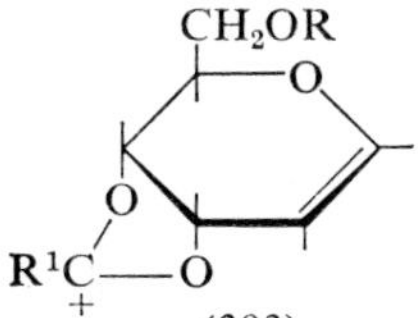

(303)

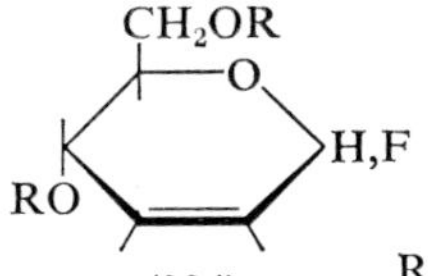

(304)

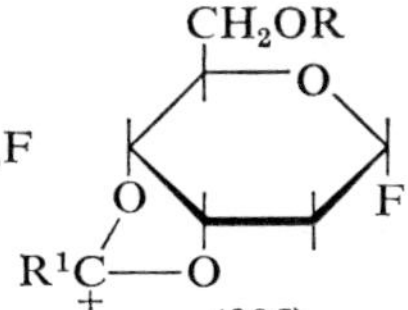

(305)

R = Ac or Bz, R^1 = Me or Ph

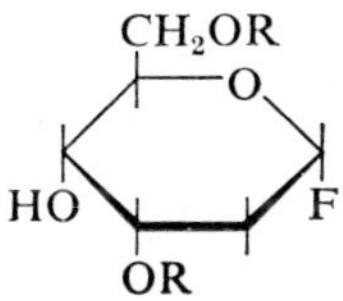

(306)

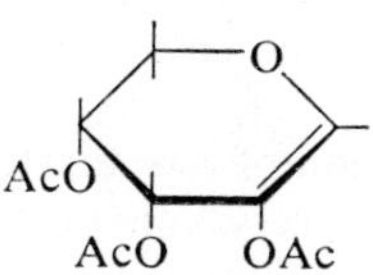

(307)

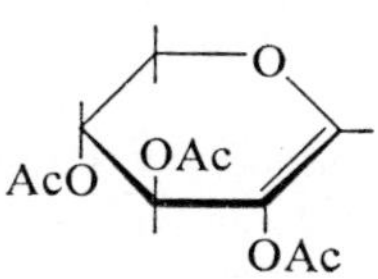

(308)

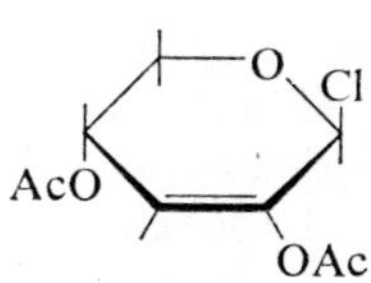

(309)

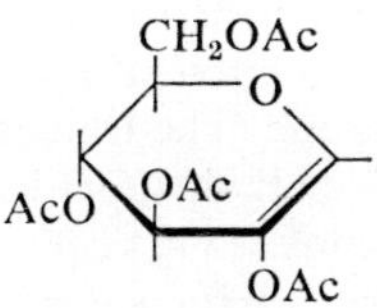

(310)

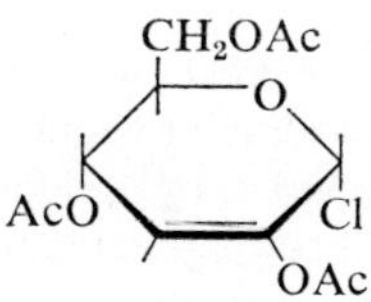

(311)

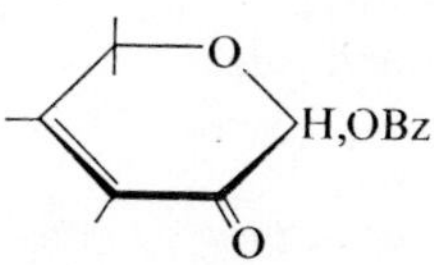

(312)

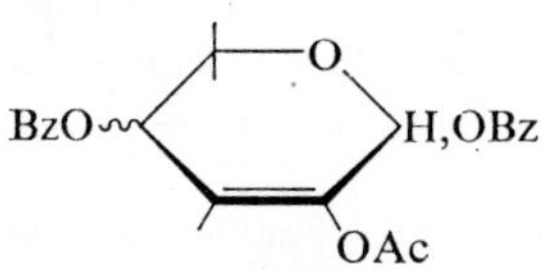

(313)

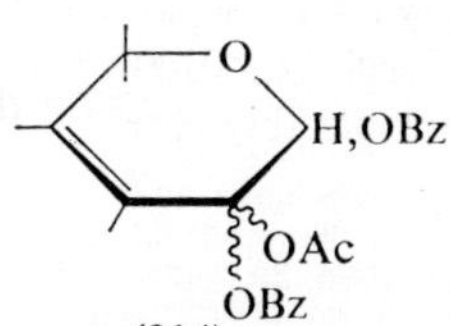

(314)

Other Unsaturated Compounds

Another method of introducing a carbon–carbon double bond into carbohydrates involved the treatment of cyclic carbonates with potassium thiocyanate. Methyl 2,3-dideoxy-α-D-*erythro*-hex-2-enopyranoside was obtained in this way and its formation was rationalised by the reactions outlined in Scheme 63.[384]

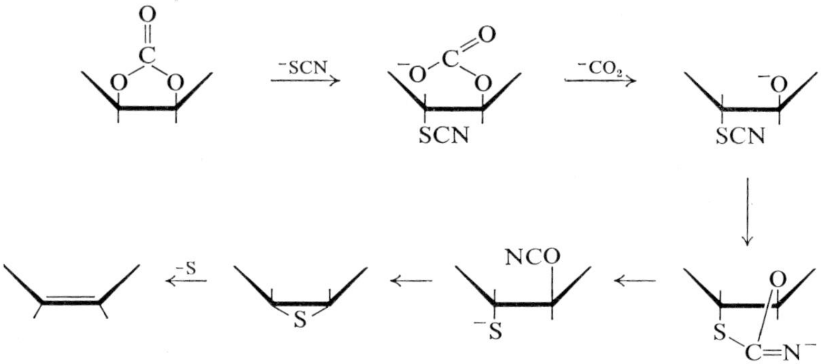

Scheme 63

Selective oxidation of the diol (315; R = CH_2OH) with manganese dioxide occurred at the allylic position to give the crystalline enone (316; R = CH_2OH) in 82% yield; the corresponding pentose derivative (315; R = H) oxidised similarly but the β-anomer was unaffected. Reduction of the enone (316; R = CH_2OH) with lithium aluminium hydride gave the starting diol exclusively. The n.m.r. spectra of the enones were described in detail.[385]

An oxidation of a different type has led to the formation of unsaturated sugars. Treatment of the phenylthio-derivative (317) with acetic anhydride–DMSO afforded a mixture of the enol acetate (318) and the 2-methylthiomethyl ether (319). Alternatively, when the oxidation was effected with DMSO–DCC in the presence of pyridinium trifluoroacetate, the 2-ulose derivative (320) was formed and was converted into (318) on treatment with acetic anhydride–pyridine at room temperature. The isomeric enol acetate (321) was also prepared by oxidation of methyl 4,6-*O*-benzylidene-2-*S*-phenyl-2-thio-α-D-altropyranoside to the 3-ulose followed by acetylation.[386]

A preliminary report has appeared on the rearrangement of the phenylthio-derivative (322) into (323). These compounds yielded the unsaturated sugars (324) and (325), respectively, when treated with base (1,5-diazabicyclo[4,3,0]-5-nonene) in DMSO.[387] The unsaturated sugar (324) is presumably formed by a double elimination.

[384] A. Klemer and G. Mersmann, *Carbohydrate Res.*, 1970, **12**, 219.
[385] B. Fraser-Reid, A. McLean, E. W. Usherwood, and M. Yunker, *Canad. J. Chem.*, 1970, **48**, 2877.
[386] S. Hanessian and A. P. A. Staub, *Chem. and Ind.*, 1970, 1436.
[387] S. Hanessian and A. P. A. Staub, *Carbohydrate Res.*, 1970, **14**, 424.

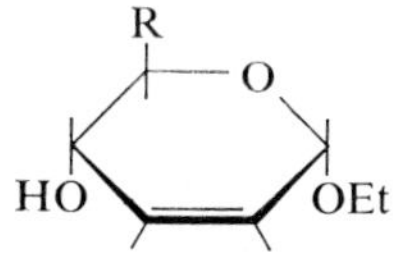

(315)

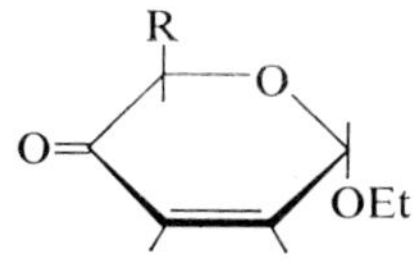

(316)

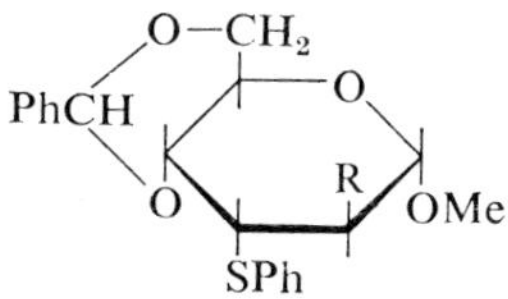

(317) R = OH
(319) R = OCH_2·SMe
(320) R = O

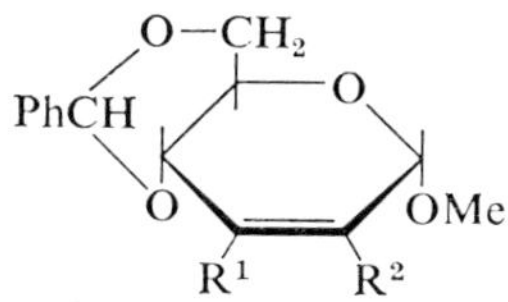

(318) R^1 = SPh, R^2 = OAc

(321) R^1 = OAc, R^2 = SPh

Epoxidations of model 2,3-unsaturated pyranosides have been mentioned earlier (see p. 27).

A series of allylic isomerisations of 2,3-unsaturated hexopyranosides to 3,4-unsaturated compounds have been reported which offer a new approach to the synthesis of a variety of modified sugars. Compounds with allylic

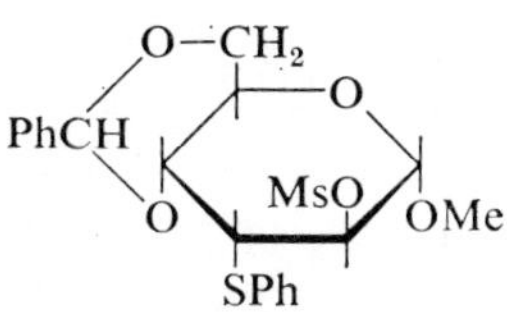

(322)

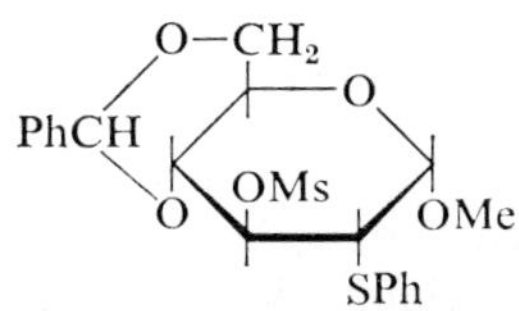

(323)

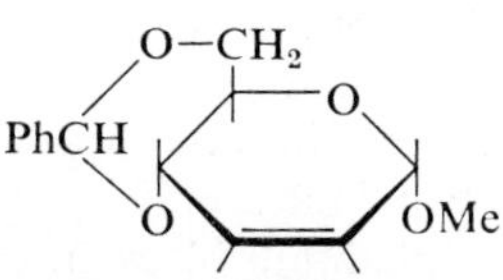

(324)

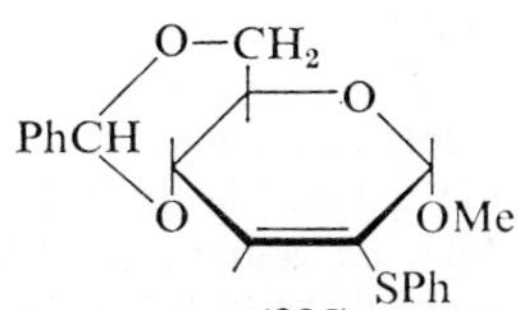

(325)

azido, thiocyanate, vinyloxy, and (methylthio)thiocarbonyloxy groups were investigated; both *erythro*- and *threo*-derivatives were isomerised on heating, the reaction proceeding more readily with *threo*-derivatives. *N*-Acetylamino, *C*-formylmethyl, and methylthio(carbonylthio) functions were introduced at C-2 by this procedure; examples are given in Schemes 64 and 65.[388] Mass spectrometry can be used to locate the positions of double bonds in such compounds (see Chapter 24).

[388] R. J. Ferrier and N. Vethaviyasar, *Chem. Comm.*, 1970, 1385.

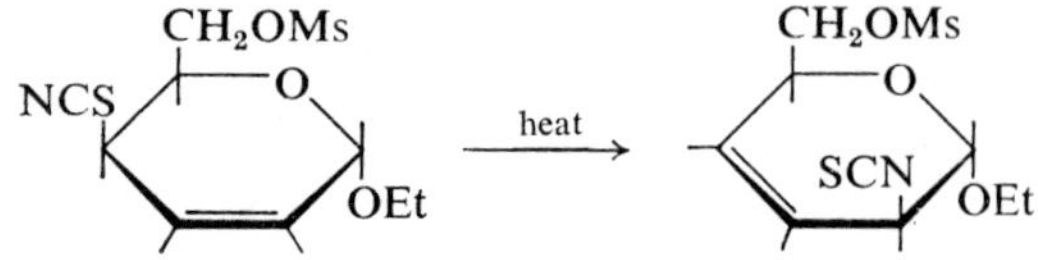

Scheme 64

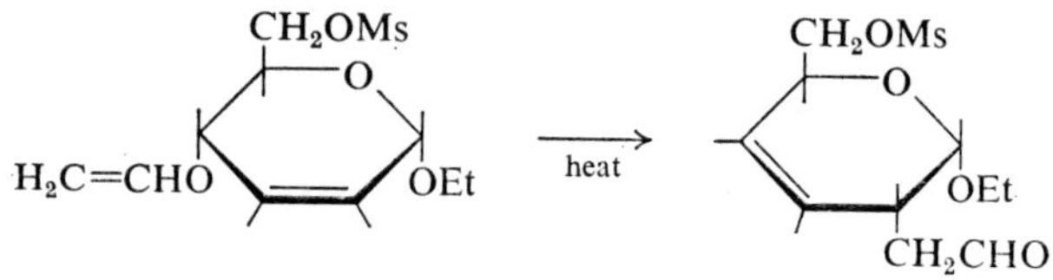

Scheme 65

Treatment of the D-*lyxo*-sulphonates (326) and (327) with sodium benzoate in DMF gave (328), whereas the analogous methyl β-D-ribofuranoside triester gave only products of nucleophilic displacement. It was

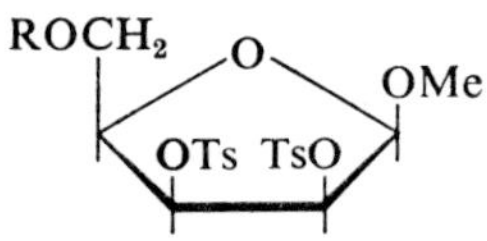

(326) R = Ts
(327) R = Bz

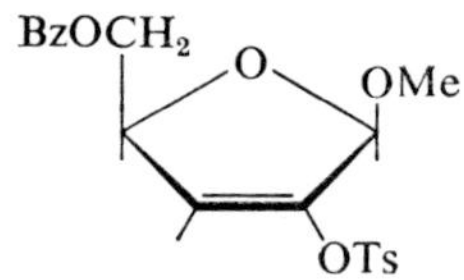

(328)

proposed that steric compression and the availability of H-2 to the base facilitate elimination with the D-*lyxo*-compounds.[389] Enones related to (328) are mentioned in Chapter 17. Products of both elimination and substitution reactions were observed on azidolysis of methyl 6-deoxy-4-*O*-methanesulphonyl-2,3-di-*O*-methyl-α-L-talopyranoside (105).[209]

Treatment of the four 1,2:5,6-di-*O*-isopropylidene-D-hexofuranos-3-uloses with acetic anhydride in pyridine yielded the two enol acetates (329) and (330), which on reduction with hydrogen over platinum black in ethanol gave almost exclusively the D-*gulo*- and D-*manno*-products as a result of *exo*-hydrogenation. However, isomerisation took place during the reductions and led to the isolation of (331) and (332). Hydrogenolysis of the ester groups also occurred during reduction with the formation of 3-deoxy products. The 3-ulose (333) did not yield (329) with isopropenyl acetate but instead gave (334), which is formed by partial hydrolysis and acylation of the hemiacetal.[390]

3′,4′-Unsaturated furanosyl pyrimidine nucleosides have been synthesised as shown in Scheme 66. Eliminations occurred mainly by a direct (*E*1cb)

[389] J. Hildesheim, A. Gaudemer, and S. D. Géro, *Chem. and Ind.*, 1970, 94.
[390] K. N. Slessor and A. S. Tracey, *Canad. J. Chem.*, 1970, **48**, 2900.

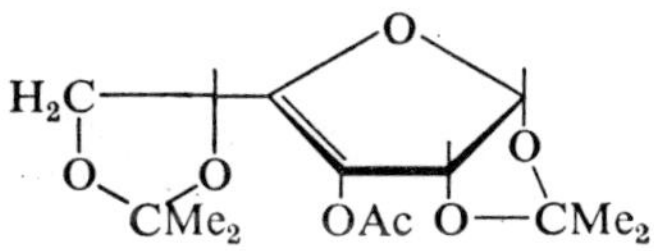

(329)

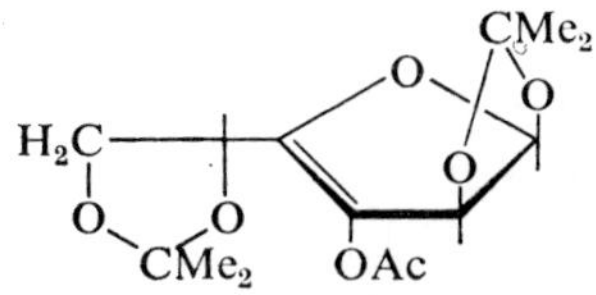

(330)

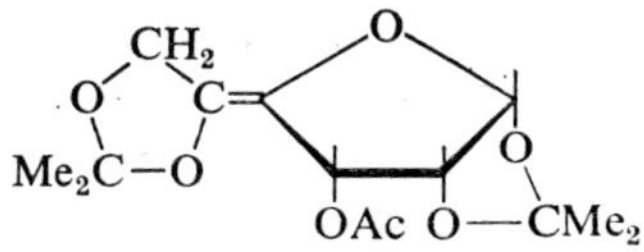

(331)

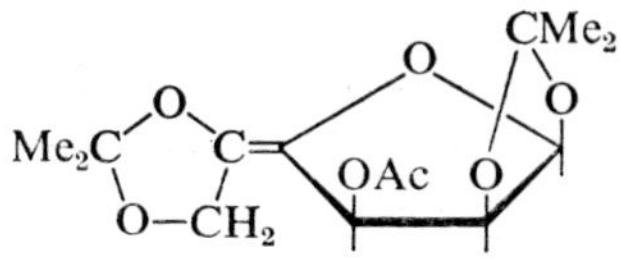

(332)

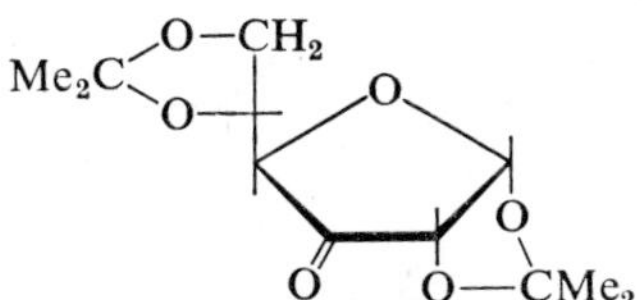

(333)

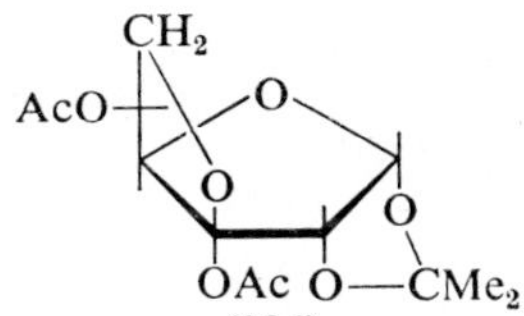

(334)

route rather than by way of a 2,3′-anhydride.[391] Other uronic acid derivatives of nucleosides are referred to in Chapter 17.

Other base-catalysed eliminations have been reported in the pyranuronic acid series. Thus, the α,β-unsaturated ester (335) was prepared in two ways

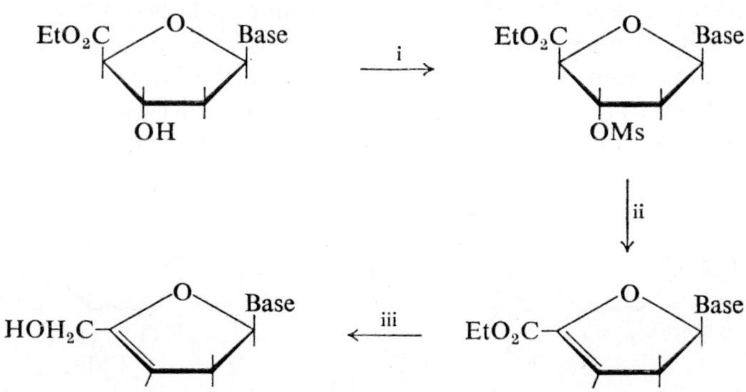

Reagents: i, MsCl, py; ii, Et_3N; iii, $(MeOCH_2CH_2O)_2AlH_2Na$

Scheme 66

as shown in Scheme 67; surprisingly, axial–axial elimination from the D-*galacto*-derivative was only slightly favoured. The n.m.r. and o.r.d. spectra of the unsaturated esters were discussed.[392] The disaccharides (336)

[391] J. Žemlička, R. Gasser, and J. P. Horwitz, *J. Amer. Chem. Soc.*, 1970, **92**, 4744.
[392] J. Kiss and F. Burkhardt, *Helv. Chim. Acta*, 1970, **53**, 1000.

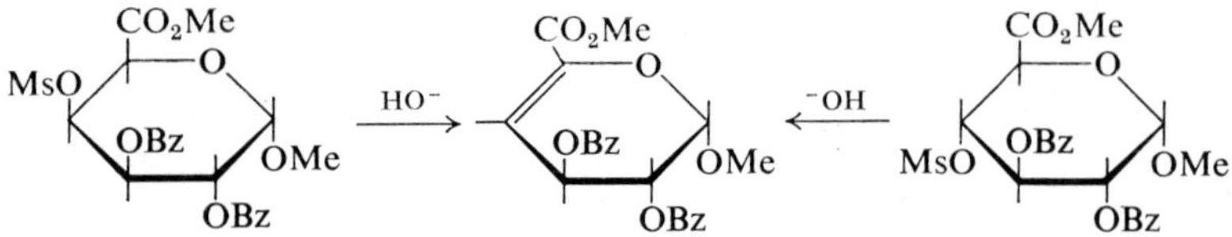

(335)

Scheme 67

and (337) have been prepared and degraded to (338) by treatment with sodium methoxide in methanol–benzene. These disaccharides served as models for heparin sulphate, the chondroitin sulphates, and hyaluronic

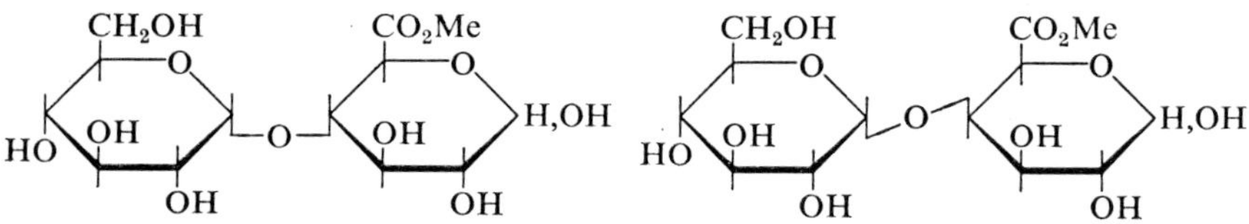

(336) (337)

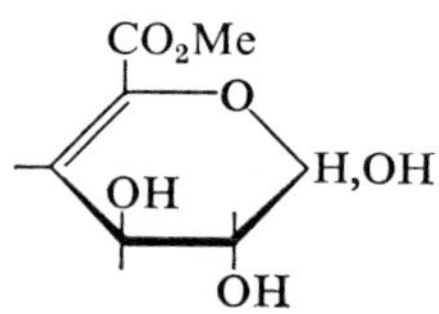

(338)

acid which are degraded enzymically to terminally-linked 4,5-unsaturated uronic acid residues. Introduction of a mesyl group at C-2 in the D-glucopyranosyl moiety of both disaccharides facilitated the β-elimination.[393]

The 4,5-unsaturated furanoid compound (339) underwent photochemically-induced addition of thioacetic acid and benzyl mercaptan as shown in Scheme 68.[375]

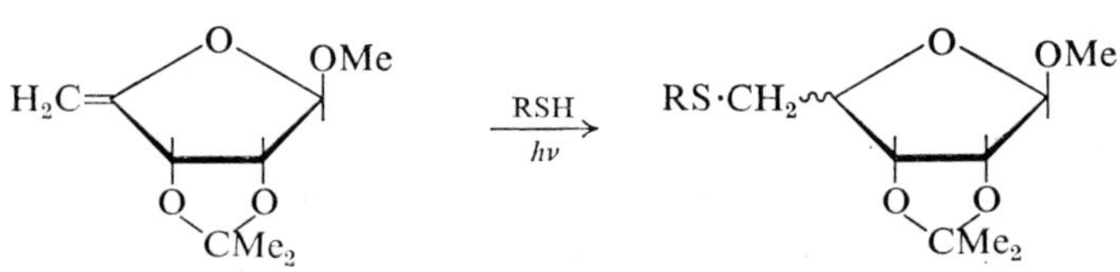

(339) R = Ac or Bn

Scheme 68

The formation of 1,2-*O*-isopropylidene-α-D-glucofuranose 3,5,6-orthoformate (85) as a by-product of the conversion of the thionocarbonate (84) into the 5,6-olefin by heating with trimethyl phosphite (Corey–Winter

[393] J. Kiss, *Tetrahedron Letters*, 1970, 1983.

reaction) was attributed to intramolecular insertion of the carbene (86) into the OH bond of the hydroxy-group at C-3, thus providing good evidence that the reaction proceeds by a carbene mechanism.[178] The nitro-olefin (218) related to the aforementioned unsaturated compound afforded a crystalline Diels–Alder adduct (340) in modest yield on treatment with boiling cyclopentadiene.[394]

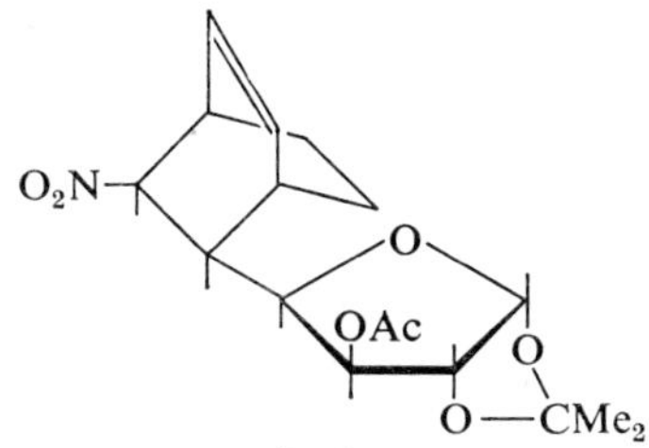

(340)

o-Nitrophenyl 6-deoxy-α-L-*arabino*-hex-5-enopyranoside has been synthesised for use as a substrate for β-galactosidase. When the enzymic reaction was conducted in the presence of glycerol, the corresponding 1′-D-glyceryl glycoside was obtained, indicating that the sugar component retains its cyclic structure during glycosyl transfer.[395]

Methylthiomethylene(triphenyl)phosphorane has been used to extend the chains of the aldehydo-compounds (341), (342), and (343) by one carbon

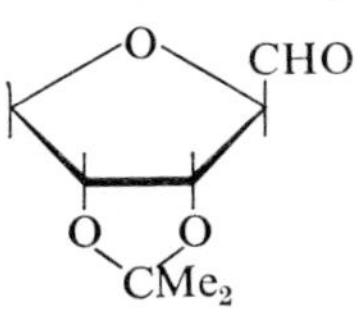

(341)

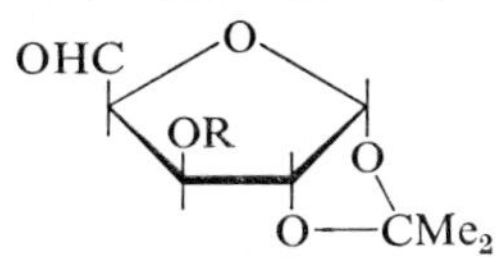

(342) R = Me
(343) R = Bn

atom.[396] A further application of the Wittig reaction to 2,3:4,5-di-*O*-isopropylidene-*aldehydo*-L-arabinose has led to a new synthesis of 2-deoxy-L-*arabino*-hexose.[364]

Methyl 4-amino-2,3,4-trideoxy-α-D-*erythro*-hex-2-enopyranosiduronic acid, a saturated sugar component of blasticidin S, has been synthesised by a route already described (see Scheme 35).[261] Syntheses of 2-amino-2-deoxy-D-allose [254] and 2,4-diamino-1,6-anhydro-2,4-dideoxy-D-talose (175) [277] *via* unsaturated precursors are reported in Chapter 8.

The additions of anthranilic acid [321] and the isomeric *m*- and *p*-aminobenzoic acids [322] to methyl 4,6-*O*-benzylidene-2,3-dideoxy-3-nitro-β-D-*erythro*-hex-2-enopyranoside (209) have been discussed in Chapter 10.

[394] W. A. Szarek and J. S. Jewell, *Canad. J. Chem.*, 1970, **48**, 1030.
[395] J. Lehmann and H. Reinshagen, *Annalen*, 1970, **732**, 112.
[396] J. M. J. Tronchet, N. Le-Hong, and F. Perret, *Helv. Chim. Acta*, 1970, **53**, 154.

15

Branched-chain Sugars

Investigations of the biosynthesis of D-aldgarose have shown that the hydroxyethyl side-chain originates from pyruvate.[397] Many nucleosides containing branched-chain sugars have been prepared (see Chapter 21).

Compounds with an R^1—C—OR^2 Branch

Syntheses of 6-deoxy-4-*C*-vinyl-L-talose and 5-deoxy-3-*C*-vinyl-D-lyxose have been described. Methyl 6-deoxy-2,3-*O*-isopropylidene-α-L-*lyxo*-hexopyranosid-4-ulose was treated with vinyl magnesium bromide and then degraded in appropriate ways.[398] Methyl 4,6-*O*-benzylidene-3-*C*-cyanomethyl-2-deoxy-α-D-*arabino*-hexopyranoside (344) was obtained by treatment of (345) with acetonitrile and lithium amide in liquid ammonia.[399] Debenzylidenation of (344) followed by reduction in the presence of ammonia with 5% rhodium–alumina as catalyst afforded methyl 3-*C*-aminoethyl-2-deoxy-α-D-*arabino*-hexopyranoside (346). A HO·C·CH_2NO_2

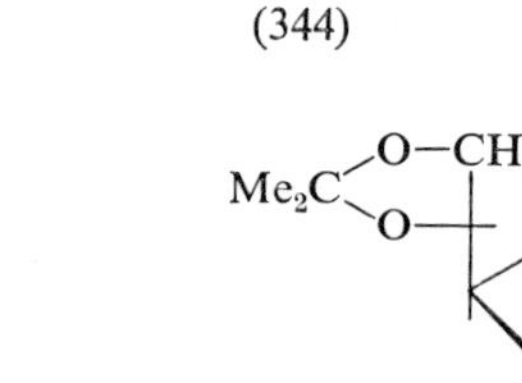

(344)

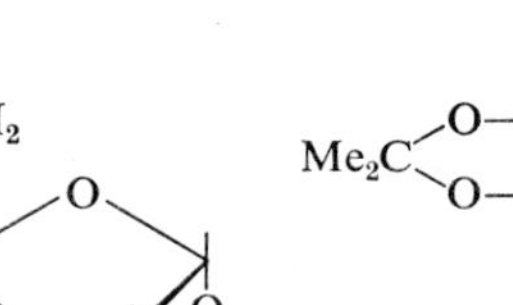

(345)

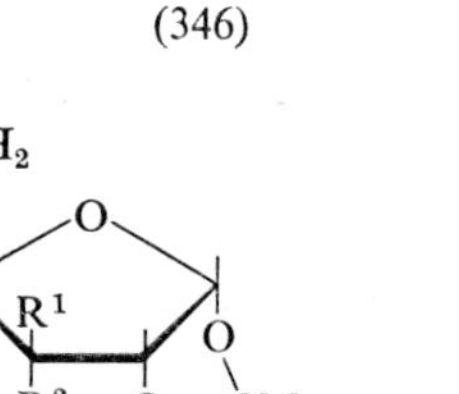

(346)

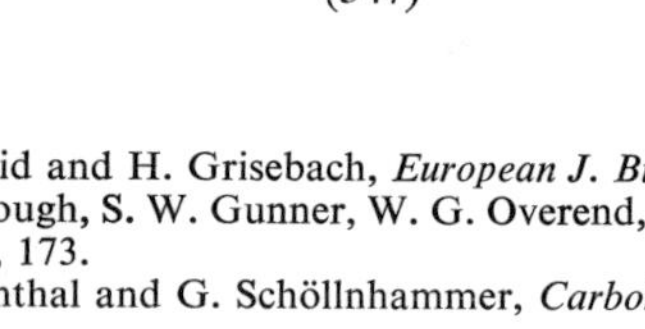

(347)

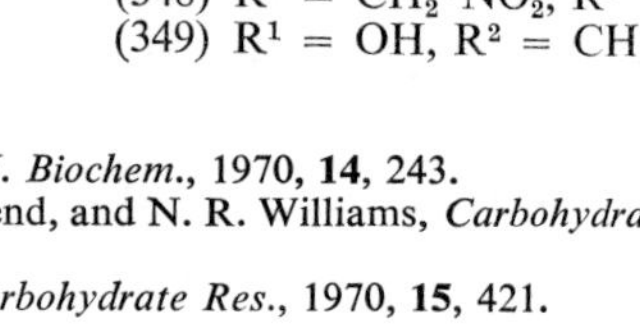

(348) $R^1 = CH_2 \cdot NO_2$, $R^2 = OH$
(349) $R^1 = OH$, $R^2 = CH_2 \cdot NO_2$

[397] R. Schmid and H. Grisebach, *European J. Biochem.*, 1970, **14**, 243.
[398] B. M. Gough, S. W. Gunner, W. G. Overend, and N. R. Williams, *Carbohydrate Res.*, 1970, **14**, 173.
[399] A. Rosenthal and G. Schöllnhammer, *Carbohydrate Res.*, 1970, **15**, 421.

branch has also been formed by the reaction of keto sugars with nitromethane. It was claimed[400] that the D-*allo*-isomer (348) was the major product from the reaction between (347) and nitromethane but another investigation has demonstrated that the D-*gluco*-isomer (349) preponderated.[401] Addition of an excess of nitromethane to (350) afforded the

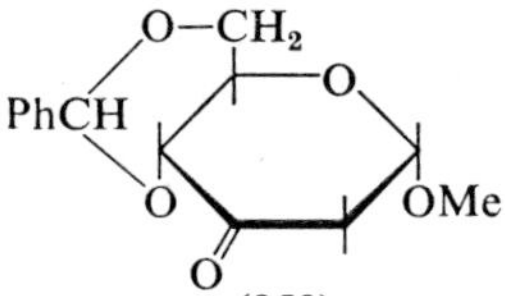

(350)

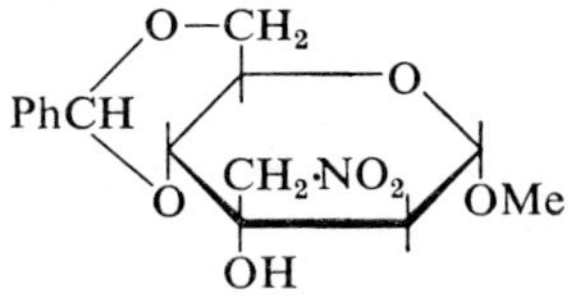

(351)

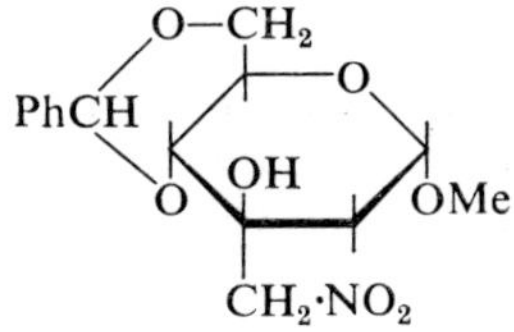

(352)

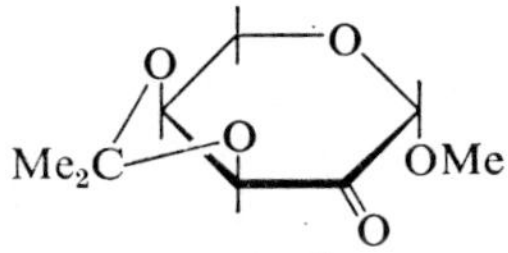

(353)

D-*ribo*-[(351), 63%] and the D-*arabino*-[(352), 22%] isomers. Compounds (351) and (352) were converted into the corresponding methyl 3-*C*-aminomethyl-2-deoxy-α-D-hexopyranosides either by acid hydrolysis and hydro-

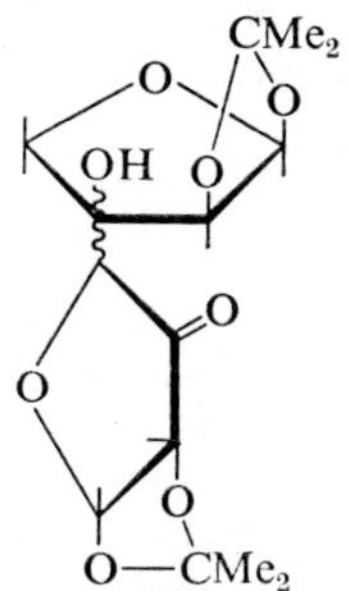

(354)

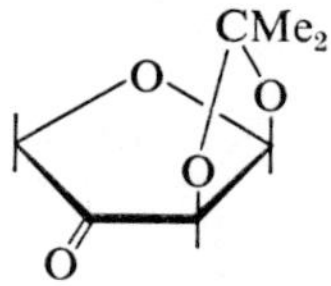

(355)

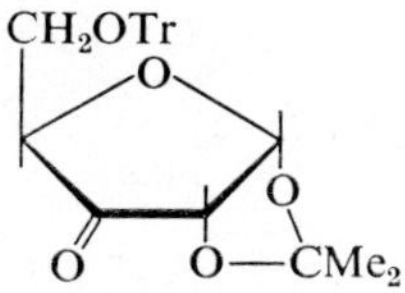

(356)

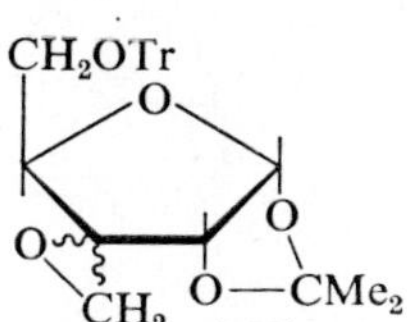

(357)

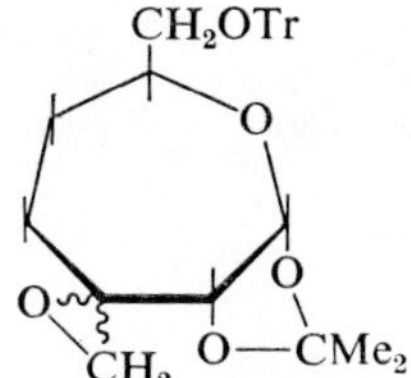

(358)

Tr = Ph_3C

[400] H. P. Albrecht and J. G. Moffatt, *Tetrahedron Letters*, 1970, 1063.
[401] A. Rosenthal, K.-S. Ong, and D. Baker, *Carbohydrate Res.*, 1970, **13**, 113.

genation or simply by hydrogenation.[402] Similar additions of nitromethane to (353) have been described,[403] and the derived amines gave the expected epoxides on diazotisation.

A synthesis of L-apiose has been described.[404] D-Apiose phenylosotriazole has proved to be a useful crystalline derivative of D-apiose in radiochemical, biosynthetic studies.[292]

The branched-chain dimer (354) was formed by distillation of (355).[405] Treatment of (356) with an excess of diazomethane in methanol–ether afforded a mixture of the epoxides (357) and (358) rather than a simple mixture of epimeric epoxides.[406]

The sugar component of the antibiotic gentamycin has been synthesised, thereby confirming its identity as 3-deoxy-4-*C*-methyl-3-methylamino-L-arabinose.[407]

Compounds with an R—C—H Branch

A useful general procedure for the synthesis of branched-chain deoxy sugars has been developed. It has been shown that sugar epoxides reacted with alkyl magnesium chlorides to give branched-chain sugars, whereas the reactions between sugar epoxides and alkylmagnesium bromides and iodides gave halohydrins. Thus, (360) was formed in good yield on treatment of (359) with ethyl magnesium chloride.[127] Compound (361), formed by oxidation of (360) with DMSO–Ac_2O, and (362), formed by base-catalysed epimerisation of (361), have been shown to give preponderantly spiro epoxides [*e.g.* (363) and (364)] with diazomethane. However, the

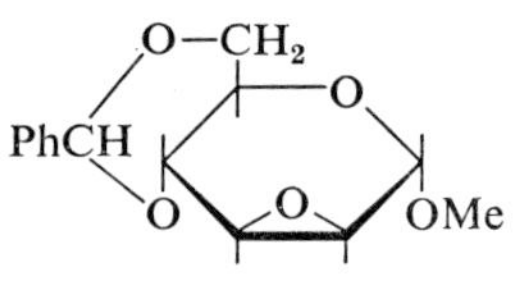

(359)

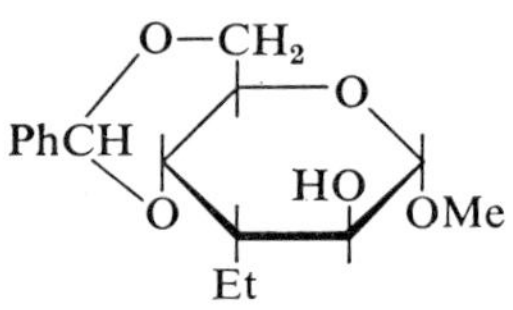

(360)

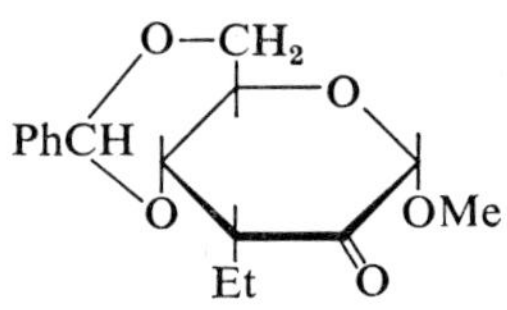

(361)

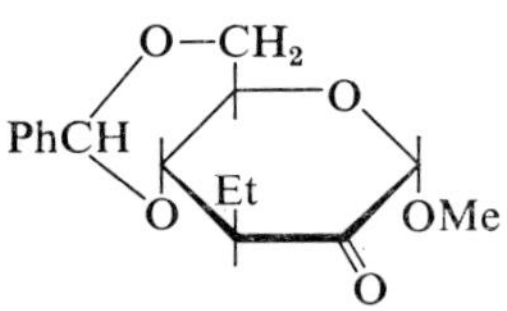

(362)

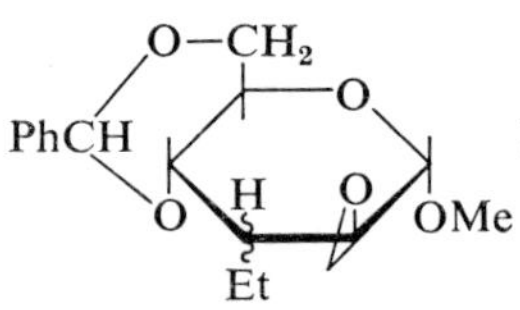

(363)

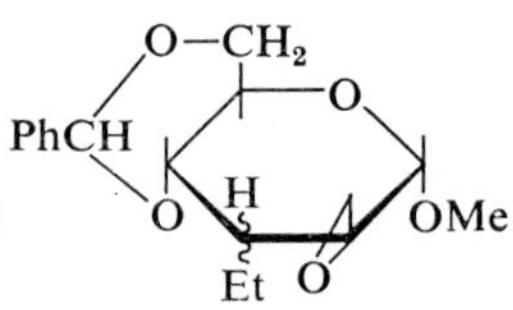

(364)

[402] A. Rosenthal and K.-S. Ong, *Canad. J. Chem.*, 1970, **48**, 3034.
[403] S. W. Gunner, R. D. King, W. G. Overend, and N. R. Williams, *J. Chem. Soc.* (*C*), 1970, 1954.
[404] W. G. Overend, A. C. White, and N. R. Williams, *Carbohydrate Res.*, 1970, **15**, 185.
[405] J. M. J. Tronchet and J. Tronchet, *Helv. Chim. Acta*, 1970, **53**, 1174.
[406] J. P. Horwitz, N. Mody, and R. Gasser, *J. Org. Chem.*, 1970, **35**, 2335.
[407] W. Meyer zu Reckendorf and E. Bischof, *Tetrahedron Letters*, 1970, 2475.

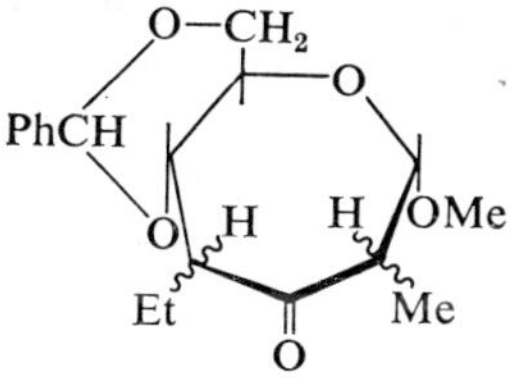

(365)

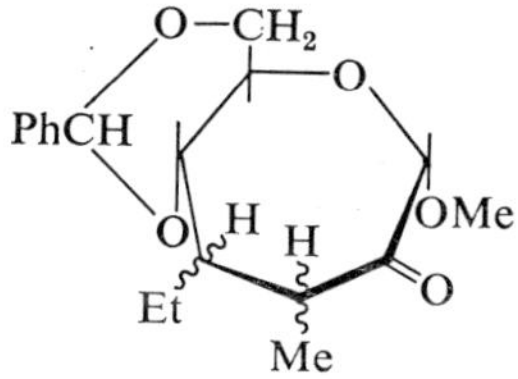

(366)

reactions with diazoethane gave preponderantly compounds (365) and (366).[408]

A number of papers have described the formation of sugars containing a CH_3—C—H branch point. Thus, (347) was converted into (367) by means of a Wittig reaction and subsequently into (368) by Raney nickel desulphurisation.[409] Similarly, the keto sugar (369) was converted into (370) and thence into (371) and (372).[409a]

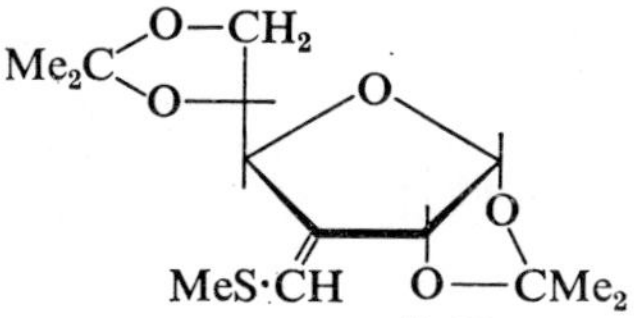

(367)

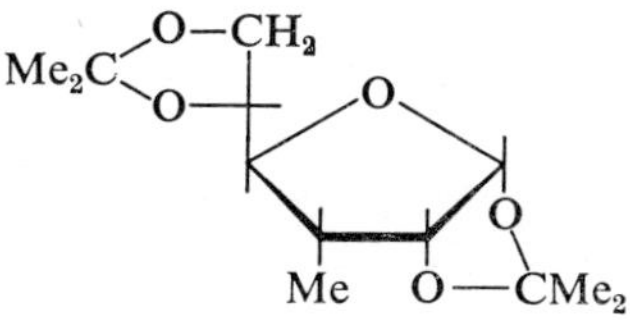

(368)

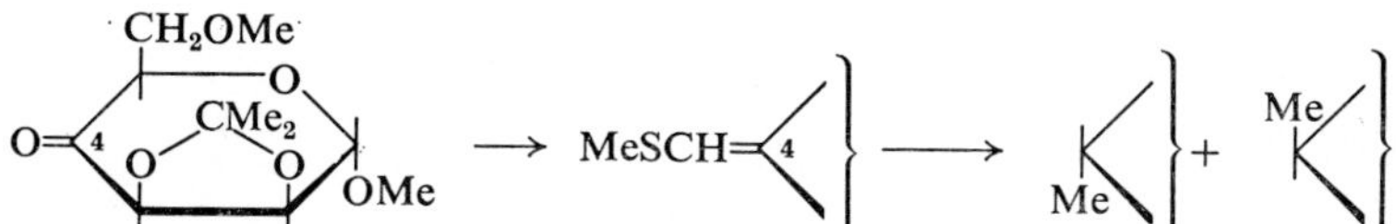

(369) (370) (371) (372)

The reaction sequence illustrated in Scheme 69 has been used to prepare some 2-deoxy-2-*C*-methylpentopyranosides.[410] Attempts to incorporate the 2-deoxy-2-*C*-methylpentopyranosides into nucleosides were unsuccessful.

Compounds (374) and (375) have been prepared by reaction of (373) with methylthiomethylene triphenylphosphorane.[411] The epimeric keto sugar (347) afforded a mixture of five compounds, *viz*. (374), (375), (376), (377), and (378), with this Wittig reagent. A variety of 3-deoxy-3-*C*-methyl

[408] T. D. Inch, G. J. Lewis, R. P. Peel, and N. Williams, *Chem. Comm.*, 1970, 1549.
[409] J. M. J. Tronchet, J. M. Bourgeois, R. Graf, and J. Tronchet, *Compt. rend.*, 1969, **296C**, 420.
[409a] J. M. J. Tronchet and J. M. Chalet, *Helv. Chim. Acta*, 1970, **53**, 364.
[410] A. Rosenthal and M. Sprinzl, *Canad. J. Chem.*, 1970, **48**, 3253.
[411] J. M. J. Tronchet and J. M. Bourgeois, *Helv. Chim. Acta*, 1970, **53**, 1463.

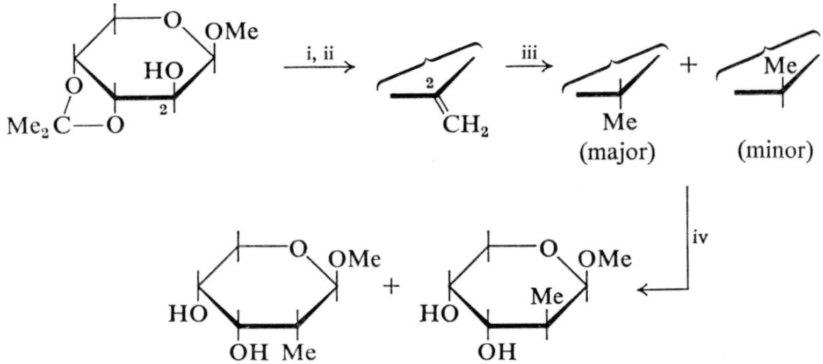

Reagents: i, RuO_4; ii, $Ph_3P\cdot MeBr$, BuLi; iii, H_2;
iv, separation of isomers followed by MeOH, H^+

Scheme 69

derivatives were obtained from these unsaturated derivatives by hydrogenation, desulphurisation and, in some cases, chain-shortening reactions.

The nitro-sugars (348) and (349) were converted into the nitro-olefin (379) either by treatment with acetic anhydride in DMSO [400] or by treatment of the acetate derivative of (349) with potassium carbonate in benzene [401] (Schmidt–Rutz reaction). One report [400] claimed that (379) was

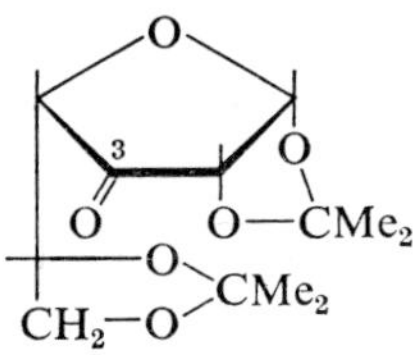

(373)

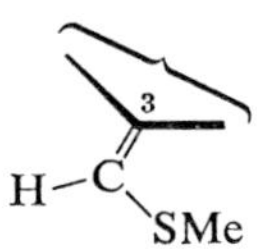

(374)

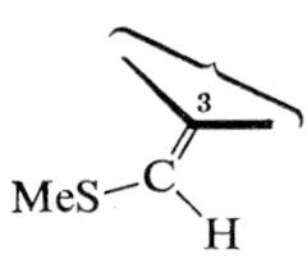

(375)

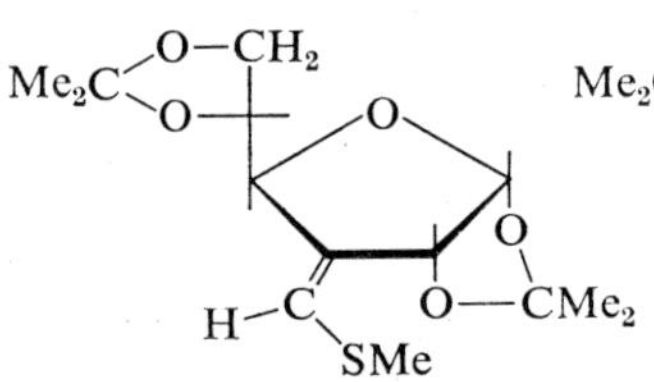

(376)

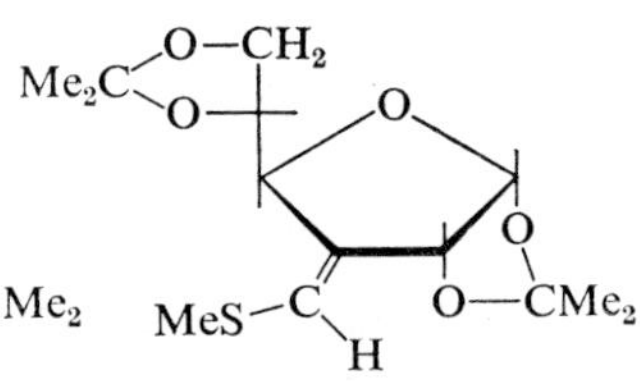

(377)

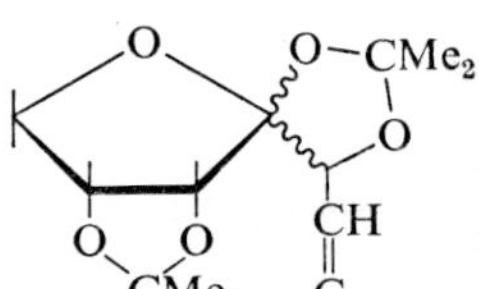

(378)

converted into (380) by borohydride reduction whereas another report [401] favoured the D-*gluco*-configuration (381) for the product.

Branched-chain sugar derivatives have been synthesised by Wittig reactions on aldehydo-derivatives of L-sorbose and L-arabinose.[412]

[412] Yu. A. Zhdanov, L. A. Uzlova, L. P. Laskina, and O. A. Gavrilenko, *Zhur. obschchei Khim.*, 1970, **40**, 666.

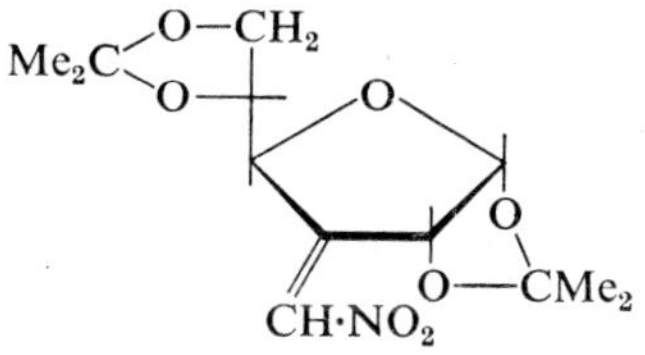

(379)

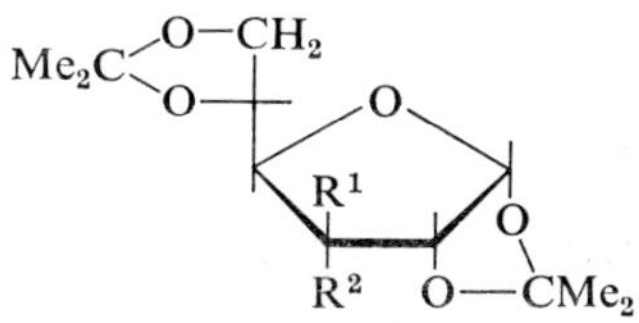

(380) $R^1 = H, R^2 = CH_2{\cdot}NO_2$
(381) $R^1 = CH_2{\cdot}NO_2, R^2 = H$

The 4-*C*-cyclopropyl-D-*ribo*-tetrofuranose derivative (382) has been prepared from the isomeric compound (383) (vol. 2, p. 145) by an oxidation–reduction sequence; it was also obtained from (384) following a Simmons–Smith reaction and hydrogenolysis of the protecting group.[413] Other cyclopropyl sugar derivatives are described in Chapter 14.

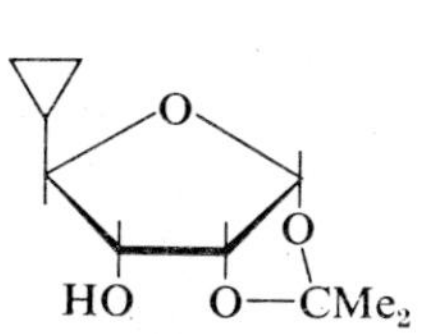

(382)

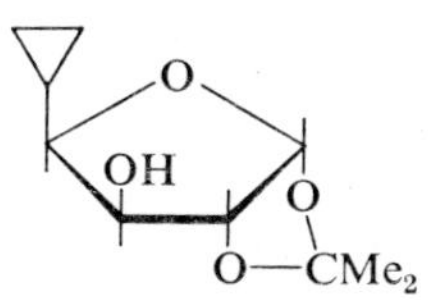

(383)

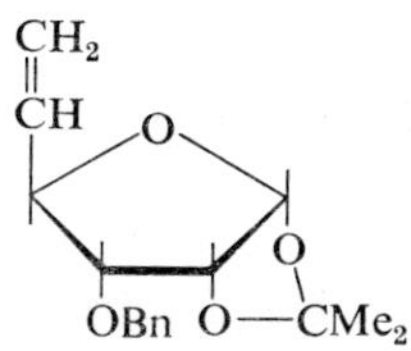

(384)

A novel branched-chain sugar (340) was obtained by treatment of (218) with cyclopentadiene.[394]

Some interesting branched-chain sugar phosphonates have been prepared as illustrated in Scheme 70.[414] The 3-phosphonate derivatives were converted into nucleoside analogues and, by condensation with DCC, into analogues of cyclic phosphates [415] (see also Chapter 18).

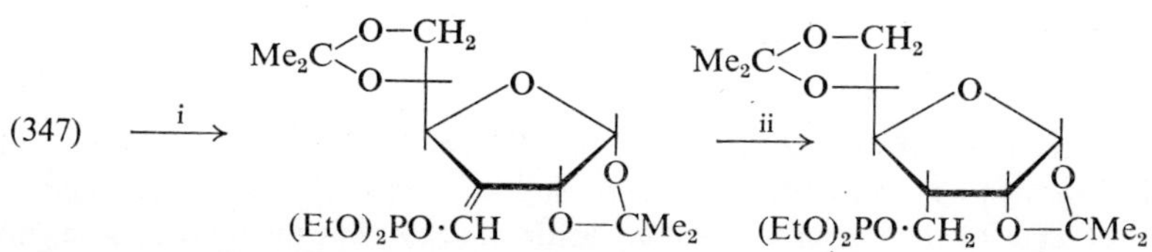

Reagents: i, $CH_2[PO(OEt)_2]_2$, BuLi; ii, H_2, Pd

Scheme 70

[413] D. Horton and C. G. Tindall, jun., *Carbohydrate Res.*, 1970, **15**, 215.
[414] H. P. Albrecht, G. H. Jones, and J. G. Moffatt, *J. Amer. Chem. Soc.*, 1970, **92**, 5511.
[415] G. H. Jones, H. P. Albrecht, N. P. Damodaran, and J. G. Moffatt, *J. Amer. Chem. Soc.*, 1970, **92**, 5510.

16
Alduloses, Dialduloses, and Diuloses

The isolation of a *talo*-heptulose and an *allo*-heptulose from avocado have been described together with new chromatographic data on heptuloses.[416] 5-Keto-D-fructose, as well as γ-pyrone derivatives, were formed on treatment of D-glucitol with *Gluconobacter suboxydans*.[417] 3-Deoxy-D-*erythro*-hexos-2-ulose (385), obtained from D-glucose by prolonged heating with *p*-toluidine, was isolated from the reaction mixture following conversion into its highly crystalline bis(benzoylhydrazone) (386). Pure (385) was regenerated from (386) by transhydrazonation with benzaldehyde. The procedure (Scheme 71) is claimed as an improved method for the preparation of 3-deoxyaldos-2-uloses.[418]

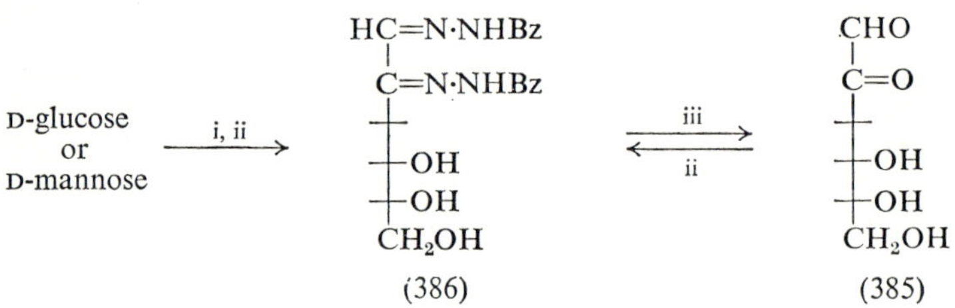

Reagents: i, *p*-toluidine; ii, $BzNH \cdot NH_2$; iii, PhCHO

Scheme 71

A synthesis of methyl 2,3-*O*-isopropylidene-β-L-*erythro*-pentopyranosid-4-ulose has been described.[404] Keto sugars [*e.g.* (207)] were obtained on photolysis of the corresponding azides [*e.g.* (206)] but the method does not appear to be of preparative significance.[317, 419] It has been demonstrated that oxidation of cyclic acetal derivatives of hexitols [*e.g.* (60)] with chromium trioxide in acetic acid afforded hex-3-ulose derivatives [*e.g.* (61)].[140]

Two distinct reaction pathways are evident in the alkaline degradation of 3-deoxyaldosuloses.[420] For example, 3-deoxy-D-*erythro*-hexosulose was

416 I. Johansson and N. K. Richtmyer, *Carbohydrate Res.*, 1970, **13**, 461.
417 K. Sato, Y. Yamada, K. Aida, and T. Uemura, *Agric. and Biol. Chem.* (*Japan*), 1969, **33**, 1606, 1612.
418 H. El Khadem, D. Horton, M. H. Meshreki, and M. A. Nashed, *Carbohydrate Res.*, 1970, **13**, 317.
419 R. L. Whistler and L. W. Doner, *J. Org. Chem.*, 1970, **35**, 3562.
420 R. M. Rowell and J. Green, *Carbohydrate Res.*, 1970, **15**, 197.

isomerised to the saccharinic acids 3-deoxy-D-*ribo*- and D-*arabino*-hexonic acids by a benzilic acid type of rearrangement, and it also suffered oxidative carbon–carbon bond cleavage to form 2-deoxy-D-*erythro*-pentonic acid. The preponderant pathway, even under oxidative conditions, was found to be the rearrangement.

The structure of 3-keto-lactose[4-*O*-(β-D-*xylo*-hexopyranosyl-3-ulose)-D-glucopyranose] has been established by n.m.r. spectroscopy;[421] the compound crystallised from methanol as a hemiacetal. ^{1}H N.m.r., i.r., and u.v. spectral data for the four 1,6-anhydro-2,4-di-*O*-*p*-tolylsulphonyl-β-D-hexopyranos-3-uloses (387)—(390) have been reported.[422] Isomerisations

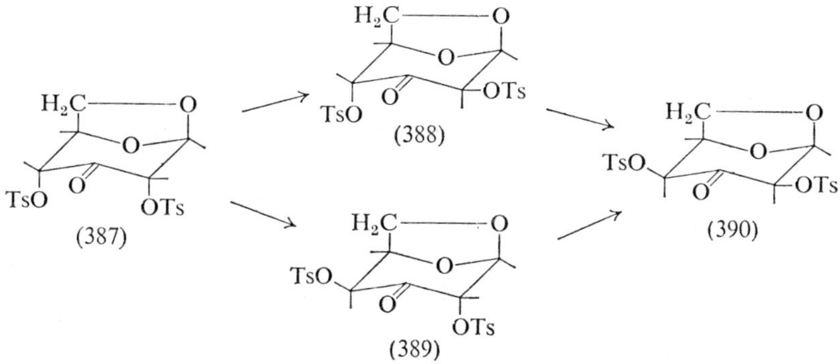

occurred in chloroform containing 5% pyridine according to the sequence (387) → [(388) and (389)] → (390). Hemiacetal formation was favoured in the compounds containing axial α-tosyloxy-groups but was inhibited when equatorial tosyloxy groups were present.

The anomeric 2-uloses of methyl 4,6-*O*-benzylidene-3-*O*-methyl-D-glucopyranoside and methyl 4,6-*O*-benzylidene-3-deoxy-D-*xylo*-hexopyranoside have been prepared. The anomers were separated by fractional crystallisation and it was suggested that the procedure might be generally useful for separating mixtures of anomeric glycosides.[423]

Reactions involving acetal rings on ulose derivatives have been described. Thus, the *gem*-diol (391) was rearranged to the hemiacetal (392) by heating at 60 °C for 3 h, whereas the keto form of (391) was rearranged to (393) on distillation. The acyclic diacetal (394) was transformed into the corresponding pyranose diacetal (395) on heating.[424]

The reactions of diazoalkanes with diuloses have already been mentioned in Chapter 15. The formation of products from the addition of diazoalkanes

[421] M. Anteunis, J. Van Beeumen, A. De Bruyn, and J. De Ley, *Bull. Soc. chim. belges*, 1969, **78**, 651.
[422] M. Černý, J. Pacák, and J. Staněk, *Carbohydrate Res.*, 1970, **15**, 379.
[423] K. Antonakis, *Compt. rend.*, 1970, **270C**, 1598.
[424] J. M. J. Tronchet, J. M. Bourgeois, and B. Baehler, *Helv. Chim. Acta*, 1970, **53**, 368.

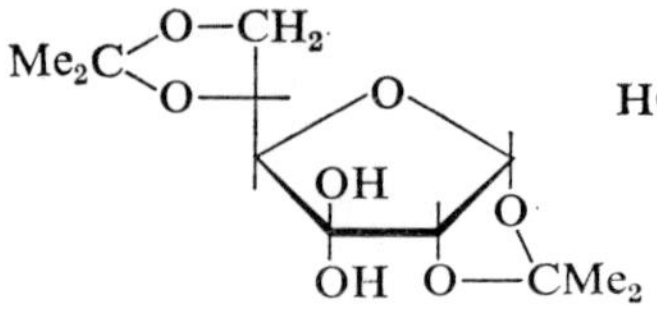
(391)

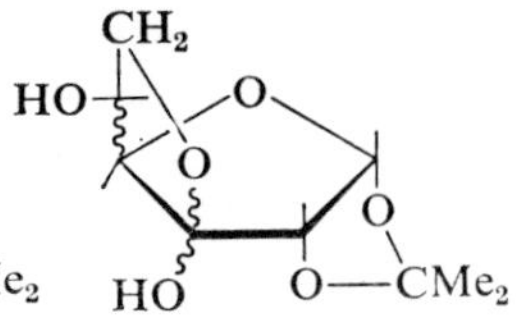
(392)

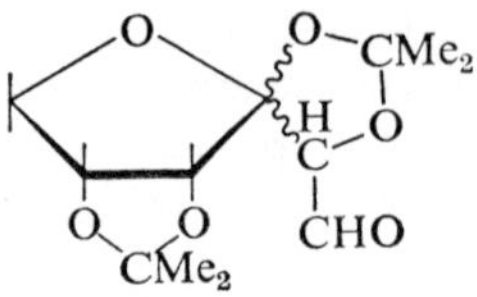
(393)

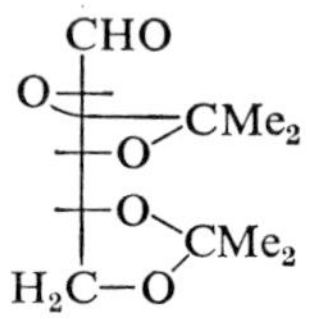
(394)

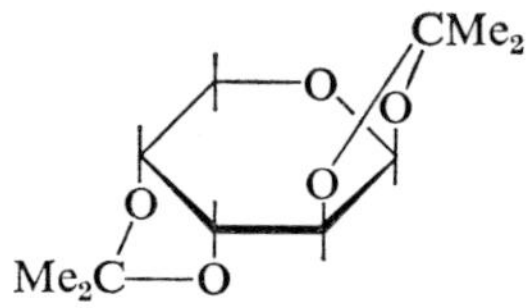
(395)

to the compound (362) was rationalised in terms of the preferred geometry of the zwitterion intermediate. Product formation appeared to be controlled by steric rather than by electronic factors.[408]

Dialdoses [*e.g.* (280)] were able to undergo Reformatsky reactions with zinc and ethyl bromoacetate to give products such as (281),[370] and were useful intermediates in the synthesis of uronic acids (see Chapter 17).[425]

D-Ribose-3-^{18}O and -2-^{18}O were conveniently synthesised for incorporation into nucleosides by reduction of the hydrates prepared from (396) and (397) by an uncatalysed exchange reaction in ^{18}O-enriched water.[426]

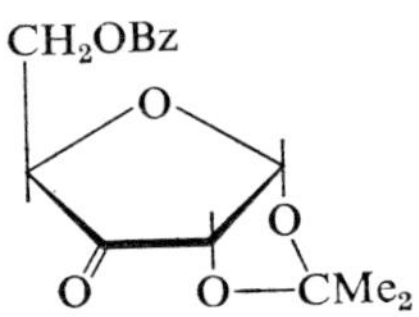
(396)

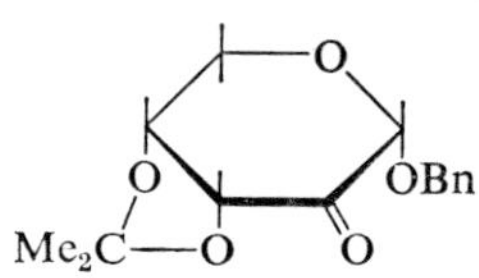
(397)

[425] D. Horton and F. O. Swanson, *Carbohydrate Res.*, 1970, **14**, 159.
[426] H. Follmann and H. P. C. Hogenkamp, *J. Amer. Chem. Soc.*, 1970, **92**, 671.

17

Sugar Acids and Lactones

Increasing interest is being shown in the chemistry of carboxylic acids derived from carbohydrates and this year ulosonic acids have received appreciable attention.

Aldonic Acids

6-Deoxy-L-gulono-1,4-lactone has been prepared from D-glucofuranurono-6,3-lactone by the route shown in Scheme 72. Reduction of the tosylhydrazone gave the lactone, presumably by way of an ammonium salt intermediate.[427]

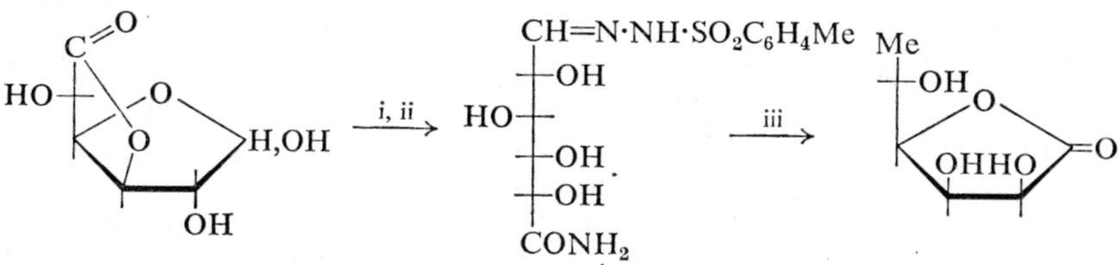

Reagents: i, $NH_2NHSO_2C_6H_4Me$; ii, NH_3; iii, KBH_4, MeOH

Scheme 72

The γ- and δ-lactones of 2-acetamido-2-deoxy-D-gluconic acid have been isolated but only the latter inhibited the action of *N*-acetyl-β-D-glucosaminidase.[427a]

The isomeric 2-acetamido-2,3-dideoxyhex-2-enono-1,4-lactones (398) and (399) have been prepared by treating 2-acetamido-2-deoxy-D-glucose, -D-mannose, and -D-galactose with unbuffered bromine water followed by alkali.[428]

A continuous process for the preparation of potassium D-arabonate from D-glucose has been described,[429] and the cobalt D-gluconate system has been studied. Several species were formed depending upon the pH of

[427] T. T. Galkowski, R. W. Kocon, and W. C. Griffiths, *Carbohydrate Res.*, 1970, **13**, 187.
[427a] T. E. Couling and R. Goodey, *Biochem. J.*, 1970, **119**, 303.
[428] N. Pravdić and H. G. Fletcher jun., *Carbohydrate Res.*, 1970, **12**, 471.
[429] A. A. Shereshevskii and E. M. Guseinov, *Khim.-Farm. Zh.*, 1969, **3**, 45 (*Chem. Abs.*, 1970, **72**, 67216*e*).

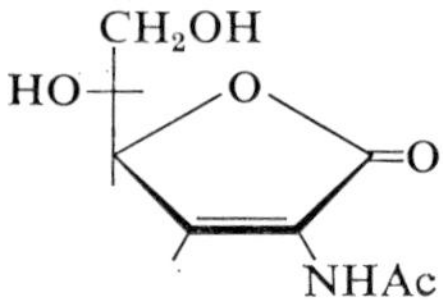

(398)

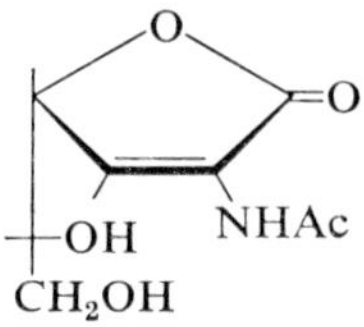

(399)

the medium.[430] Modifications to the standard preparation of D-glucono-lactone from D-glucose have been described.[431]

Tetra-*O*-benzoyl derivatives of D-glucono-γ- and -δ-lactone have been reported,[432] and the 2,3:4,5-di-*O*-isopropylidene derivatives of D-xylonic and D-arabonic acid have been prepared.[145]

'α'-D-Isosaccharinic acid has the *erythro*-structure, as revealed by an *X*-ray crystallographic study[433] (*cf.* vol. 2, p. 201; vol. 3, p. 180).

Ulosonic Acids

Several papers have dealt with 3-deoxyoctulosonic acids because of their biological significance. 3-Deoxy-D-*manno*-octulosonic acid has been synthesised by converting D-mannose into 2-deoxy-D-*manno*-heptose, followed by application of the cyanohydrin method and selective oxidation of the C-2 hydroxy-groups of the resulting epimeric 3-deoxyoctonic acids. All 3-deoxyaldulosonic acids gave rise to a molar proportion of 3-*C*-formylpyruvate when oxidised with periodate in dilute sulphuric acid, and this may be estimated by the thiobarbituric acid method.[434]

The 3-deoxyoctulosonic acids, obtained by condensation of D-arabinose with oxalacetic acid, on treatment with ethane thiol and concentrated hydrochloric acid at room temperature gave (400) and (401) in the ratio 5 : 1.[435]

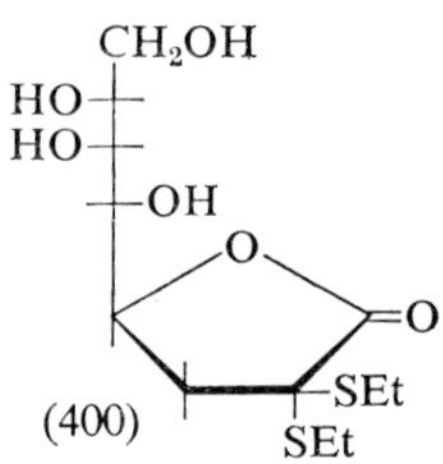

(400)

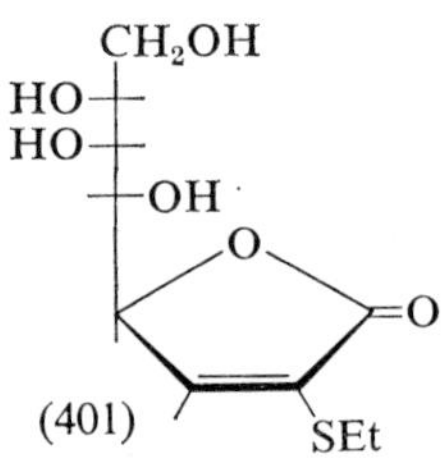

(401)

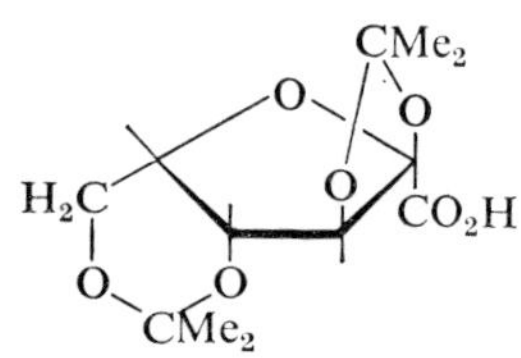

(402)

[430] J. F. Ashton and W. F. Pickering, *Austral. J. Chem.*, 1970, **23**, 1367.
[431] I. Z. Sergienko and B. N. Stepanenko, *Priklad. Biokhim. i Mikrobiol.*, 1969, **5**, 715.
[432] R. M. De Lederkremer, A. F. Cirelli, and J. O. Deferrari, *Carbohydrate Res.*, 1970, **13**, 9.
[433] D. L. Hughes, J. Trotter, and J. Howard, *J. Chem. Soc.* (*B*), 1970, 983.
[434] D. Charon, R. S. Sarfati, D. R. Strobach, and L. Szabo, *European J. Biochem.*, 1969, **11**, 364.
[435] B. A. Dmitriev and L. V. Backinowsky, *Carbohydrate Res.*, 1970, **13**, 293.

3-Deoxyheptulosonic acids with the D-*arabino*- and D-*lyxo*-configurations have been prepared from D-glucose and D-galactose, respectively.[436]

An interesting use of the 2-keto-L-gulonic acid derivative, 2,3:4,6-di-*O*-isopropylidene-α-L-*xylo*-hexofuranulosonic acid (402), was in the resolution of chiral bases.[437]

A preliminary communication has reported an interesting application of the Reformatsky reaction to give a 2-deoxy-3-octulosonic acid derivative (Scheme 73).[438]

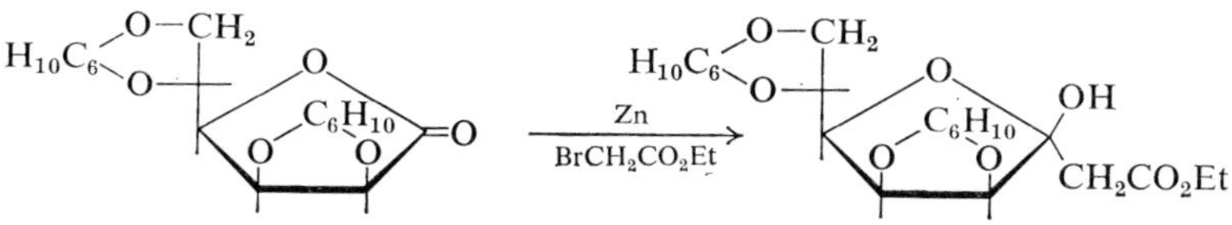

Scheme 73

5-Ulosonic acids have also received attention. In particular, a full paper has appeared on a reaction previously reported briefly (vol. 3, p. 133) whereby acetylated methyl β-D-hexopyranosides were converted into the corresponding methyl 5-hexulosonates with chromium trioxide in acetic acid (Scheme 74). The reaction appears to be specific for glycosides

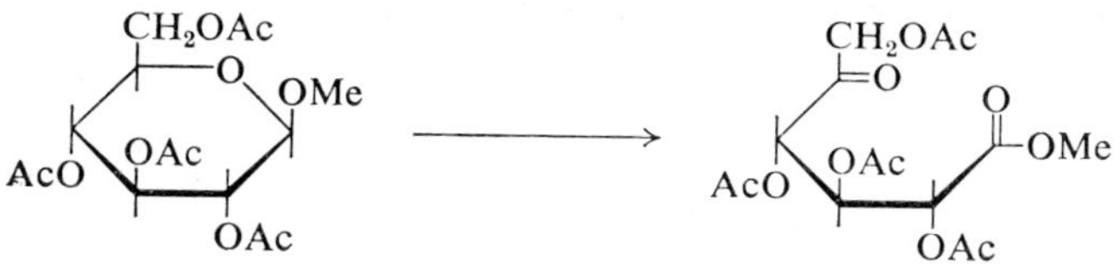

Scheme 74

having equatorial aglycones since methyl α-D-glycopyranosides, with the exception of the D-*ido*-compound which exists to some degree in the 1*C*-conformation, did not react. Hydrogenation of the resulting ketones led to selective syntheses of derivatives of L-aldonic acids. Both α- and β-anomers of acetylated furanosides gave the corresponding 4-ketones (Scheme 75), and the lactose peracetates gave the corresponding ketones

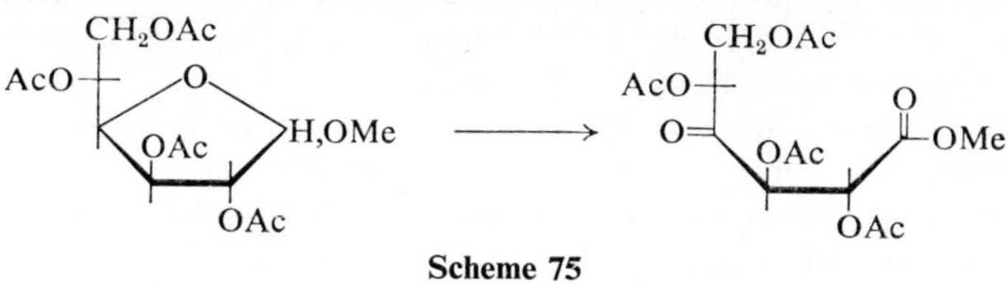

Scheme 75

[436] G. B. Paerels and H. W. Geluk, *Rec. Trav. chim.*, 1970, **89**, 813.
[437] A. Brossi and S. Teitel, *J. Org. Chem.*, 1970, **35**, 3559.
[438] Yu. A. Zhdanov, Yu. E. Alexeev, and Ch. A. Khourdanov, *Carbohydrate Res.*, 1970, **14**, 422.

with acyclic residues derived from the D-galactose moieties. Surprisingly, the branched-chain glycoside (403) was oxidised as shown in Scheme 76.[439]

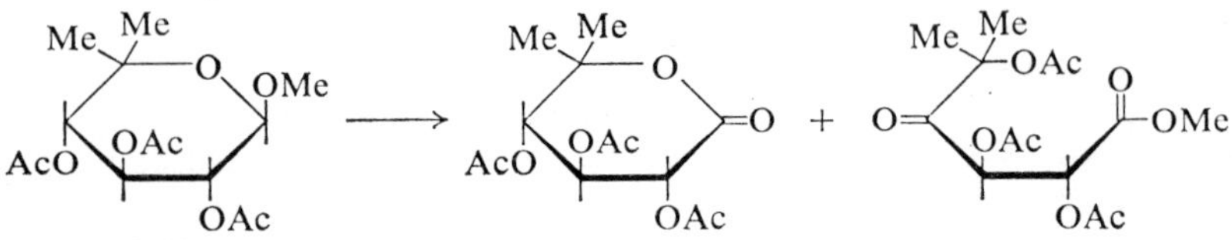

(403)

Scheme 76

The catalytic hydrogenation of calcium D-*xylo*-5-hexulosonate on transition-metal catalysts has been studied with the view to developing conditions for obtaining the maximum yield of calcium L-idonate.[440]

Studies using u.v., i.r., and n.m.r. spectroscopy suggested that sodium 5-keto-D-gluconate exists in aqueous solution as a mixture of the hemi-acetal and keto forms.[441]

Uronic Acids

Various uronoside derivatives are described in Chapter 3, and the flavone glucuronosyl glucuronoside (404) has also been reported.[442]

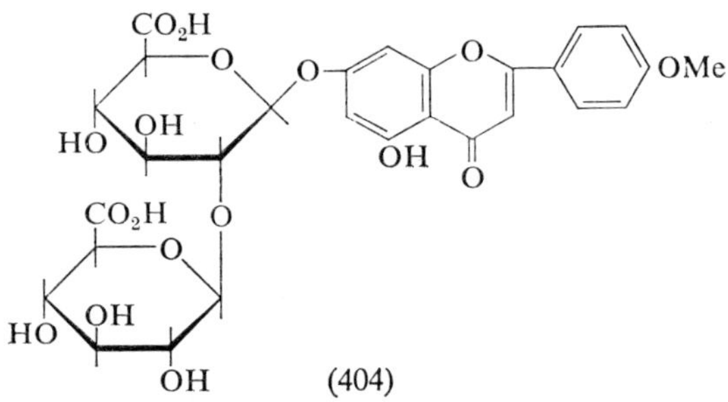

(404)

A synthetic route to uronic acids is illustrated in Scheme 77. It was suggested that this procedure offers a useful general method which has a number of advantages over the approach involving the cyanohydrin reaction applied to ω-aldehydo-sugar derivatives.[425]

A related route involved application of the Wittig reaction to these derivatives, but compound (405) on treatment with (ethoxycarbonylmethylene)triphenyl phosphorane did not give the expected product (406),

439 S. J. Angyal and K. James, *Austral. J. Chem.*, 1970, **23**, 1209.
440 C.-Y. Chen, H. Yamamoto, and T. Kwan, *Chem. and Pharm. Bull.* (*Japan*), 1969, **17**, 2349.
441 C.-Y. Chen, H. Yamamoto, and T. Kwan, *Chem. and Pharm. Bull.* (*Japan*), 1970, **18**, 815.
442 M. Okigawa, H. Hatanaka, N. Kawano, I. Matsunaga, and Z. Tamura, *Tetrahedron Letters*, 1970, 2935.

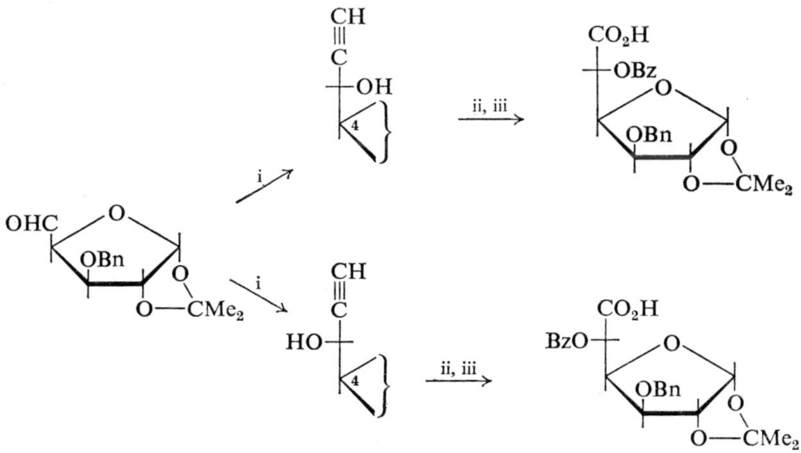

Reagents: i, HC≡CMgBr; ii, BzCl, py; iii, O_3

Scheme 77

but instead (407) and (408) were formed. Compound (406) was required as an intermediate in the synthesis of the nucleotide analogue (409), which was finally obtained in low yield as the methyl ester by anodic coupling of (410) with methyl hydrogen succinate.[443]

It was further reported in the nucleoside series that oxidation of 2′,3′-*O*-isopropylidene-inosine with chromium trioxide in acetic acid gave the

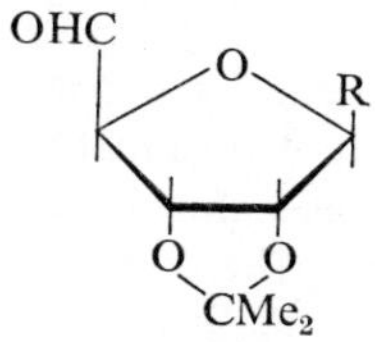

(405)

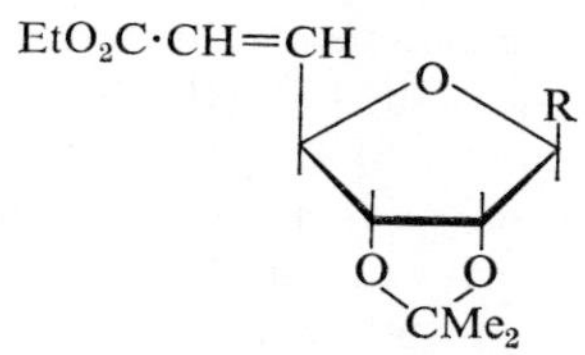

(406)

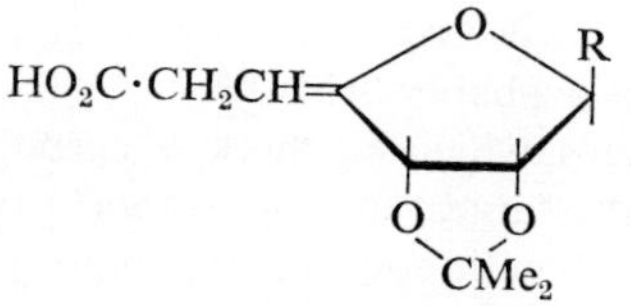

(407)

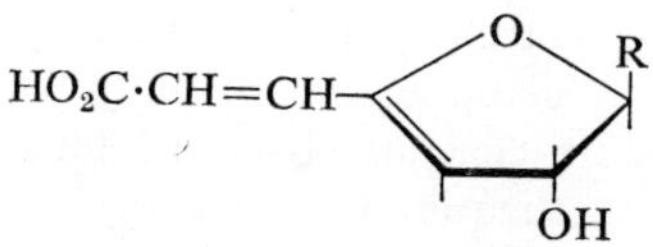

(408)

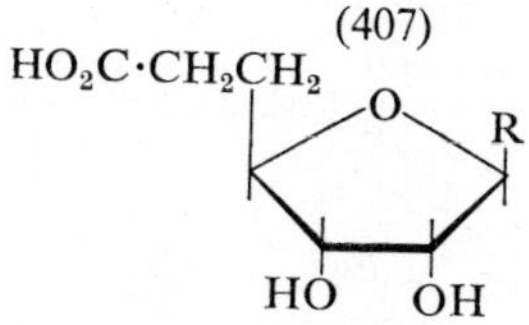

(409)

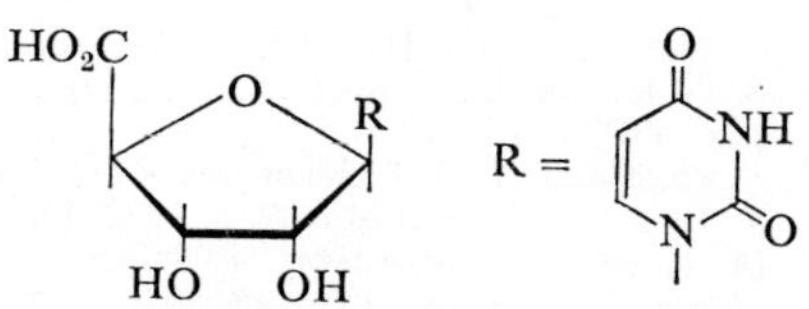

(410)

[443] P. Howgate, A. S. Jones, and J. R. Tittensor, *Carbohydrate Res.*, 1970, **12**, 403.

uronic acid analogue in good yield. Various derivatives of the product were also described.[444] Other specific compounds which have been reported include 2-amino-2-deoxy-D-mannuronic acid (prepared as in Scheme 78),[445]

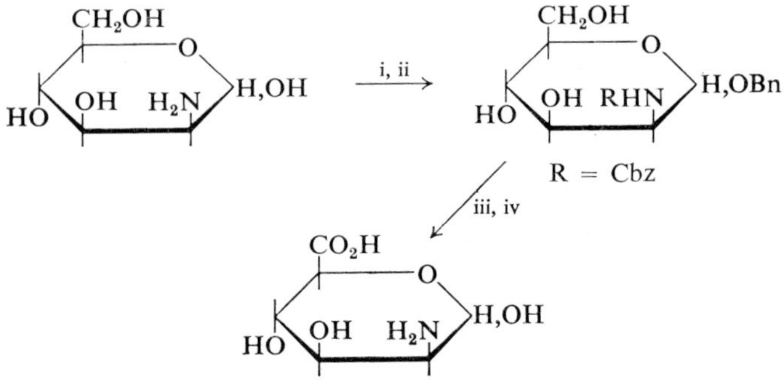

Reagents: i, RCl; ii, BnOH, HCl; iii, O_2–Pt; iv, H_2–Pd/C

Scheme 78

benzyl 2,3,4-tri-*O*-benzyl-β-D-glucopyranosiduronic acid,[446] and [^{14}C]-methyl D-glucuronate.[447] 1,2:3,5-Di-*O*-benzylidene-α-D-glucofuranuronic acid was prepared by treatment of D-glucuronic acid with zinc chloride and benzaldehyde, and gave [^{14}C]methyl D-glucuronate following reaction with [^{14}C]methyl iodide in DMF and hydrogenolysis of the protecting groups.[447] [6-^{14}C]-L-Iduronic acid has been synthesised.[250]

D-Glucuronolactone has been transformed into the synthetically useful derivative 3,5-di-*O*-acetyl-1,2-*O*-isopropylidene-5-*O*-*p*-tolylsulphonyl-α-D-glucofuranose (96) (see p. 45).[207] The yellow product formed when D-glucuronolactone is treated with alkali has been shown to be the reductone (411) which exists in the ene–diol form.[448] The tautomerism of (411) was studied by polarographic and u.v. methods.[449] In addition to 2-furaldehyde and reductic acid, the degradation of D-glucuronic acid at pH 3.5 gave catechol, 2,3-dihydroxytoluene, 2,3-dihydroxybenzoic acid, 2,3-dihydroxyacetophenone, (412), (413), and probably (414).[450]

A method for cleaving glycuronoside linkages, which Kochetkov's group have been investigating for some years, was the subject of a new report. In this reaction the derived amides were treated with sodium hypochlorite

[444] R. R. Schmidt and H.-J. Fritz, *Chem. Ber.*, 1970, **103**, 1867.
[445] N. G. Kundu, J. F. Crawford, B. Prajsnar, E. J. Reed, and S. Rosenthal, *Carbohydrate Res.*, 1970, **12**, 225.
[446] E. Zissis and H. G. Fletcher jun., *Carbohydrate Res.*, 1970, **12**, 361.
[447] R. H. Shah, *Carbohydrate Res.*, 1970, **12**, 43.
[448] M. Kawata, Y. Mizutani, N. Shinriki, M. Kimura, and M. Ishidate, *Chem. and Pharm. Bull.* (*Japan*), 1970, **18**, 50.
[449] M. Kawata, Y. Mizutani, N. Shinriki, M. Kimura, and M. Ishidate, *Chem. and Pharm. Bull.* (*Japan*), 1970, **18**, 55.
[450] T. Popoff and O. Theander, *Chem. Comm.*, 1970, 1576.

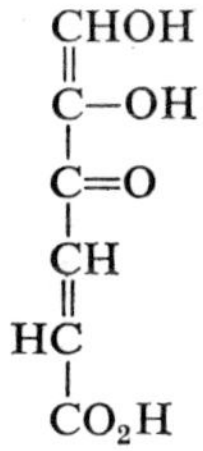

(411)

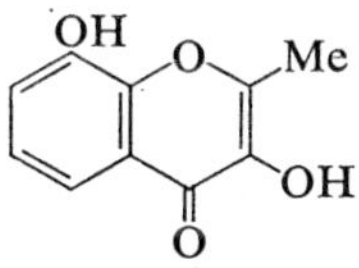

(412)

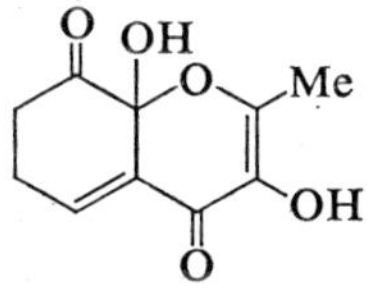

(413)

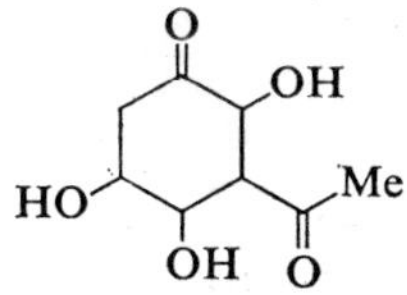

(414)

followed by mild hydrolysis with acid (Scheme 79). Details of studies with model hexopyranosiduronic acids were reported; the dialdehydes produced were characterised as pentitol acetates and the procedure has been applied to a xylan from which 94% of the terminal uronic acid residues were removed without cleavage of other glycosidic linkages.[451]

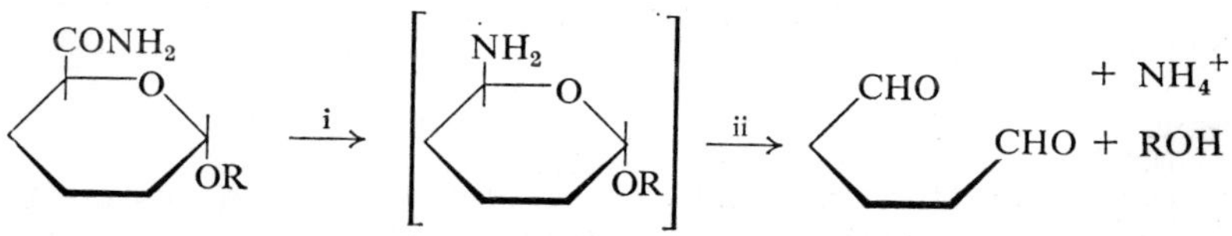

Reagents: i, NaOCl; ii, H_3O^+

Scheme 79

Aldaric Acids

An enzyme which catalysed the oxidation of D-galacturonic acid to galactaric acid and of D-glucuronic acid to D-glucaric acid has been isolated from *Pseudomonas syringae* and purified.[452]

The unsaturated acid (415) was obtained on Wolff degradation of the 1,8-bisdiazo-*galacto*-octodiulose derivative (416). Elimination also occurred

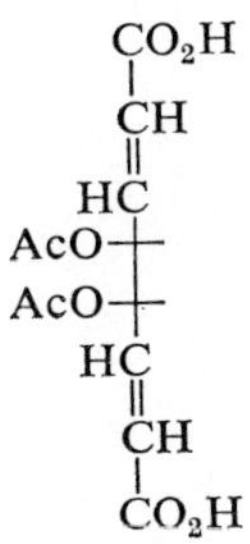

(415)

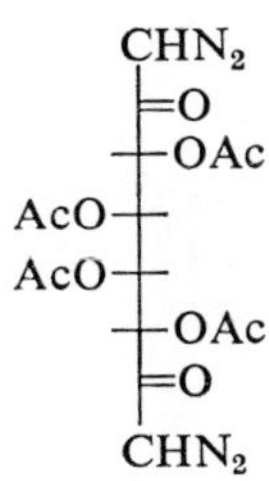

(416)

[451] N. K. Kochetkov, O. S. Chizhov, and A. F. Svirdov, *Carbohydrate Res.*, 1970, **14**, 277.

[452] D. F. Bateman, T. Kosuge, and W. W. Kilgore, *Arch. Biochem. Biophys.*, 1970, **136**, 97.

during an analogous degradation of (233) and the unsaturated aldonic acid (234) was the product.[341]

Ascorbic Acid

The browning reaction of L-ascorbic acid with a variety of amines has been studied,[453] and the kinetics of its oxidation with osmium(VIII) have been reported.[454] The specifically-labelled acids, [6-^{14}C]-L-ascorbic and [5-^{14}C]-L-ascorbic acids, have been prepared.[455]

A simple procedure for the preparation of dehydroascorbic acid has been reported, and spectroscopic measurements indicated that the product is dimeric.[456] Procedures have been devised for the determination of the dimer.[457]

Muramic Acid

The absolute configuration of the lactyl side-chain of muramic acid [2-amino-3-*O*-(D-1-carboxyethyl)-2-deoxy-D-glucose] has been confirmed by two methods.[457a]

[453] T. Ozawa and Y. Nakamura, *J. Pharm. Soc. Japan*, 1970, **90**, 93.
[454] U. S. Mehrotra and S. P. Mushran, *Canad. J. Chem.*, 1970, **48**, 1148.
[455] D. B. Karr, E. M. Baker, and B. M. Tolbert, *J. Labelled Compounds*, 1970, **6**, 155.
[456] W. Müller-Mulot, *Z. physiol. Chem.*, 1970, **351**, 52, 56.
[457] W. Müller-Mulot, *Z. Analyt. Chem.*, 1970, **252**, 20.
[457a] A. Veyrières and R. W. Jeanloz, *Biochemistry*, 1970, **9**, 4153.

18
Inorganic Derivatives

Interesting nucleoside derivatives containing carbon–phosphorus bonds have been reported. Compounds (417), (418), and (419) were synthesised

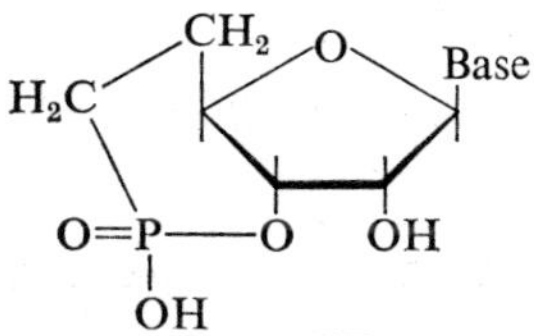

(417)

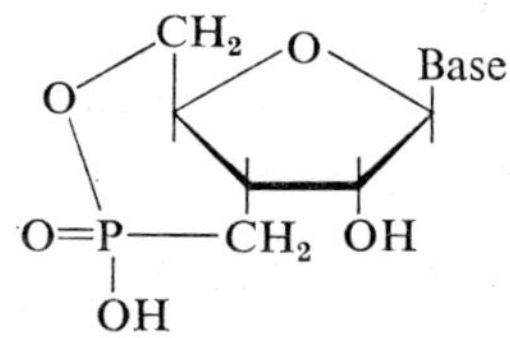

(418)

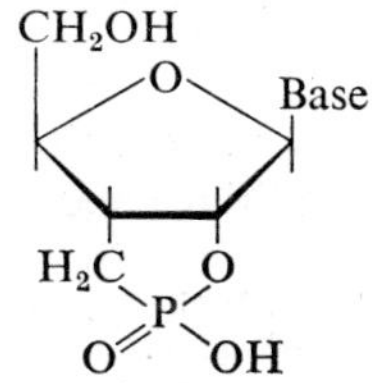

(419)

as analogues of nucleoside cyclic phosphates by cyclisation of the appropriate phosphonates using DCC. Various dinucleotide analogues were also prepared as possible enzyme substrates.[415] 3′-Deoxy-3′-(dihydroxyphosphinyl)nucleosides (420), which are isosteric analogues of nucleoside 3′-phosphates, were synthesised as shown in Schemes 70 and 80.[414] In

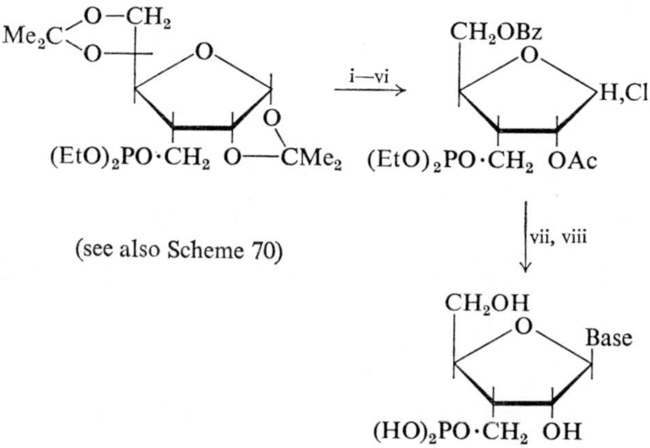

(420)

Reagents: i, H_3O^+; ii, IO_4^-; iii, $NaBH_4$; iv, BzCl, py; v, Ac_2O, AcOH, H^+; vi, HCl; vii, base derivative; viii, HO^-

Scheme 80

related work, the Michaelis–Arbuzov reaction has been used to prepare such sugar derivatives as (421) and (422), with carbon–phosphorus bonds, from halogenated precursors.[458]

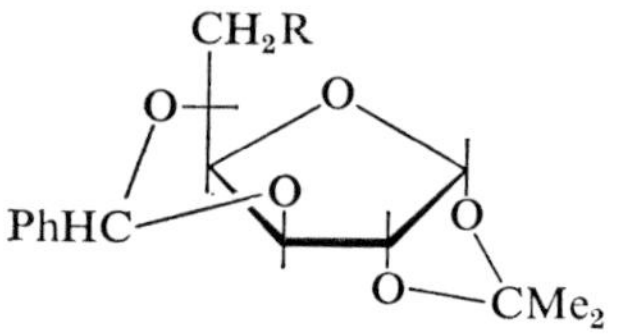

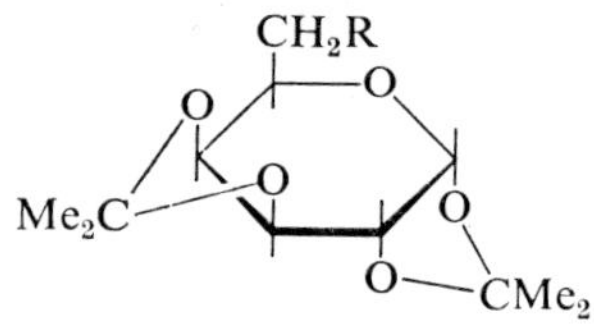

R = $\cdot$P(O)(OEt)$_2$ or $\cdot$P(O)(NEt$_2$)$_2$

(421) (422)

Some cobalt(II) D-gluconates have been prepared under different pH conditions and have been studied.[430]

[458] S. Inokawa, Y. Tsuchiya, H. Yoshida, and T. Ogata, *Bull. Chem. Soc. Japan*, 1970, **43**, 3224.

19
Cyclitols

A system of nomenclature based on stereospecific numbering has been proposed for chiral, asymmetrically-substituted *myo*-inositols.[459]

1,4/2,5-Cyclohexanetetraol, 2-deoxy-*myo*-inositol, and *myo*-inositol have been isolated from extracts of *Mycobacterium lutheri*;[460] the latter compound has also been isolated from roasted *Coffea arabica*.[461]

Two papers have discussed the γ-radiolysis of inositols. γ-Irradiation of *myo*-inositol in water gave *myo*-2-inosose as the major product under a variety of conditions. The by-products varied, but idaric acid was always present in the mixtures.[462] The structures of the primary radicals resulting from γ-radiolysis of *meso*-inositol have been considered in detail,[463] and it was shown that certain primary radicals underwent loss of water.

X-Ray crystallographic studies on the various hydrates of dodecasodium *myo*-inositol hexaphosphate (phytic acid) have established the degree of hydration.[464] The structures of *epi*- and *myo*-inositols have been compared by crystallography.[465] It was suggested that the primary effects controlling conformations are intramolecular *O*–*O*-forces, but that in rigid cyclitol molecules the effects of these forces are much less marked than, for example, in alditols.

A method has been developed for assaying the enzyme responsible for the conversion of D-glucose 6-phosphate into *myo*-inositol 1-phosphate. The method was based on the observation that only the cyclitol phosphate liberated inorganic phosphate on oxidation with periodate.[466]

A number of inositol esters have been prepared. Inositol hexasulphates can be prepared conveniently by direct interaction of *myo*-inositol with either fuming sulphuric acid or chlorosulphonic acid.[467] The latter reagent caused less oxidation and was more suitable for more sensitive inositols.

459 B. A. Klyashchitskii, V. I. Shvets, and N. A. Preobrazhenskii, *Chem. and Phys. Lipids* 1969, **3**, 393.
460 M. V. Laycock and J. S. Craigie, *Canad. J. Biochem.*, 1970, **48**, 699.
461 A. R. Mishkin, R. S. Bower, and L. E. Anderson, *Carbohydrate Res.*, 1970, **13**, 170.
462 W. J. Criddle and E. Ward, *J. Chem. Soc.* (*B*), 1970, 40.
463 I. V. Nikitin, V. A. Sharpatii, L. I. Kudryashov, N. K. Kochetkov, and N. M. Emanuel, *Doklady Akad. Nauk S.S.S.R.*, 1970, **190**, 635.
464 M. R. Truter and M. E. Tate, *J. Chem. Soc.* (*B*), 1970, 70.
465 G. A. Jeffrey and H. S. Kim, *Carbohydrate Res.*, 1970, **15**, 310.
466 J. E. G. Barnett, R. E. Brice, and D. L. Corina, *Biochem. J.*, 1970, **119**, 183.
467 A. J. Fatiadi, *Carbohydrate Res.*, 1970, **12**, 293.

1,2,4,5,6-Penta-*O*-*p*-tolylsulphonyl-*myo*-inositol has been synthesised from (±)-bornesitol (423) by *p*-tolylsulphonylation and subsequent demethylation with boron tribromide.[468] 2,3,4,5,6-Penta-*O*-*p*-tolylsulphonyl-*myo*-inositol has also been synthesised.[469]

The syntheses of 1-*O*-methyl- and 5-*O*-methyl-DL-[^{14}C]*myo*-inositol (methyl [^{14}C]bornesitol and methyl [^{14}C]sequoyitol, respectively) have been

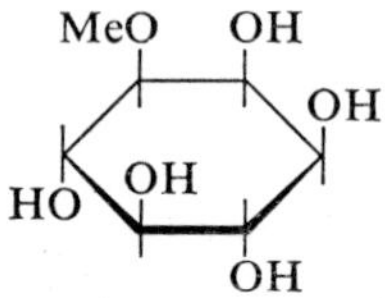

(423)

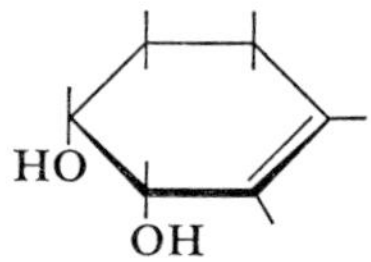

(424)

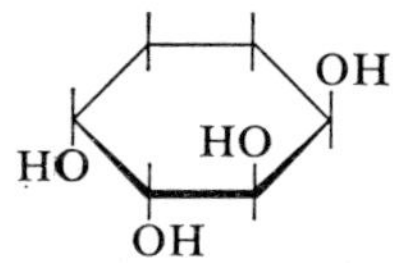

(425)

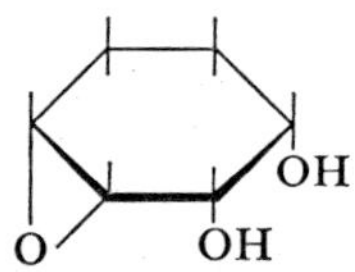

(426)

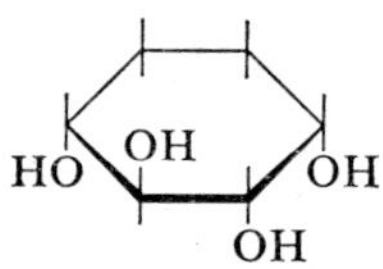

(427)

carried out.[470] Various isomers and derivatives of quinic acid have been prepared and characterised stereochemically by n.m.r. methods.[471] Tetraols have been prepared by treatment of cyclohexa-1,3-diene and cyclopenta-1,3-diene with alkaline or neutral permanganate.[472] From cyclohexadiene the normal *cis*-hydroxylation products (424) and (425) were obtained but, in addition, the abnormal products (426) and (427) were also formed. In the cyclopentadiene series, the abnormal pathway predominated.

A synthesis of a diastereoisomeric mixture of phosphatidylinositols has been described.[473] Reaction of racemic (428) with optically-active (429) afforded (430), which on treatment with sodium iodide in dry acetone followed by hydrogenolysis gave a mixture of diastereoisomeric phosphatidylinositols (431).

The syntheses of 1- and 3-*O*-α-D-mannopyranosyl-*myo*-inositol by the orthoester route have been described.[474] Resolution of the orthoesters prepared from 1,4,5,6-tetra-*O*-benzyl-*myo*-inositol and D-mannose orthoacetate afforded (432) and (433). Benzylation and hydrolysis of (432) afforded (434), and similar treatment of (433) yielded the enantiomeric

[468] E. G. Zhelvakova, V. I. Shvets, and N. A. Preobrazhenskii, *Zhur. org. Khim.*, 1970, **6**, 62.
[469] E. G. Zhelvakova, O. I. Ulyanova, V. I. Shvets, and N. A. Preobrazhenskii, *Khim. prirod. Soedinenii*, 1970, **2**, 163.
[470] R. H. Shah and F. Loewus, *J. Labelled Compounds*, 1970, **6**, 333.
[471] J. Corse and R. E. Lundin, *J. Org. Chem.*, 1970, **35**, 1904.
[472] H. Z. Sable, K. A. Powell, H. Katchian, C. B. Niewoehner, and S. B. Kadlec, *Tetrahedron*, 1970, **26**, 1509.
[473] P. A. Gent, R. Gigg, and C. D. Warren, *Tetrahedron Letters*, 1970, 2575.
[474] B. A. Klyashchitskii, V. I. Shvets, S. D. Sokolov, and N. A. Preobrazhenskii, *Zhur. obshchei Khim.*, 1970, **40**, 1148.

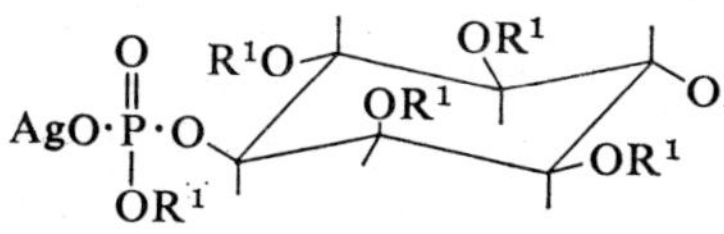

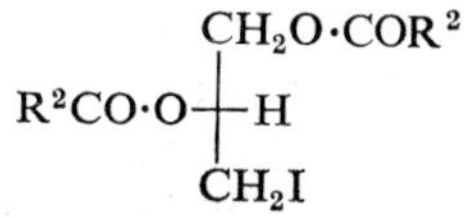

(428) R^1 = Bn (429) R^2 = $\cdot(CH_2)_{14}Me$

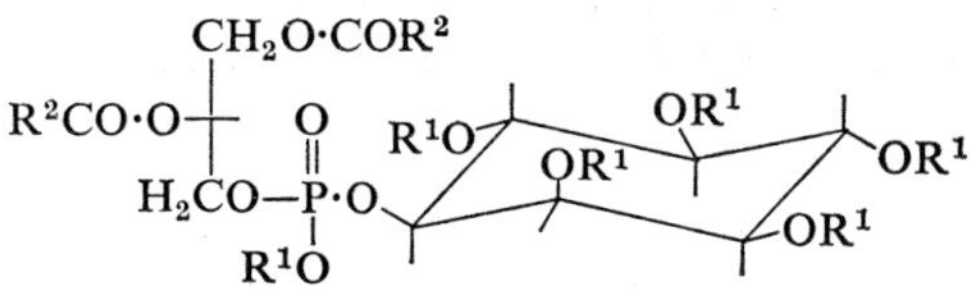

(430) R^1 = Bn
(431) R^1 = H

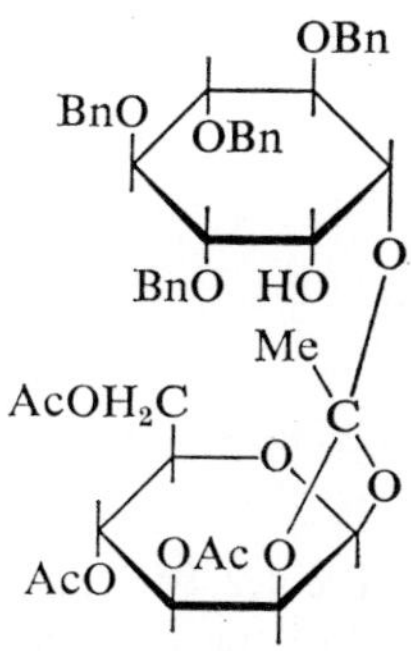

(432)

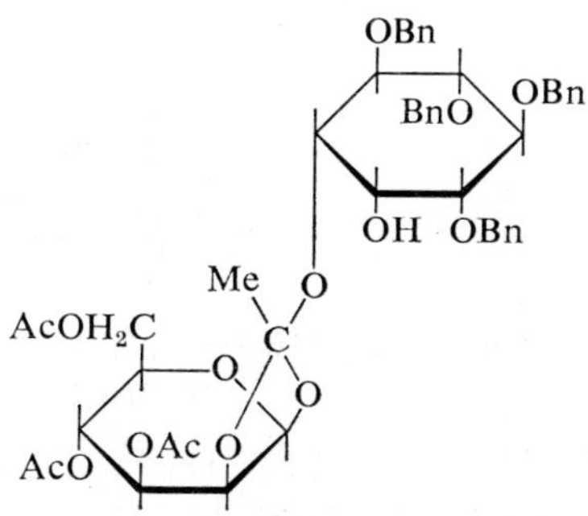

(433)

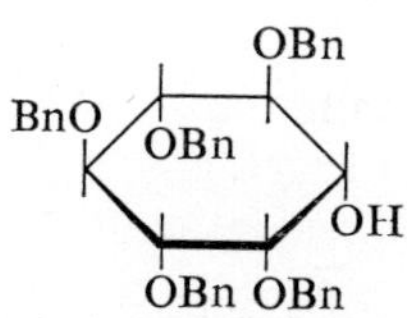

(434)

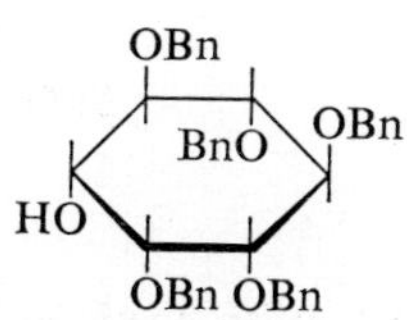

(435)

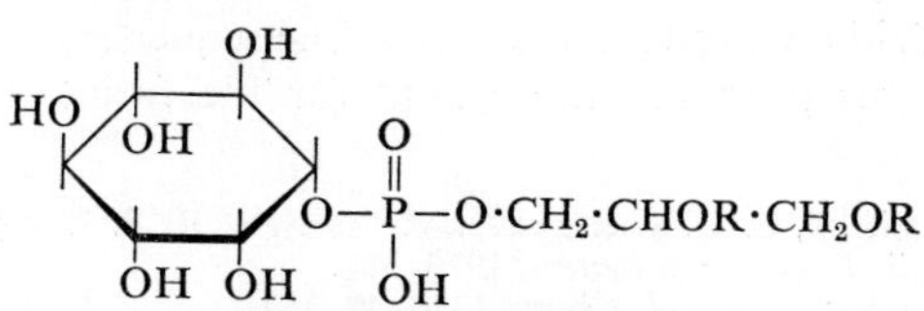

(436) R = palmitoyl

compound (435). Treatment of (434) with dichlorophenyl phosphate and 1,2-dipalmitoylglycerol followed by hydrogenolysis gave 1-*O*-[1′,2′-dipalmitoylglyceryl-3′-phosphoryl]-*myo*-inositol (436), which had physical and spectral characteristics similar to those of natural monophosphoinositides.[475]

Other reactions of the cyclitols to be described include application of the Corey–Winter procedure to (437) to provide a new synthesis of conduritol B (Scheme 81).[476] Aromatisation occurred when 1,2-*O*-isopropylidene-*myo*-inositol 3,4,5,6-tetra-*O*-alkyl ethers were treated with potassium

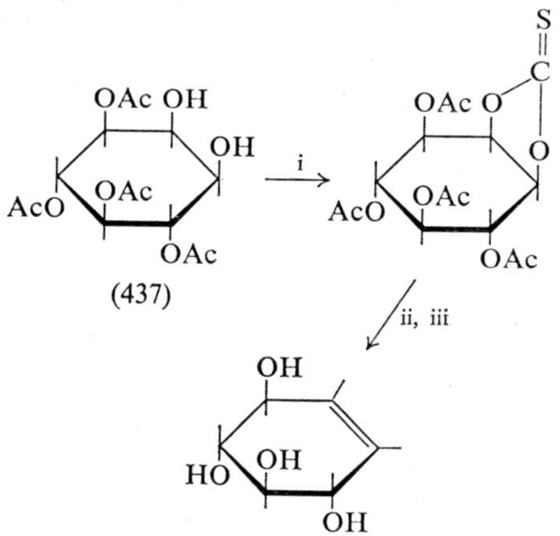

Reagents: i, *NN*′-thiocarbonyldiimidazole; ii, $(MeO)_3P$; iii, Et_3N

Scheme 81

tertiary butoxide in DMSO.[477] The products were 1,2,4-trialkoxybenzenes and in the case of the tetra-*O*-benzyl-*myo*-inositol the intermediate (438) was isolated. Aromatisation also occurred when the triketoinositol (439)

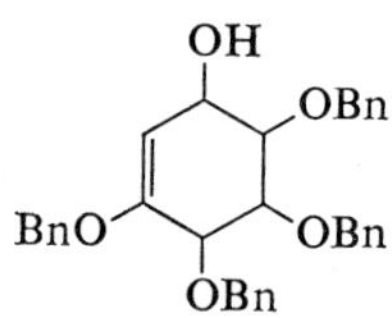

(438)

was acetylated.[478] With acetic anhydride in pyridine, the hexa-acetate of (440) was the major product, whereas with acetic anhydride in the presence of phosphoric acid the product was (441). The sequence of reactions shown in Scheme 82 was proposed to account for the formation of the latter product.

[475] B. A. Klyashchitskii, E. G. Zhelvakova, V. I. Shvets, R. P. Evstigneeva, and N. A. Preobrazhenskii, *Tetrahedron Letters*, 1970, 587.
[476] T. L. Nagabhushan, *Canad. J. Chem.*, 1970, **48**, 383.
[477] P. A. Gent and R. Gigg, *J. Chem. Soc.* (*C*), 1970, 2253.
[478] A. J. Fatiadi, *Carbohydrate Res.*, 1970, **12**, 131.

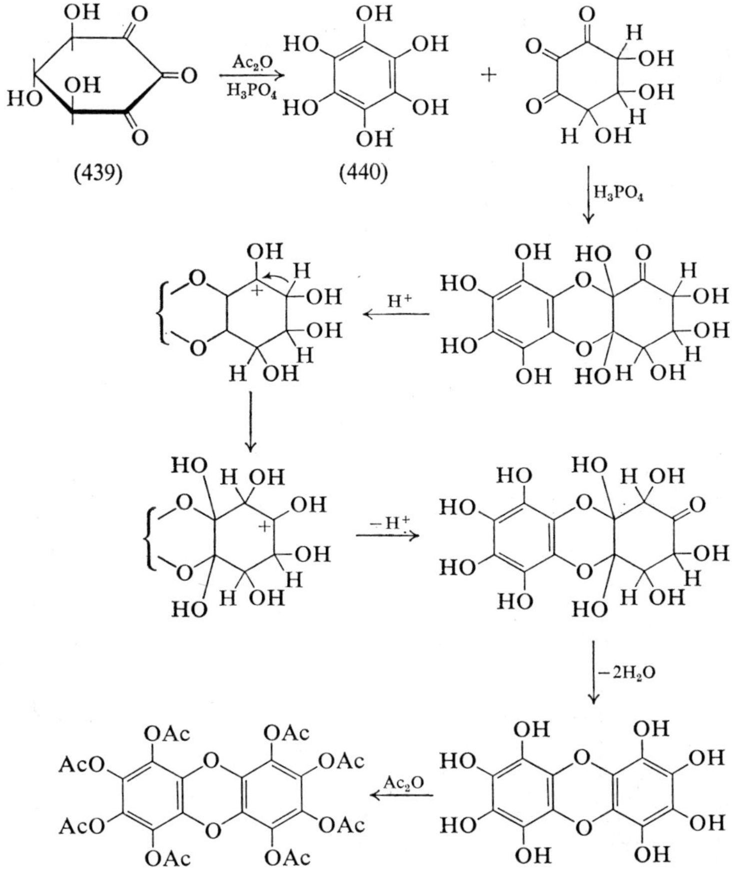

(439) (440) (441)

Scheme 82

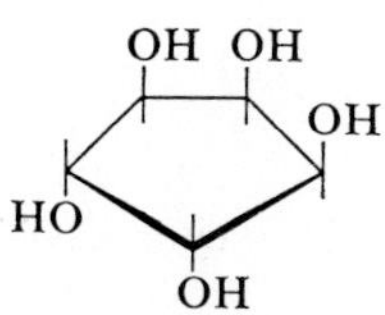

(442)

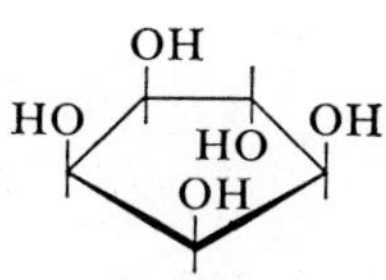

(443)

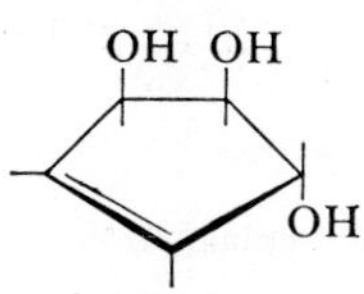

(444)

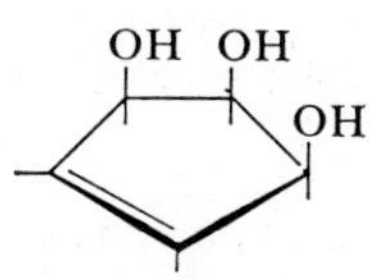

(445)

Two new cyclopentanepentaols (442) and (443) resulted from hydroxylation of suitable 3,4,5-trihydroxy-1-cyclopentene derivatives,[479] and the preparation of the previously unknown cyclopentenetriols (444) and (445) has been described.[480]

Nitrogen-containing Derivatives

Treatment of compound (446), which exists as a cyclic hydrate, with nitromethane afforded a mixture of isomeric nitrocyclopentane derivatives,

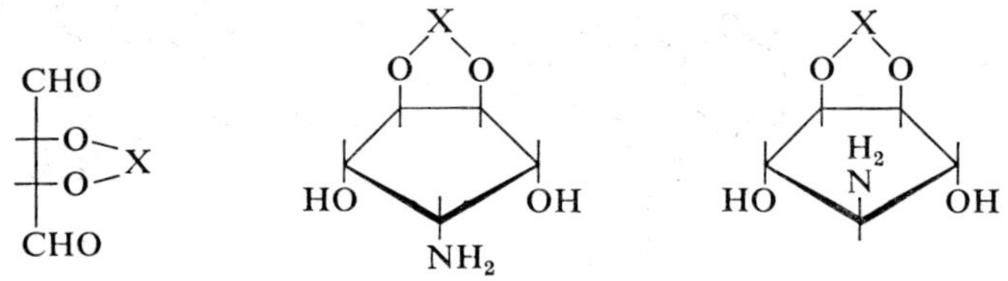

(446) (447) (448)

X = cyclohexylidenyl

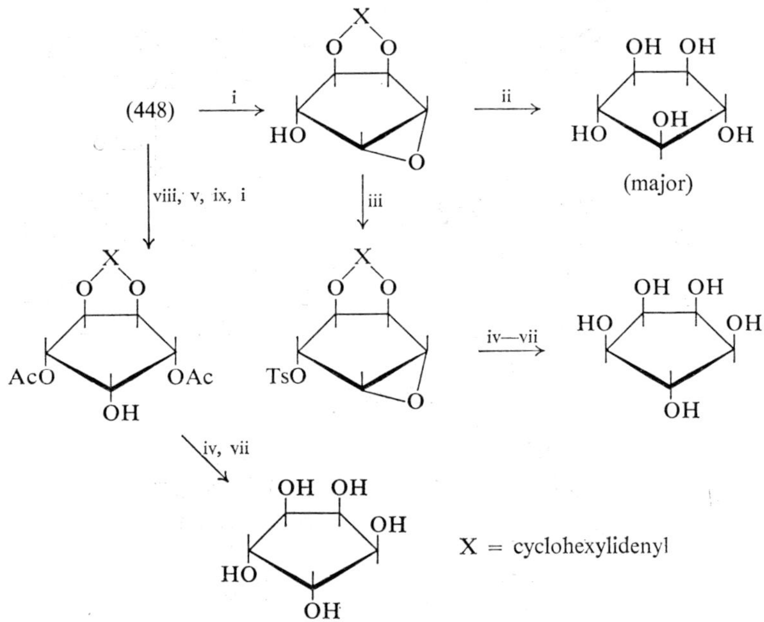

Reagents: i, HNO_2; ii, AcOH; iii, TsCl, py; iv, dil. H_2SO_4; v, Ac_2O, py; vi, DMF,NaOAc; vii, NaOMe; viii, $NO_2\cdot C_6H_4\cdot CHO$; ix, HCl, 4-dioxan

Scheme 83

[479] G. Wolczunowicz, L. Bors, F. Cocu, and T. Posternak, *Helv. Chim. Acta*, 1970, **53**, 2288.

[480] G. Wolczunowicz, F. G. Cocu, and T. Posternak, *Helv. Chim. Acta*, 1970, **53**, 2275.

which on hydrogenation gave the corresponding aminocyclopentane derivatives, two of which [(447) and (448)] were isolated.[481] Three of the four possible cyclopentanepentaols were synthesised from (448) as illustrated in Scheme 83.[482]

2-Deoxystreptamine has been converted into (449) by treatment with *p*-nitrophenoxycarbonyl chloride in the presence of basic resin.[271]

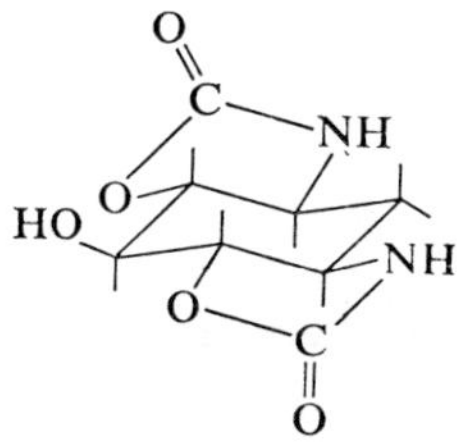

(449)

[481] R. Ahluwalia, S. J. Angyal, and B. M. Luttrell, *Austral. J. Chem.*, 1970, **23**, 1819.
[482] S. J. Angyal and B. M. Luttrell, *Austral. J. Chem.*, 1970, **23**, 1831.

20
Antibiotics

Methyl α- and β-glycosides [483] of streptozotocin have been synthesised and also the analogues (450), (451), and (452).[484] The diamycins have been shown to contain 2-amino-2-deoxy-D-glucose as the only sugar component.[485] The

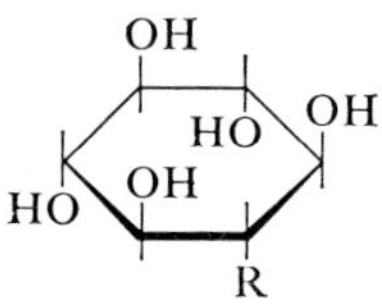

(450)

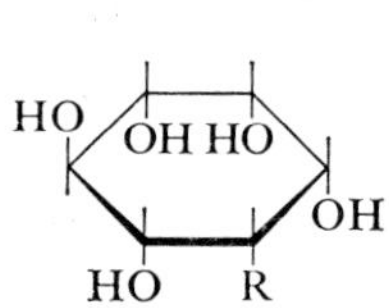

(451)

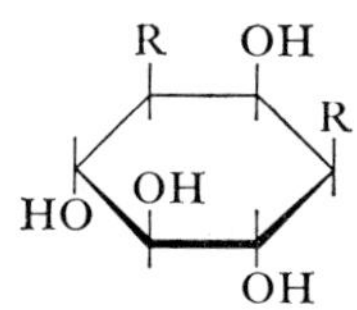

(452)

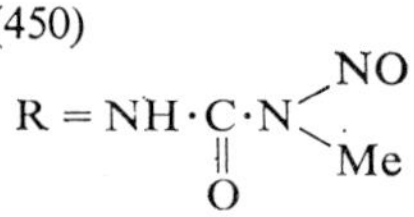

sugar component of antibiotics LL-AC541 and LL-AB644 [streptothricin-like compounds] was identified as 2-deoxy-2-methylamino-D-gulosamine.[486] Mycosamine has been isolated from antibiotic DJ400.[487] A synthesis of perosamine, a component of perimycin, has been described.[246a] Nojirimycin has been shown to be a potent inhibitor of glucosidase.[488] The sugar moiety of aranciamycin has been identified as 2-*O*-methyl-L-rhamnose.[489]

Neomycins D, E, and F have been shown to be identical with paromamine, paromomycin I, and paromomycin II, respectively.[490] A new species of *Micromonospora* (*M. pallida*) has yielded mannosidostreptomycin.[491]

483 T. Suami and T. Machinami, *Bull. Chem. Soc. Japan*, 1970, **43**, 3013.
484 T. Suami and T. Machinami, *Bull. Chem. Soc. Japan*, 1970, **43**, 2953.
485 E. Meyers, D. M. Slusarchyk, J. L. Bouchard, and F. L. Weisenborn, *J. Antibiotics*, 1969, **22**, 490.
486 D. B. Borders, K. J. Sax, J. E. Lancaster, W. K. Hausmann, L. A. Mitscher, E. R. Wetzel, and E. L. Patterson, *Tetrahedron*, 1970, **26**, 3123.
487 F. Bohlmann, E. V. Dehmlow, H.-J. Neuhahn, R. Brandt, and B. Reinicke, *Tetrahedron*, 1970, **26**, 2191.
488 T. Niwa, S. Inouye, T. Tsuruoka, Y. Koaze, and T. Niida, *Agric. and Biol. Chem. (Japan)*, 1970, **34**, 967.
489 W. Keller-Schierlein and A. Müller, *Experientia*, 1970, **26**, 929.
490 E. J. Hessler, H. K. Jahnke, J. H. Robertson, and K. Tsuji, *J. Antibiotics*, 1970, **23**, 465.
491 G. F. Gauze, M. G. Brazhnikova, M. A. Sveshnikova, R. S. Ukholina, N. P. Nechaeva, G. V. Gavrilina, M. F. Lavrova, I. N. Kovsharova, V. V. Proshlyakova, M. K. Kudinova, and S. P. Shapovalova, *Antibiotikii*, 1970, **15**, 99.

Reduction of streptomycin with aluminium amalgam gave a compound which was neither of the known dihydro- or dihydrodeoxy-streptomycins, but was probably produced by reduction of the tertiary hydroxy-group in the streptose moiety.[492] A new aminoglycoside antibiotic SF-733 has had its structure deduced [493] and confirmed by synthesis [494] as (453). Kanamycin

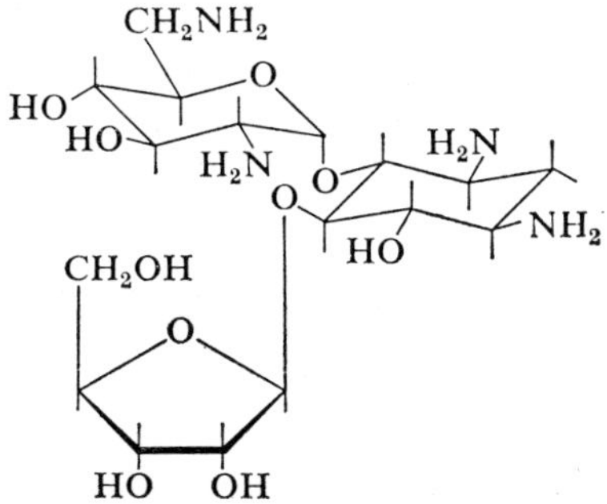

(453)

6″-uronic acid has been prepared *via* the tetra-*N*-ethoxycarbonyl-6″-*O*-trityl derivative [495] and several 'kanamycinotetracyclines' have been described.[496] The synthesis of 4-*O*- and 6-*O*-(α-D-glucopyranosyl)-2-deoxystreptamine is reported in Chapter 3.[57] Hybrimycin A and B have been hydrolysed to the neamine-like fragments (454) and (455), named hydrimycins A3 and B3, respectively.[497]

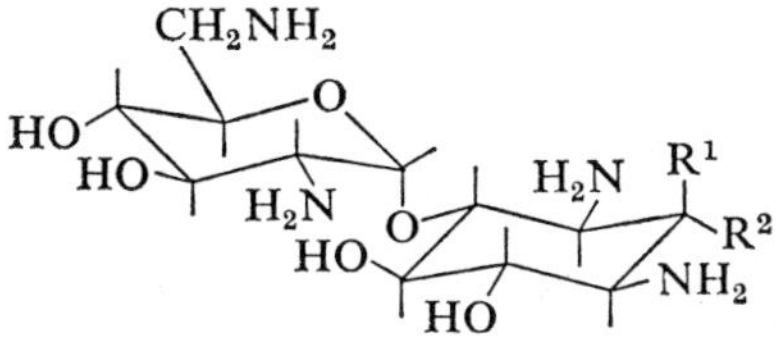

(454) R^1 = H, R^2 = OH
(455) R^1 = OH, R^2 = H

Garamine, one of the sugar components of gentamicin C, has been shown to be 3-deoxy-4-*C*-methyl-3-methylamino-L-arabinose (456).[407] Gentosamine, a component of gentamicin A, has been identified as 3-deoxy-3-methylamino-D-xylose (457), thus completing the structure of the anti-

[492] I. Fujimaki and K. Tsuji, *J. Agric. Chem. Soc. Japan*, 1970, **44**, 463, 471.
[493] E. Akita, T. Tsuruoka, N. Ezaki, and T. Niida, *J. Antibiotics*, 1970, **23**, 173.
[494] T. Ito, E. Akita, T. Tsuruoka, and T. Niida, *Agric. and Biol. Chem. (Japan)*, 1970, **34**, 980.
[495] T. Kobayashi, T. Tsuchiya, K. Tatsuta, and S. Umezawa, *J. Antibiotics*, 1970, **23**, 225.
[496] W. Sobiczewski, W. Chojnowski, and H. Cendrowska, *Acta Polon. Pharm.*, 1970, **27**, 99.
[497] W. T. Shier, K. L. Rinehart jun., and D. Gottlieb, *J. Antibiotics*, 1970, **23**, 51.

biotic.[246] Full details have been given of the synthesis of actinamine and hyosamine[498] (see Vol. 2, p. 166 and Vol. 3, p. 149).

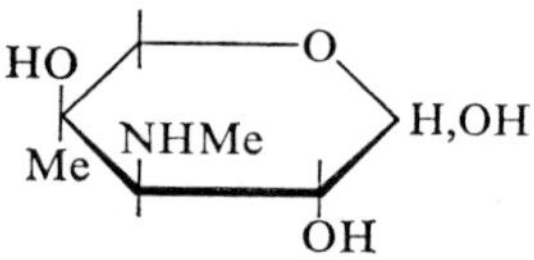

(456)

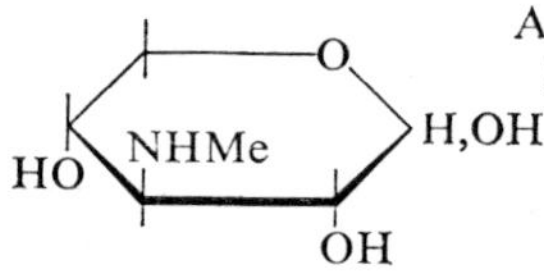

(457)

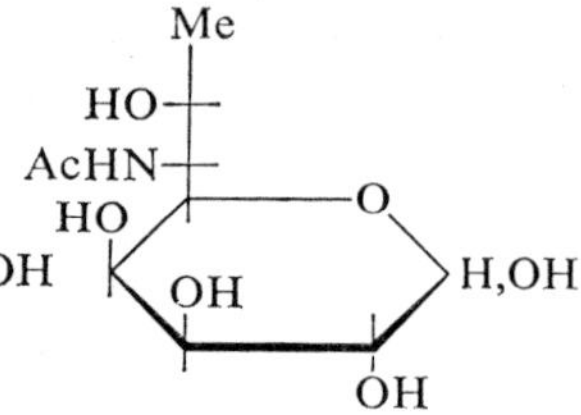

(458)

Full experimental details have now been presented on the two routes leading to the synthesis of *N*-acetyl-lincosamine (458) (*cf.* Vol. 3, p. 152). In the full paper, the conversion of (458) into methyl thiolincosaminide was also described; this is tantamount to a total synthesis of lincomycin.[499] Another synthesis of *N*-acetyl-lincosamine is shown in Scheme 84,[500] the

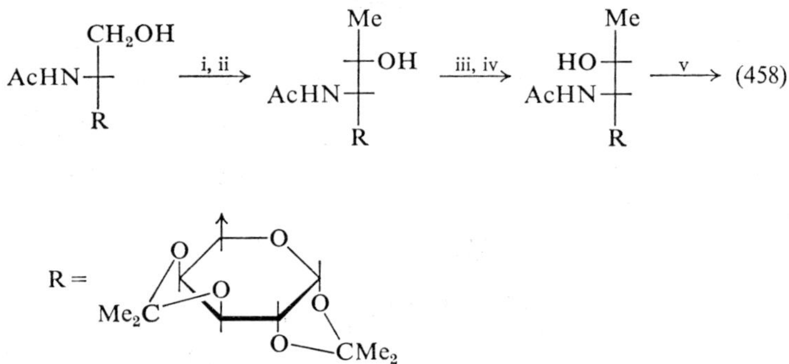

Reagents: i, DMSO, DCC; ii, MeMgI, Et_2O; iii, CrO_3, py; iv, $NaBH_4$, MeOH; v, aq AcOH

Scheme 84

starting material having been described previously (see Vol. 3, p. 91). A third approach (Scheme 85) yielded the epimeric derivatives (459), which were used in a synthesis of the antibiotic.[501]

Acidic and alkaline isomerisations and hydrolysis of lincomycin monoesters have been studied. In alkaline media, the 2-, 3-, and 4-monohexanoates were rapidly isomerised, each entering into a facile equilibrium with the other two, and with accompanying hydrolysis. In acidic media, the 3- and 4-hexanoates were isomerised to give a mixture containing only

498 T. Suami, S. Ogawa, and H. Sano, *Bull. Chem. Soc. Japan*, 1970, **43**, 1843.
499 G. B. Howarth, W. A. Szarek, and J. K. N. Jones, *J. Chem. Soc.* (*C*), 1970, 2218.
500 H. Saeki and E. Ohki, *Chem. and Pharm. Bull.* (*Japan*), 1970, **18**, 412, 789.
501 B. J. Magerlein, *Tetrahedron Letters*, 1970, 33.

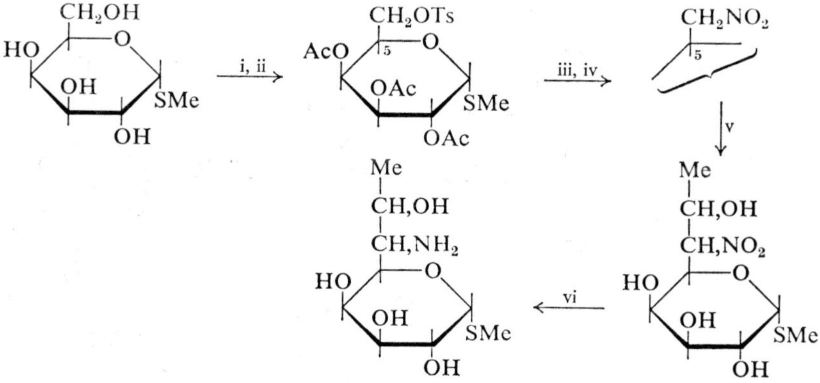

(459)

Reagents: i, TsCl, py; ii, Ac_2O, py; iii, NaI, Me_2CO; iv, $NaNO_2$, DMF; v, MeCHO, NaOMe, MeOH; vi, H_2–Pt

Scheme 85

a trace of the 2-ester; the latter ester underwent hydrolysis without re-arrangement.[502] A synthesis of lincomycin 2-phosphate has been described.[502a] Clindamycin and other analogues of lincomycin have been oxidatively *N*-demethylated.[503]

Full details of the crystal and molecular structure of showdomycin have now been given,[504] and a synthesis of the antibiotic has been reported (Scheme 86).[505]

A series of puromycin analogues have been prepared in which the 3′-amino-group was substituted with a variety of amino-acids.[506] The nucleoside 'C-substance' (459a), produced on hydrolysis of gougerotin,[507] and the analogue (460) of pyrazomycin [508] have been synthesised.

A trideoxy-sugar believed to be rhodinose has been isolated from rubomycin B.[509] Kujimycin B was found to be identical to lankamycin, whilst kujimycin A is lankamycin deacetylated at C-4 of the L-arcanose moiety.[510] A revised structure has been proposed for lankamycin having D-chalcose attached to position 5 and 4-*O*-acetyl-L-arcanose at position 3 of the macrocycle; this reverses the previous assignments.[511]

502 T. O. Oesterling, *Carbohydrate Res.*, 1970, **15**, 285.

502a W. Morozowich, D. J. Lamb, H. A. Karnes, F. A. Mackellar, C. Lewis, K. F. Stern, and E. L. Rowe, *J. Pharm. Sci.*, 1969, **58**, 1485.

503 R. D. Birkenmeyer and L. A. Dolak, *Tetrahedron Letters*, 1970, 5049.

504 Y. Tsukuda and H. Koyama, *J. Chem. Soc.* (*B*), 1970, 1709.

505 L. Kalvoda, J. Farkaš, and F. Šorm, *Tetrahedron Letters*, 1970, 2297.

506 L. V. Fisher, W. W. Lee, and L. Goodman, *J. Medicin. Chem.*, 1970, **13**, 775.

507 K. A. Watanabe, M. P. Kotick, and J. J. Fox, *J. Org. Chem.*, 1970, **35**, 231.

508 J. M. J. Tronchet and F. Perret, *Helv. Chim. Acta*, 1970, **53**, 648.

509 G. B. Fedorova, M. G. Brazhnikova, A. S. Mezentsev, and I. Kshepinsky, *Antibiotikii*, 1970, **15**, 403.

510 S. Omura, T. Muro, S. Namiki, M. Shibata, and J. Sawada, *J. Antibiotics*, 1969, **22**, 629.

511 R. S. Egan and J. R. Martin, *J. Amer. Chem. Soc.*, 1970, **92**, 4129.

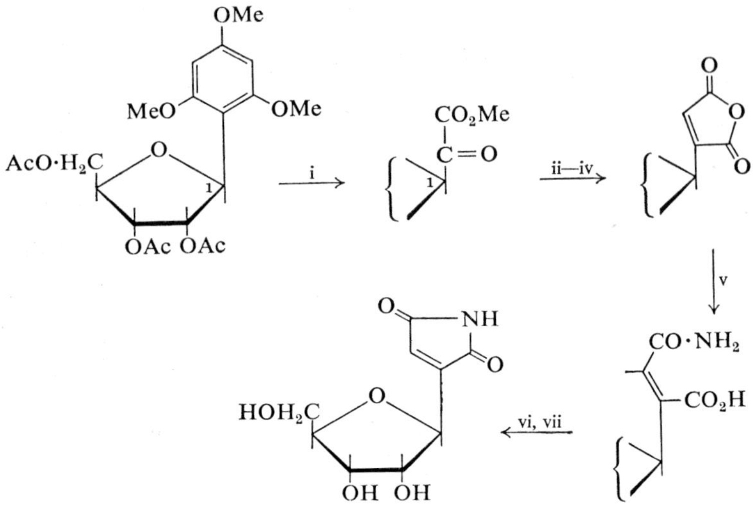

Reagents: i, O_3; ii, $Ph_3P{=}CH{\cdot}CO_2Me$; iii, ^-OH; iv, Ac_2O, CF_3CO_2H; v, NH_3, PhH; vi, ethyl polyphosphate, DMF; vii, H^+

Scheme 86

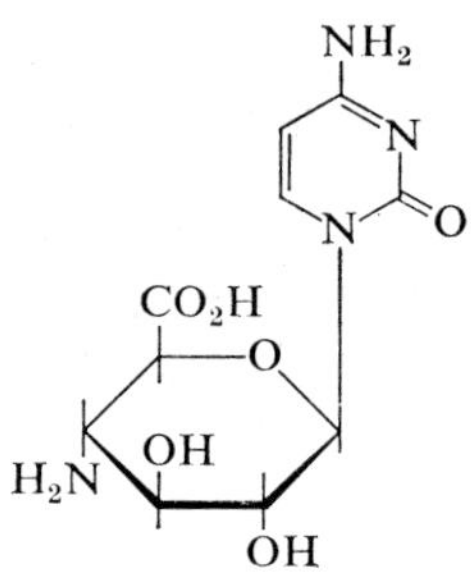

(459a)

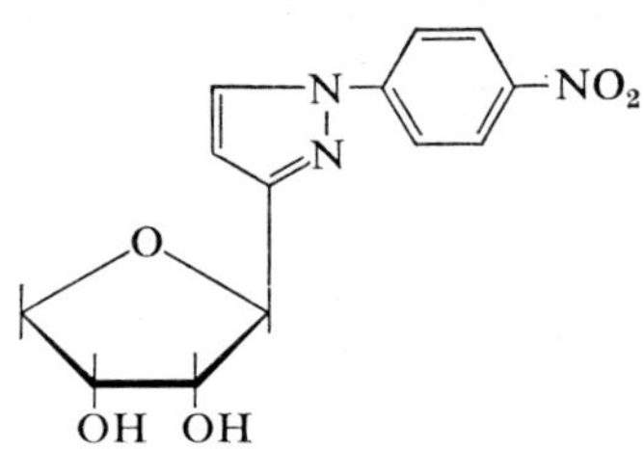

(460)

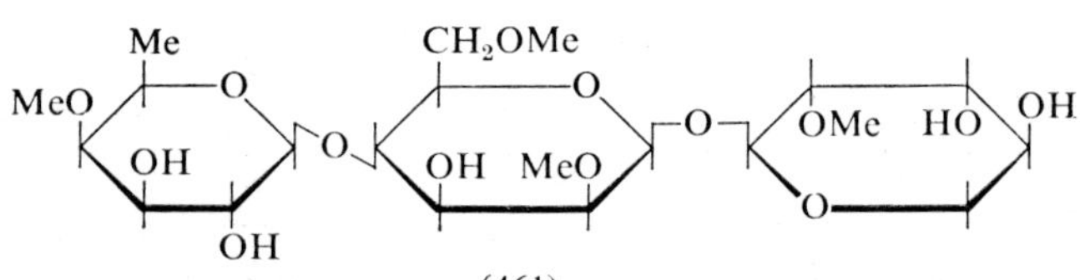

(461)

Evertriose, a trisaccharide component of everninomicin D, has been shown to have the structure (461).[512]

[512] A. K. Ganguly and O. Z. Sarre, *Chem. Comm.*, 1970, 911.

21
Nucleosides

Synthesis

Recent developments in the synthesis of nucleosides have been reviewed.[513] There have been many reports of the synthesis of new nucleosides and their derivatives, most of which have involved conventional procedures. Thus, 1-(2-deoxy-D-*arabino*-hexopyranosyl)-5-bromo-6-azauracil,[514] 2′,3′-dideoxyuridine,[515] 3-(2-deoxy-β-D-ribofuranosyl)uracil,[516] 8-aza-adenosine,[517] some (D-ribofuranosyl)indazoles,[518] some 4-methyl-2-pyrimidinoneribonucleosides,[519] various D-ribofuranosyl and 2-deoxy-D-ribofuranosyl 1,2,4-triazole nucleosides,[519a] and the α- and β-anomers of 1-(2-deoxy-D-ribofuranosyl)-2-pyridone[520] have been prepared. The syntheses and some properties of nucleotides containing 4-thio-D-ribofuranose have been described[521] and 3-(β-D-ribofuranosyl)-5,7-dihydroxy-1*H*-pyrazolo[4,3-*d*]-pyrimidine has been synthesised as part of a study of nucleoside antibiotics.[522] 6-Chloro-9-(4-thio-β-D-ribofuranosyl)purine has been prepared from 2,3,5-tri-*O*-acetyl-4-thio-D-ribofuranosyl chloride,[523] and a number of new 2-fluororibonucleosides have been examined for cytotoxicity.[524] 2′(3′)-*O*-Aminoacyl and 2′,3′-*O*-bis(aminoacyl) derivatives of adenosine,[525] and 9-(3-*O*-hexyl-α-D-xylofuranosyl)adenine, its β-anomer, and 5-deoxy-analogues have all been synthesised.[526]

[513] W. W. Zorbach, *Synthesis*, 1970, 329.
[514] G. J. Durr and S. Hammond, *J. Heterocyclic Chem.*, 1970, **7**, 743.
[515] Y. Furukawa, Y. Yoshioka, K.-I. Imai, and M. Honjo, *Chem. and Pharm. Bull. (Japan)*, 1970, **18**, 554.
[516] M. W. Winkley, *J. Chem. Soc. (C)*, 1970, 1365.
[517] J. A. Montgomery, H. J. Thomas, and S. J. Clayton, *J. Heterocyclic Chem.*, 1970, **7**, 215.
[518] G. R. Revankar and L. B. Townsend, *J. Heterocyclic Chem.*, 1970, **7**, 117.
[519] R. S. Klein, I. Wempen, K. A. Watanabe, and J. J. Fox, *J. Org. Chem.*, 1970, **35**, 2330.
[519a] J. T. Witkowski and R. K. Robins, *J. Org. Chem.*, 1970, **35**, 2635.
[520] M. P. Mertes, *J. Medicin. Chem.*, 1970, **13**, 149.
[521] D. J. Hoffman and R. L. Whistler, *Biochemistry*, 1970, **9**, 2367.
[522] M. Bobek, J. Farkaš, and F. Šorm, *Tetrahedron Letters*, 1970, 4611.
[523] M. Bobek, R. L. Whistler, and A. Bloch, *J. Medicin. Chem.*, 1970, **13**, 411.
[524] J. A. Montgomery and K. Hewson, *J. Medicin. Chem.*, 1970, **13**, 427.
[525] S. Chládek, P. Pulkrábek, J. Sonnenbichler, J. Žemlička, and I. Rychlik, *Coll. Czech. Chem. Comm.*, 1970, **35**, 2296.
[526] D. H. Murray and J. Prokop, *J. Pharm. Sci.*, 1970, **59**, 344.

Syntheses of 1-(3-deoxy-3-fluoro- and 2-deoxy-2-fluoro-β-D-xylofuranosyl)cytosines and 1-(2-deoxy-2-fluoro-α- and -β-D-arabinofuranosyl)cytosines have been accomplished by reaction of suitably protected fluoroglycosyl bromides with bis(trimethylsilyl)cytosine. Attempts to react the fluoro-sugars with cytosine by the mercuric cyanide–nitromethane method were unsuccessful.[527] A variation of the Hilbert–Johnson procedure using trimethylsilylated derivatives of pyrimidines and 1-halogeno- or 1-*O*-acylribofuranose triesters with catalysts such as stannic chloride gave excellent yields (60—95%) of nucleosides.[528] Several D-ribofuranosyltriazine derivatives have been prepared by treatment of tri-*O*-acetyl-D-ribofuranosyl bromide with appropriate trimethylsilylated derivatives. The antibiotic nucleoside 5-azacytidine has been synthesised in this way.[529]

An important observation has been made concerning the stereochemistry of the coupling reaction between (462) and, for example, trimethylsilylated derivatives of 5-mercaptouracil.[530] The presence of trimethylsilyl chloride in the reaction mixture gave mainly α-nucleosides, whereas the β-anomers preponderated under conditions where the chloride ion was rapidly removed. During control experiments, it was observed that compound (462) rearranged to (463) and (464) on standing in benzene solution.

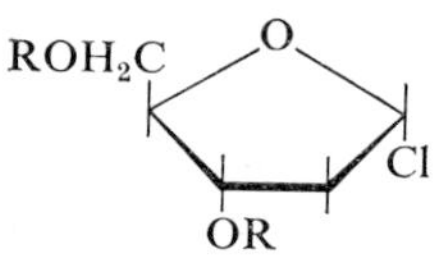

(462)

R = p-Cl·C_6H_4·CO

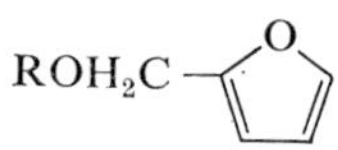

(463)

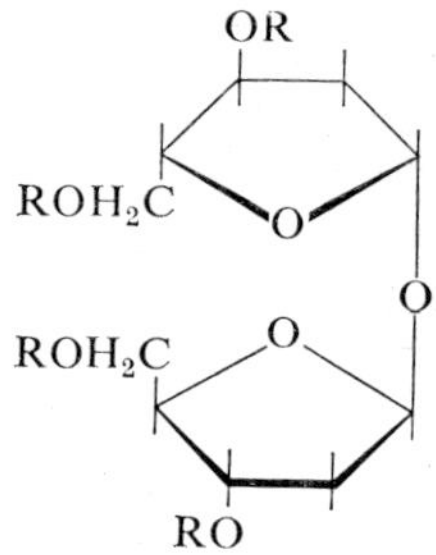

(464)

1-*O*-Acylfuranoses in which the other hydroxy-groups were protected by isopropylidene groups have been converted into adenine nucleosides by coupling with 6-benzamidomercuripurine by the titanium tetrachloride method.[531] Yields obtained with L-rhamnose derivatives were better than those with D-mannose derivatives. Great care was required in the hydrolysis of the isopropylidene residues to avoid cleavage of the glycosidic bond.

A new route to *C*-nucleosides has been provided by way of the 1-diazo-sugar (465). Compound (465) was treated with dimethylacetylene dicarboxylate to give (466), which was then converted into (467).[532] 'Double-

[527] J. A. Wright, D. P. Wilson, and J. J. Fox, *J. Medicin. Chem.*, 1970, **13**, 269.
[528] U. Niedballa and H. Vorbrüggen, *Angew. Chem. Internat. Edn.*, 1970, **9**, 461.
[529] M. W. Winkley and R. K. Robins, *J. Org. Chem.*, 1970, **35**, 491.
[530] M. P. Kotick, C. Szantay, and T. J. Bardos, *J. Org. Chem.*, 1969, **34**, 3806.
[531] L. M. Lerner and Y. Y. Cheng, *Carbohydrate Res.*, 1970, **14**, 297.
[532] E. M. Acton, K. J. Ryan, and L. Goodman, *Chem. Comm.*, 1970, 313.

headed' nucleosides (*e.g.* 468) have been synthesised by nucleophilic displacements of 5′-sulphonic esters using sodium adenide.[533]

D-Ribose, its 5-phosphate, and D-arabinose were each treated with cyanamide in aqueous solution to produce amino-oxazoline derivatives.[534]

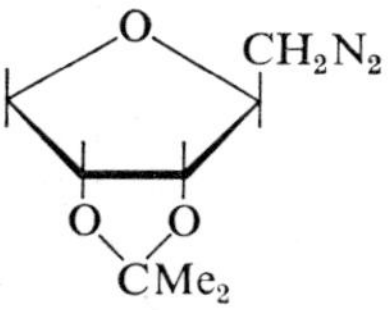

(465)

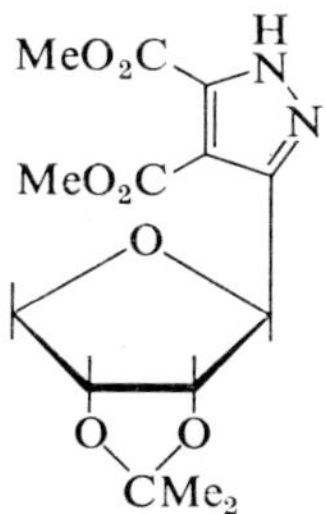

(466)

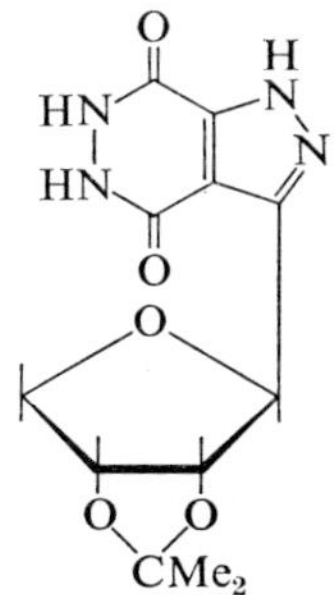

(467)

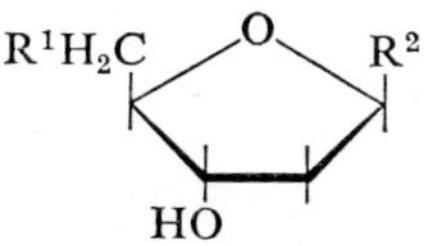

R^1 = adenine
R^2 = adenine or thymine

(468)

The products were then allowed to react with cyanoacetylene to give α-5′-cytidylic acid, α-cytidine, and (β-D-arabinosyl)cytosine, respectively, which were anomerised and epimerised on irradiation.

The first report of the use of a 2-deoxyglycoside in nucleoside synthesis has appeared. Thus, (469) was converted into a mixture of (470) and its anomer using dichloroacetic acid as a catalyst.[535]

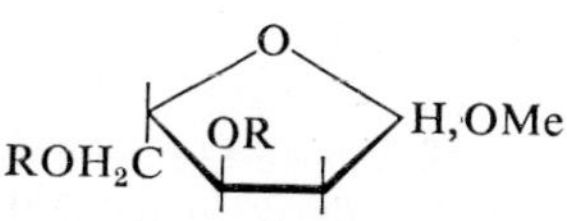

(469)

R = *p*-Me·C_6H_4·CO

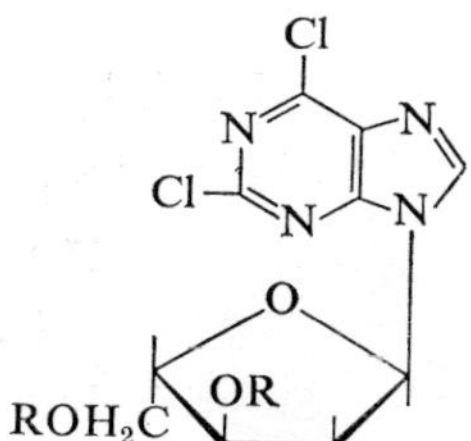

(470)

[533] R. Fecher, K. H. Boswell, J. J. Wittick, and T. Y. Shen, *J. Amer. Chem. Soc.*, 1970, **92**, 1400.
[534] R. A. Sanchez and L. E. Orgel, *J. Mol. Biol.*, 1970, **47**, 531.
[535] M. J. Robins, T. A. Khwaja, and R. K. Robins, *J. Org. Chem.*, 1970, **35**, 636.

A convenient synthesis of 1-(β-D-arabinofuranosyl)cytosine used N^4-acetylcytosine and 2-*O*-acetyl-5-*O*-ethoxycarbonyl-3-*O*-*p*-tolylsulphonyl-α-D-xylofuranosyl chloride as starting materials.[536] Various approaches have been considered for the synthesis of D-ribo- and deoxy-D-ribo-nucleotides labelled with deuterium at C-2′.[536a] 2′- and 3′-Ketocytidine derivatives were obtained by oxidation of the appropriately protected nucleoside with DCC–DMSO.[537] Reductions with sodium borotritide then afforded D-*arabino*-, D-*xylo*-, and D-*ribo*-nucleosides with tritium labels at the 2′ and 3′ positions.

A number of nucleosides containing 3-deoxy-3-methylamino-D-ribofuranose,[538] and branched-chain amino-sugars [265] [*e.g.* (160)] have been described.

Several papers have been devoted to the synthesis and reactions of anhydronucleosides. Thus, treatment of cytidine with thionyl chloride in DMF gave (471), from which 2,2′-cyclocytidine was obtained.[539] 2′,5′-Dideoxy-5′-fluororibonucleosides have been prepared by treating the 5′-*O*-tosyl derivatives with an excess of tetrabutylammonium fluoride in

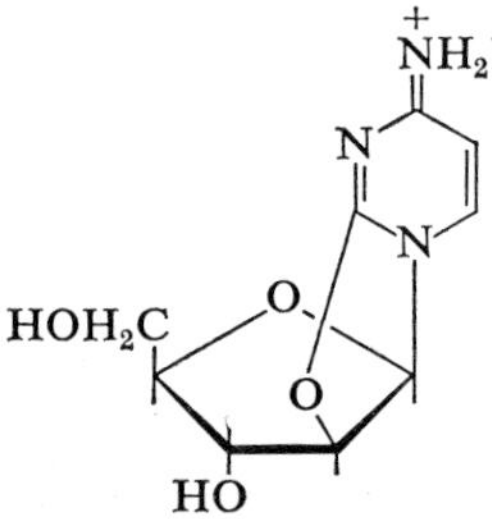

(471)

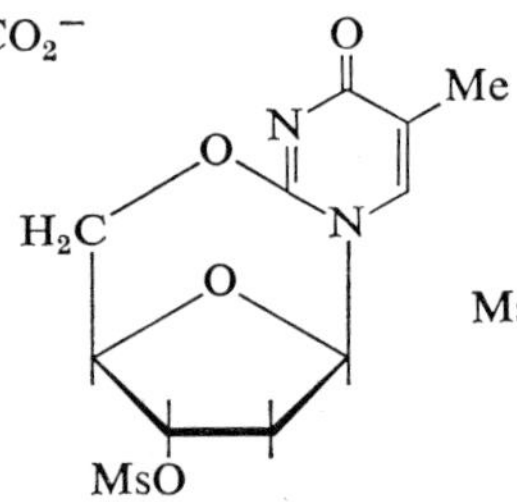

(472)

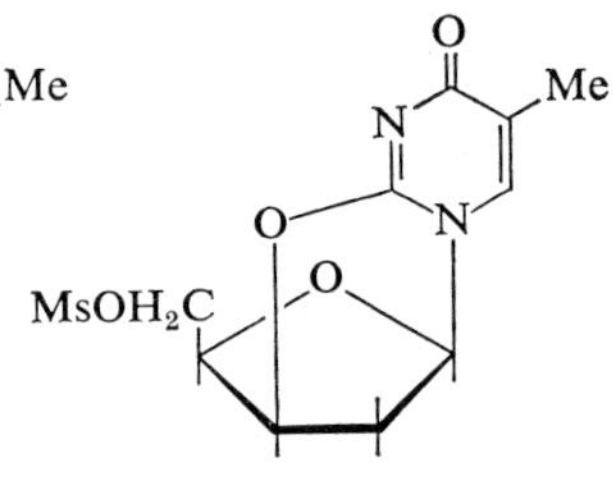

(473)

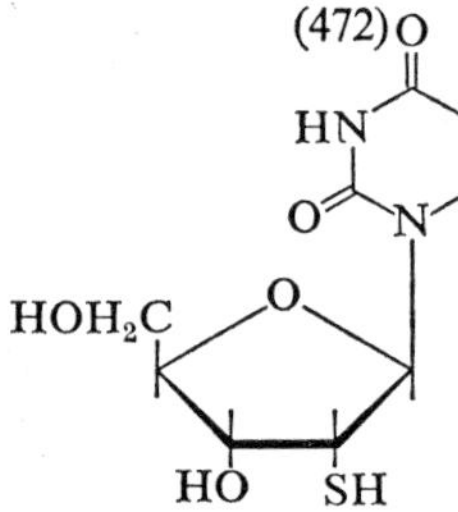

(474)

DMF at 50 °C. Surprisingly, treatment of 3′,5′-di-*O*-methanesulphonylthymidine with this reagent gave the anhydro-derivatives (472) and (473) rather than products containing fluorine.[540]

[536] B. Shimizu and F. Shimizu, *Chem. and Pharm. Bull.* (*Japan*), 1970, **18**, 1060.
[536a] S. David, J. Eustache, and C. Rouzeau, *Compt. rend.*, 1970, **270C**, 1821.
[537] U. Brodbeck and J. G. Moffatt, *J. Org. Chem.*, 1970, **35**, 3552.
[538] W. W. Lee, G. L. Tong, R. W. Blackford, and L. Goodman, *J. Org. Chem.*, 1970, **35**, 3808.
[539] K. Kikugawa and M. Ichino, *Tetrahedron Letters*, 1970, 867.
[540] G. Kowollik, K. Gaertner, G. Etzold, and P. Langen, *Carbohydrate Res.*, 1970, **12**, 301.

2′-Deoxy-2′-mercapto-uridine (474) resulted from treatment of 2,2′-anhydro-3′,5′-di-*O*-acetyluridine with thiolacetic acid in DMF and de-acetylation of the product.[541] 8,2′-Anhydro-8-oxy-β-D-arabinofuranosyl-adenine was converted into the corresponding 8,5′-cyclonucleoside by treatment with warm, dilute sodium hydroxide.[542] Thymidine derivatives substituted at the 5′-position gave 3′-deoxy-3′-iodo-compounds on reaction with triphenylphosphite methiodide by way of O^2,3′-cyclonucleosides. 5′-Subsituted-uridines did not afford iodinated products with this reagent but gave instead mixtures of 2′- and 3′-*O*-methylphosphonates. 2′,5′-Di-*O*-trityluridine gave the expected 3′-deoxy-3′-iodo-D-xylose derivative, and also underwent selective loss of the trityl groups, depending on the solvent used.[242] N^3,5′-Anhydronucleosides were formed in high yield by the re-action of adenosine and 2′,3′-*O*-isopropylidene derivatives of purine nucleosides with triphenylphosphite methiodide in DMF.[241]

A series of 2′,6-anhydropyrimidine nucleosides and their reactions have been described. The nucleosides underwent facile ring-openings in aqueous base to give D-arabinose derivatives but they were stable in dilute aqueous acid.[543]

Several nucleosides have been oxidised with sodium periodate and the products were characterised as bis(arylhydrazones).[544] Oxidation of 2′,3′-*O*-isopropylidene-inosine with chromium trioxide in acetic acid afforded the corresponding carboxylic acid in good yield.[444]

The factors involved in the browning reaction of 5′-ribonucleotides in the presence and in the absence of D-glucose have been investigated.[545]

4-Methoxytetrahydropyran-4-yl has proved to be a useful symmetrical alternative to the tetrahydropyranyl protecting-group in nucleoside transformations,[152] and it has been used in the synthesis of deoxyribo-nucleotides.[546]

Compounds such as (475) have been condensed with polyalanine-polylysine to give potential antigens.[547] Two sarcosyl-D-seryl-amino-sugar analogues of gangerotin have been synthesised.[548] Nitrogen analogues [*e.g.* (476)] of adenosine 3′,5′-cyclophosphate have been prepared.[549] Adeno-sine 5′-bis(dihydroxyphosphinylmethyl)phosphinate, the bis-methylene analogue of ATP, has been synthesised.[550]

541 M. Imazawa and T. Ueda, *Tetrahedron Letters*, 1970, 4807.
542 M. Ikehara, M. Kaneko, and Y. Ogiso, *Tetrahedron Letters*, 1970, 4673.
543 E. A. Falco, B. A. Otter, and J. J. Fox, *J. Org. Chem.*, 1970, **35**, 2326.
544 V. P. Chernetskii, E. A. Ponomareva, and V. V. Stavitskii, *Khim. geterotsikl. Soedinenii*, 1970, 557.
545 M. Fujimaki, N. van Chuyen, T. Matsumoto, and T. Kurata, *J. Agric. Chem. Soc. Japan*, 1970, **44**, 275.
546 D. P. L. Green, T. Ravindrannathan, C. B. Reese, and R. Saffhill, *Tetrahedron*, 1970, **26**, 1031.
547 J. P. Coat and S. David, *Carbohydrate Res.*, 1970, **12**, 335.
548 F. W. Lichtenthaler, G. Trummlitz, and P. Emig, *Tetrahedron Letters*, 1970, 2061.
549 A. Murayama, B. Jastorff, and J. Hettler, *Angew. Chem. Internat. Edn.*, 1970, **9**, 640.
550 D. B. Trowbridge and G. L. Kenyon, *J. Amer. Chem. Soc.*, 1970, **92**, 2181.

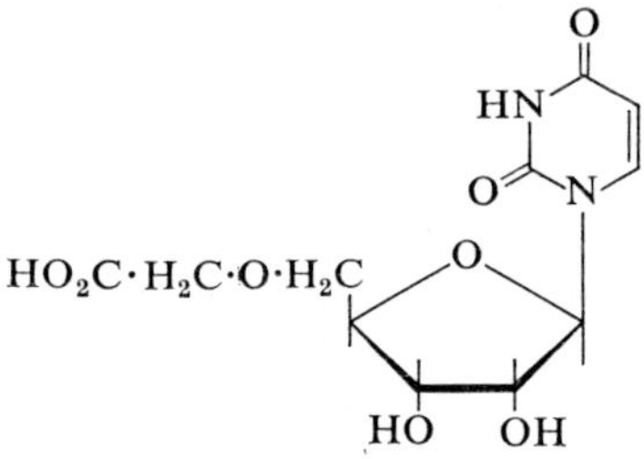

(475)

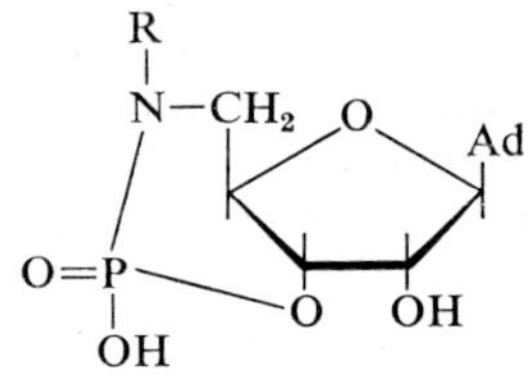

(476) R = H or Me

Rearrangements of nucleosides and their derivatives have been described. Thus, (477) was converted into (478) on treatment with methanolic hydrogen chloride.[551] Treatment of (479) with potassium fluoride dihydrate in

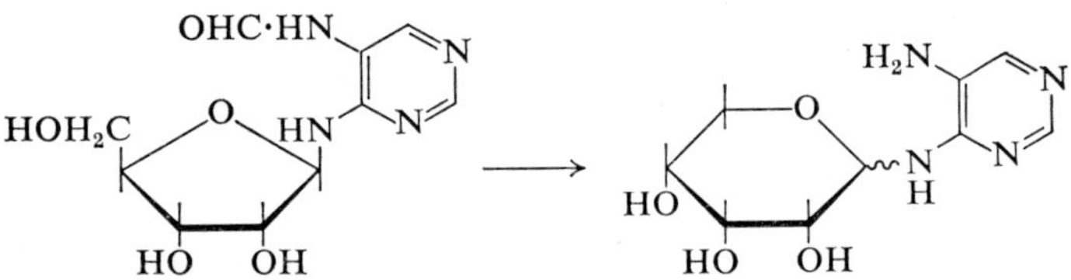

(477) (478)

methanol gave (480) (16%), (481) (7%), and the novel '5′-*O*-nucleoside' (482) (33%).[552] It has been suggested that in the reaction of glycosyl halides with the chloromercury salt of such compounds as N^6-benzoyladenine, the

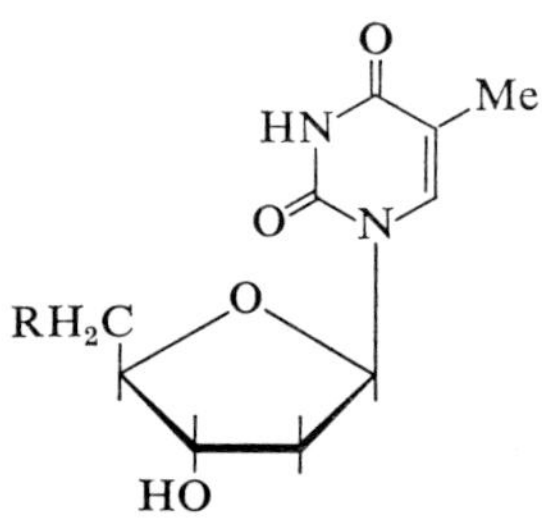

(479) R = OTs
(480) R = F
(481) R = OMe

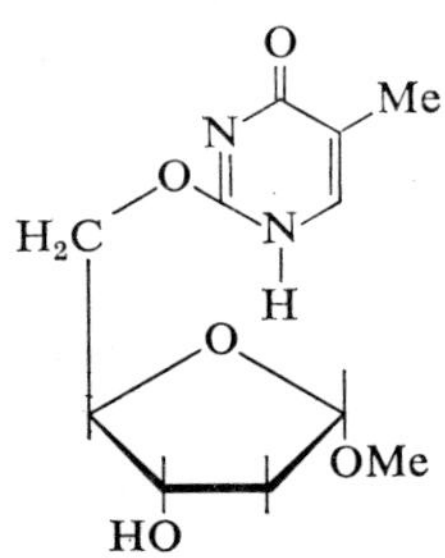

(482)

N-3-glycoside is formed first and then rearranges into the more stable *N*-9-isomer.[553]

Numerous papers have reported methods for the partial or selective esterification and etherification of nucleosides. Methods for tritylation,

[551] J. A. Montgomery and H. J. Thomas, *Chem. Comm.*, 1970, 265.
[552] G. Kowollik, P. Langen, and A. Holý, *J. prackt. Chem.*, 1970, **312**, 145.
[553] M. Miyaki and B. Shimizu, *Chem. and Pharm. Bull.* (*Japan*), 1970, **18**, 732.

benzylation, and acetylation have been reported.[554–557] 5-Chloro-8-hydroxyquinoline esters have been used for the selective acylation of nucleoside aminoacyl derivatives,[558] and syntheses of 3′(2′)-*O*-aminoacyl esters of nucleoside-5′-mono- and tri-phosphates have been described.[559] Tosylation of adenosine 5′-monophosphate in aqueous alkali gave mainly the 2′-*O*-tosyl derivative; the best yields were obtained with four moles of the reagent, suggesting that a tosylphosphate anhydride might be formed which protects the 3-hydroxy-group from attack.[560] 5′-*O*-Trityl-, 5′-*O*-benzoyl-, 5′-*O*-acetyl-, and unprotected 8-bromoadenosine have been sulphonylated using sodium hydride and tri-isopropylbenzenesulphonyl chloride in DMF solution.[561] The percentage of the 2′-*O*-sulphonate increased from 38% to 70% as the bulk of the 5′-substituent increased.

Many other reports, some of which are noted in Chapter 6, have described nucleoside esters.[168, 186, 190, 192, 197, 562]

Nucleosides containing Unsaturated Carbohydrate Components

A number of nucleosides containing unsaturated sugar residues have been reported, some of which have been mentioned elsewhere.[391, 443] Condensation of the L-rhamnose perester (483) with theophylline in the presence of *p*-tolylsulphonic acid gave (484), which was also formed when theophylline and (485) were heated with phosphorus pentoxide in DMF. Elimination occurred in the sugar moiety before condensation with

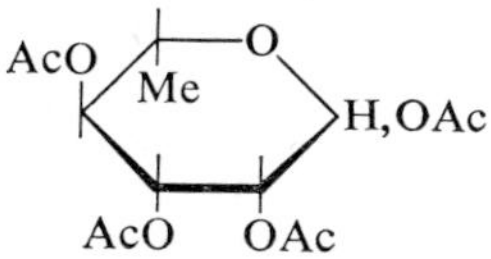

(483)

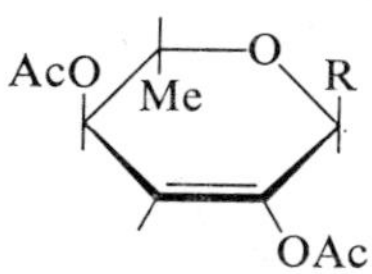

(484)

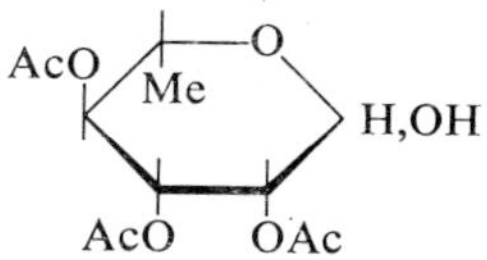

(485)

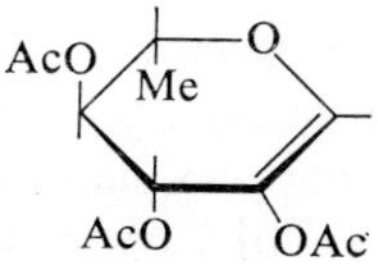

(486)

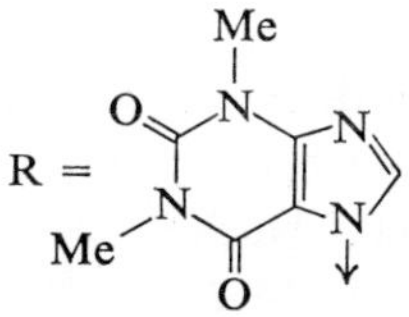

[554] H.-U. Blank, D. Frahne, A. Myles, and W. Pfleiderer, *Annalen*, 1970, **742**, 34.
[555] H.-U. Blank and W. Pfleiderer, *Annalen*, 1970, **742**, 29.
[556] H.-U. Blank and W. Pfleiderer, *Annalen*, 1970, **742**, 16.
[557] H.-U. Blank and W. Pfleiderer, *Annalen*, 1970, **742**, 1.
[558] V. Gut, S. Chládek, and J. Žemlička, *Coll. Czech. Chem. Comm.*, 1970, **35**, 2398.
[559] B. P. Gottikh, A. A. Krayevsky, N. B. Tarussova, P. P. Purygin, and T. L. Tsilvich, *Tetrahedron*, 1970, **26**, 4419.
[560] M. Ikehara and S. Uesugi, *Tetrahedron Letters*, 1970, 713.
[561] M. Ikehara and M. Kaneko, *Tetrahedron*, 1970, **26**, 4251.
[562] S. J. Benkovic and R. C. Hevey, *J. Amer. Chem. Soc.*, 1970, **92**, 4971.

theophylline, since (485) was rearranged to (486) on heating with *p*-tolylsulphonic acid.[563] Nucleosides containing unsaturated sugars were also formed by fusion of theophylline with 2-hydroxy-D-glucal, D-galactal, L-rhamnal, 2-hydroxy-D-xylal, and 1,2,4,6-tetra-*O*-acetyl-3-deoxy-D-*threo*-hex-2-enopyranose,[563] and in the acid-catalysed reaction between 3,4-*O*-acetyl-D-xylal and 6-chloropurine.[564]

Treatment of (487) with silver acetate in acetonitrile did not afford the expected anhydronucleoside but instead gave (488).[565] Three products,

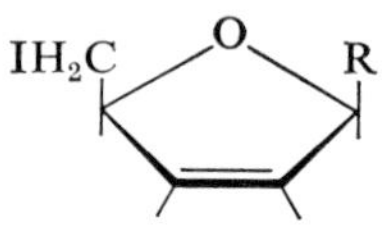

(487)

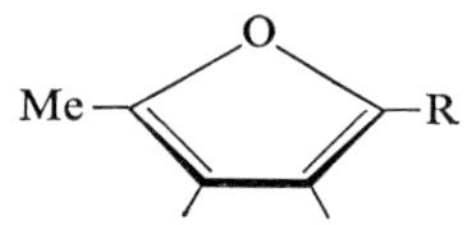

(488)

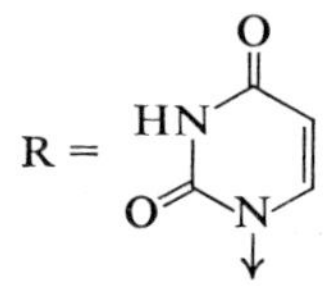

viz. (489), (490), and (491), were obtained in approximately equal proportions from the acid-catalysed fusion of 3,4,6-tri-*O*-acetyl-D-glucal and 2-methylthio-6-chloropurine. Compounds (490) and (491) represent a new class of purine nucleoside.[566]

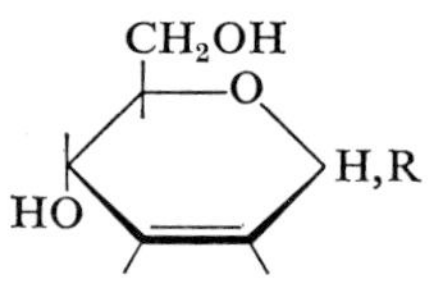

(489)

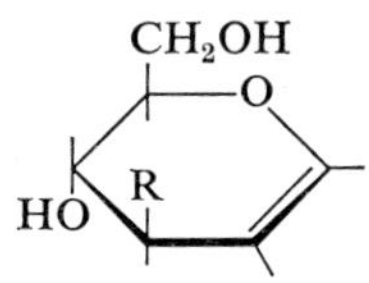

(490)

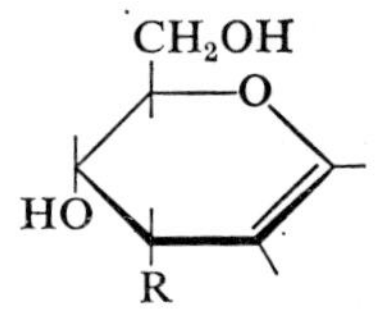

(491)

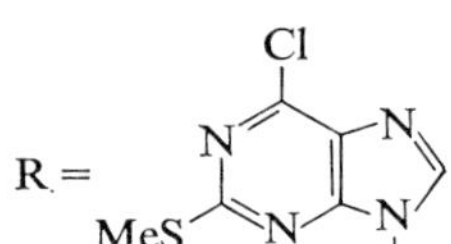

Nucleosides containing Branched-chain Sugar Components

The apiosylnucleoside (492)[567] and the 5′,5′-di-*C*-methyladenosine derivative (493)[568] have been described. Nucleosides containing phosphonate groups[414] [*e.g.* (420)] and phostonate groups[415] [*e.g.* (418)] have been prepared.

The adenine derivative (494), containing a nitromethyl group, was formed on treatment of (495) with nitromethane,[569] and branched-chain

[563] K. Onodera and T. Yajima, *Carbohydrate Res.*, 1970, **13**, 97.
[564] M. Fuertes, G. Garcia-Muñoz, T. Madroñero, M. Stud, and M. Rico, *Tetrahedron*, 1970, **26**, 4823.
[565] M. W. Winkley, *Carbohydrate Res.*, 1970, **13**, 173.
[566] E. E. Leutzinger, R. K. Robins, and L. B. Townsend, *Tetrahedron Letters*, 1970, 3751.
[567] J. M. J. Tronchet and J. Tronchet, *Helv. Chim. Acta*, 1970, **53**, 853.
[568] P. J. Harper and A. Hampton, *J. Org. Chem.*, 1970, **35**, 1688.
[569] A. Rosenthal, M. Sprinzl, and D. A. Baker, *Tetrahedron Letters*, 1970, 4233.

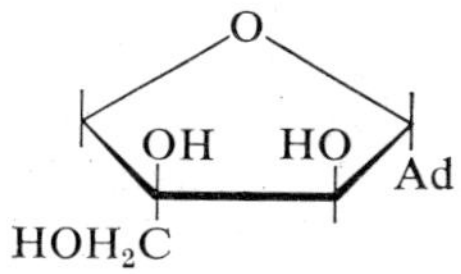

(492)

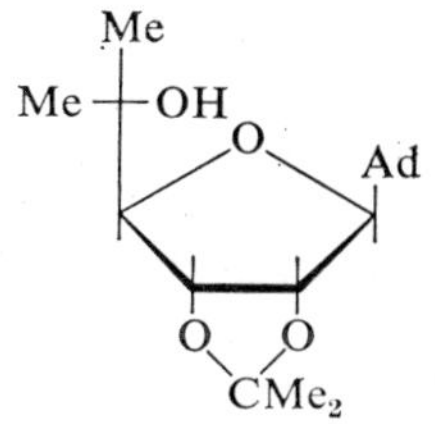

(493)

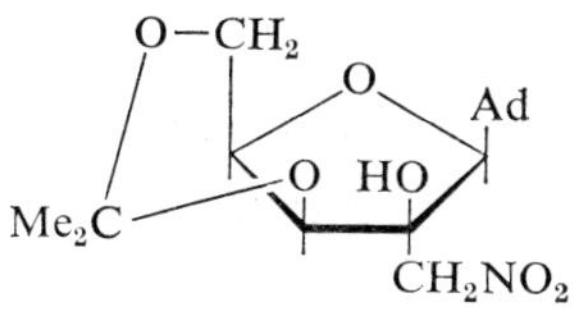

(494)

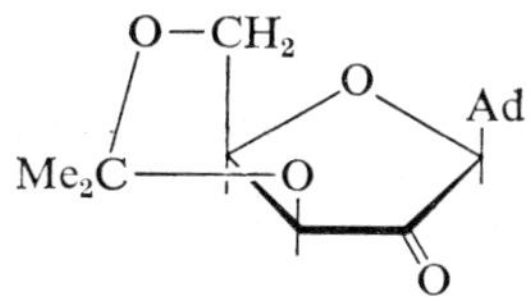

(495)

cyanomethyl derivatives resulted from condensation of pyrimidines with 1-*O*-acyl-3-cyanomethyl sugars.

Physical Measurements

The crystal structures of 3′-*O*-acetyladenosine,[570] and 8-bromoguanosine and 8-bromoadenosine[571] have been determined.

Mass spectroscopy has been shown to be useful in the sequence analysis of nucleosides[572] and for distinguishing between anomers.[573]

Pseudouridine has been examined by n.m.r. spectroscopy at 100 MHz. The preferred conformation of the hydroxymethyl group was that in which the hydroxy-group was located over the furanose ring. The orientation of the D-ribofuranose ring was such that H-6 of the uracil nucleus resided over the sugar.[574] A detailed analysis of the conformation of β-pseudouridine was also reported.[99] The following complexes have been detected in DMSO solution by n.m.r. methods: adenosine–Co–guanosine, guanosine–Zn–imidazole, cytidine–Zn–imidazole, and guanosine–Zn–adenosine.[575]

O.r.d. studies of purine and pyrimidine nucleosides have been reported.[576, 577] Polarographic reductions of adenine nucleosides and nucleotides have been shown to be similar, in principle, to that of adenine itself.[578]

570 S. T. Rao, M. Sundaralingam, S. K. Arora, and S. R. Hall, *Biochem. Biophys. Res. Comm.*, 1970, **38**, 496.

571 S. S. Tavale and H. M. Sobell, *J. Mol. Biol.*, 1970, **48**, 109.

572 J. J. Dolhun and J. L. Wiebers, *Org. Mass Spectroscopy*, 1970, **3**, 669.

573 S. J. Shaw, D. M. Desiderio, K. Tsuboyama, and J. A. McCloskey, *J. Amer. Chem. Soc.*, 1970, **92**, 2510.

574 F. E. Hruska, A. A. Grey, and I. C. P. Smith, *J. Amer. Chem. Soc.*, 1970, **92**, 214.

575 L. S. Kan and N. C. Li, *J. Amer. Chem. Soc.*, 1970, **92**, 281.

576 G. T. Rogers and T. L. V. Ulbricht, *Biochem. Biophys. Res. Comm.*, 1970, **39**, 414.

577 G. T. Rogers and T. L. V. Ulbricht, *Biochem. Biophys. Res. Comm.*, 1970, **39**, 419.

578 B. Janik and P. J. Elving, *J. Amer. Chem. Soc.*, 1970, **92**, 235.

The mechanisms of the acid-catalysed hydrolysis of D-*ribo*- and deoxy-D-*ribo*-purine nucleosides have been examined by spectrophotometric methods, and were shown to involve specific, hydronium-ion catalysis.[579] It was proposed that hydrolysis involved initial protonation of the base followed by a rate-determining cleavage of the glycosidic bond, *i.e.* the mechanism is akin to that of the hydrolysis of glycosides with acids.

Complex formation between Cu^{II} and D-*ribo*- and deoxy-D-*ribo*-nucleosides has been studied. Above pH 6, complex formation occurred only with the sugar moiety, and the deoxy-D-ribose nucleosides did not complex.[580]

579 J. A. Zoltewicz, D. F. Clark, T. W. Sharpless, and G. Grahe, *J. Amer. Chem. Soc.*, 1970, **92**, 1741.
580 H. Reinert and R. Weiss, *Z. physiol. Chem.*, 1969, **350**, 1321.

22
Oxidation and Reduction

Periodate Oxidation

A colorimetric procedure has been described for the quantitative micro-determination of glycolic acid produced by periodate oxidation of ketoses. Interference by glyoxylic acid, formaldehyde, periodate, and iodate has been eliminated so that the glycolic acid can be estimated directly in oxidation mixtures.[581] Glyoxylic or glycolic acids were formed in reactions between periodate and 2-ketoses depending on whether oxidation occurred preferentially between C-1 and C-2 or C-2 and C-3 of the ketose. Previous studies of these reactions have been complicated by a further oxidation of glyoxylic acid with periodate. However, in 0·1N sulphuric acid no such over-oxidation occurred. Careful analysis of the products of periodate-oxidised ketoses has been carried out to assess the importance of steric factors in influencing attack at the C(1)–C(2) and the C(2)–C(3) bonds.[581a] Perseulose is initially oxidised by periodic acid between C-1 and C-2 to a greater extent than between C-2 and C-3.[582]

A new method for the microdetermination of the oxidation of vicinal diols with periodate has been described.[583] The procedure was based on the quantitative precipitation of periodate by aluminium hydroxide and spectrophotometric determination of the iodate left in solution. It was claimed that the method can be used for the determination of 0·01—0·25 μmol of diol. Experimental conditions have been developed for the determination of periodate ion and formaldehyde by polarography. The technique can be applied during the periodate oxidation of carbohydrates down to the 0·04 μmol level.[584]

The periodate oxidation of methyl 2-amino-4,6-*O*-benzylidene-2-deoxy-α-D-altropyranoside and its 3-amino-3-deoxy-isomer has been studied in some detail.[585] The oxidations, which followed first-order kinetics only up to the consumption of 0·6 mol of oxidant per mol, were accompanied by over-oxidation. It was suggested that alternative oxidation mechanisms to

[581] S. R. Sarfati and P. Szabó, *Carbohydrate Res.*, 1970, **12**, 290.
[581a] S. R. Sarfati and P. Szabó, *Carbohydrate Res.*, 1970, **13**, 441.
[582] P. Fleury, J. E. Courtois, and D. Darzens-Souloumiac, *Ann. pharm. franç.*, 1970, **28**, 17.
[583] H. Eibl and W. E. M. Lands, *Analyt. Biochem.*, 1970, **33**, 58.
[584] R. D. Corlett, W. G. Breck, and G. W. Hay, *Canad. J. Chem.*, 1970, **48**, 2474.
[585] C. B. Barlow and R. D. Guthrie, *Carbohydrate Res.*, 1970, **13**, 199.

those which operate for the six 'normal' amino-sugars (Vol. 3, p. 166) must operate for the aminoaltrosides.

Oxidation of D-*glycero*-D-*gulo*-heptono-1,4-lactone with 1 mol of periodate afforded D-arabinose; this behaviour contrasts with that of other γ-lactones where the side-chain is cleaved preferentially.[586]

The oxidations of 2,4-*O*-benzylidene-D-glucitol with lead tetra-acetate and with sodium periodate have been studied.[587] With the former reagent, the anomers of (496) and the dimer (497) were isolated following acetylation,

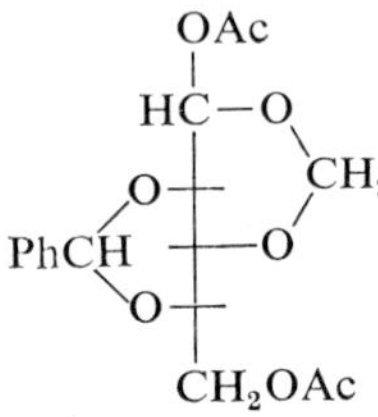

(496)

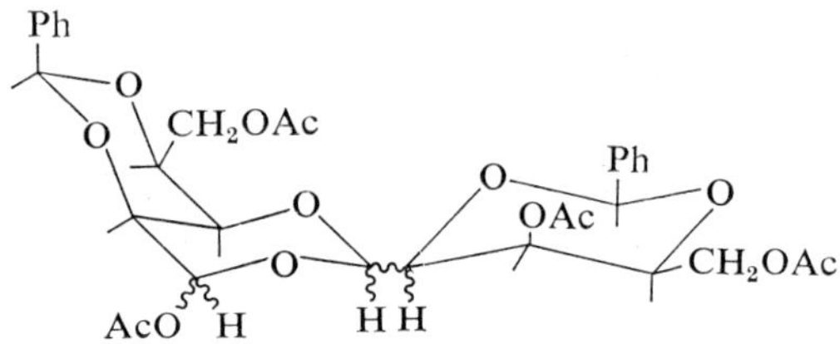

(497)

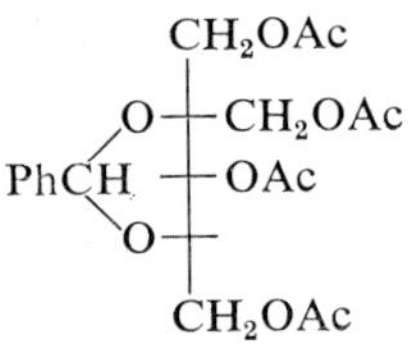

(498)

whereas with sodium periodate the only product isolated after acetylation was (498), many other products were detected in the oxidation mixtures. Oxidation of (264) with periodate also gave a dimeric product[355] (Chapter 12).

DMSO-based Oxidations

Sequences using oxidation with Ac_2O–DMSO followed by borohydride reduction afforded convenient procedures for inverting the configuration at a carbon atom in sugar derivatives. Thus, 2,3,6-tri-*O*-methyl-α-D-galactose was obtained from the corresponding methyl D-glucoside,[109] and the D-galactopyranose derivative (499) was converted into the D-gulopyranose derivative (500).[588] Similarly, the D-galactopyranose derivative (501) was transformed into the D-talopyranose derivative (502) by oxidation with P_2O_5 in DMSO followed by $LiAlH_4$ reduction, thereby providing a convenient route to D-talose. Oxidation of (501) with acetic anhydride in DMSO afforded the 2-methylthiomethyl ether and not the expected hexopyranosid-2-ulose.[588]

[586] W. C. Griffiths, T. T. Galkowski, R. W. Kocon, and K. M. Reardon, *Carbohydrate Res.*, 1970, **13**, 177.

[587] S. J. Angyal and K. James, *Carbohydrate Res.*, 1970, **15**, 91.

[588] G. J. F. Chittenden, *Carbohydrate Res.*, 1970, **15**, 101.

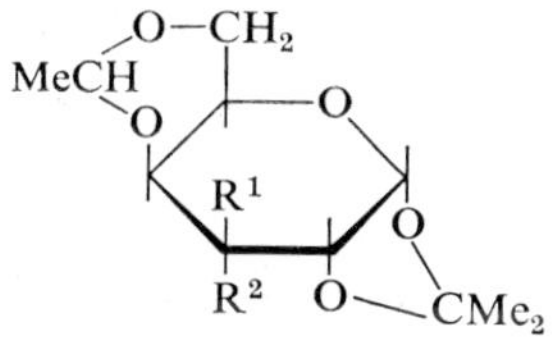

(499) R¹ = OH, R² = H
(500) R¹ = H, R² = OH

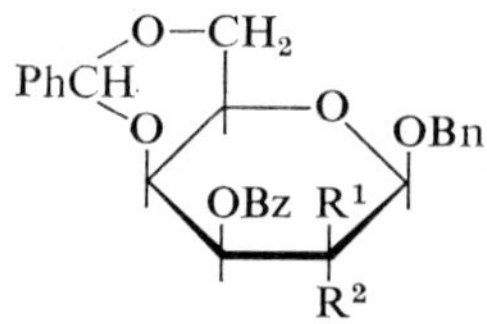

(501) R¹ = H, R² = OH
(502) R¹ = OH, R² = H

Related oxidations of methyl 4,6-*O*-benzylidene-3-*S*-phenyl-3-thio-α-D-altropyranoside (317) and the isomeric 2-*S*-phenyl-2-thio-compound [386] have been discussed in Chapter 14, and the theophylline derivative (503) has been prepared by oxidation of the corresponding alcohol with Ac_2O–DMSO.[589]

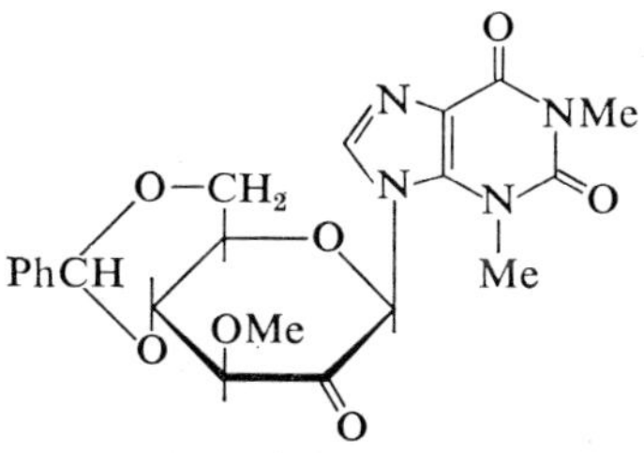

(503)

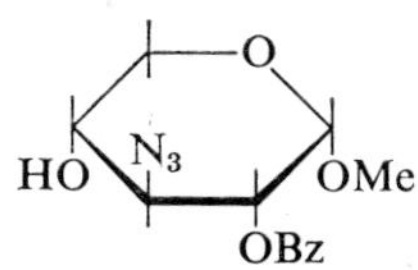

(504)

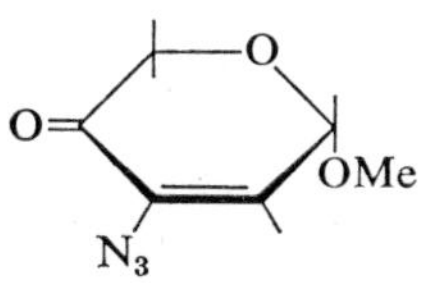

(505)

Oxidation of (504) with the Pfitzner–Moffatt reagent afforded (505) and not the expected 4-ulose.[174]

Other Oxidations

D-Glucal has been selectively oxidised to the enose (287) in 60—80% yield with silver carbonate on Celite [376] (Fétizon's reagent). Oxidation of aldohexoses and ketohexoses with Ag^{III} has been studied;[590] at 30 °C, aldohexoses required 12 equivalents and ketohexoses required 14 equivalents of the oxidant. Complete oxidation occurred at 80 °C and 24 equivalents of the oxidant were required by both classes. The complete oxidation of aldoses with Cu^{III} compounds has been reported.[591]

The oxidations of 1-*O*-benzoyl-2,3:5,6-di-*O*-isopropylidene-D-gulitol, -D-allitol, -DL-galactitol, and -D-mannitol with the chromium trioxide–

[589] M. K. Antonakis and F. Leclercq, *Compt. rend.*, 1970, **271C**, 1197.
[590] P. K. Jaiswal, *Microchem. J.*, 1970, **15**, 122.
[591] P. K. Jaiswal, *Microchem. J.*, 1970, **15**, 434.

pyridine complex have been compared.[592] Only the galactitol derivative was oxidised smoothly to the 4-ulose, and it was suggested that in the other isomers the formation of the chromate ester intermediate is impeded by a 1,4-*cis*-interaction with the 1-*O*-benzoyl substituent. The 4-hydroxy- and 1-benzoate groups are *trans*-orientated about the 2,3-dioxolan ring in the galactitol derivative so that chromate ester formation is not impeded. Attention has been drawn to the use of chromium trioxide in acetic acid as a reagent for oxidative demethylation. For example, 1,2,4,6-tetra-*O*-acetyl-3-*O*-methyl-α-D-glucopyranose was converted into 1,2,4,6-tetra-*O*-acetyl-3-*O*-formyl-α-D-glucopyranose in 64% yield, but attempts to selectively hydrolyse the formate ester were unsuccessful.[593] Benzyl ethers were similarly oxidised to benzoic esters. The use of chromium trioxide in acetic acid in the oxidation of cyclic acetals [140] has been noted in Chapter 5.

Ruthenium tetroxide has been used in the synthesis of some of the uloses described in Chapters 15 and 16. Ruthenium tetroxide oxidation of (506) gave the ring-expanded product (507) in addition to the 3-ulose derivative

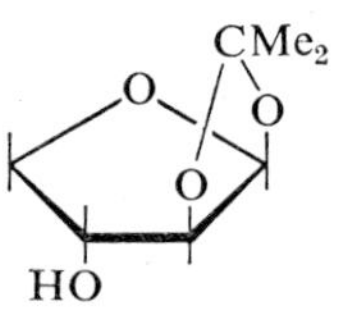

(506)

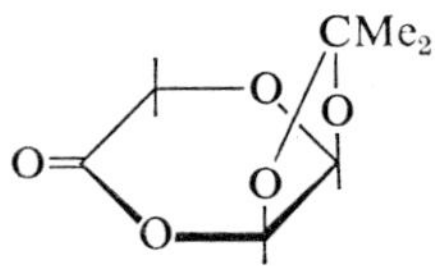

(507)

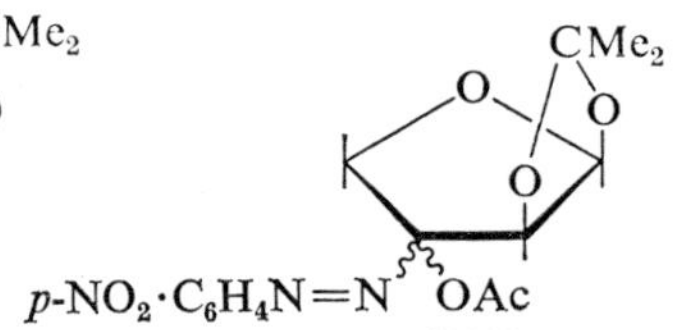

(508)

(355).[405] Lead tetra-acetate oxidation of the *p*-nitrophenylhydrazone derived from (355) afforded the *gem*-azoacetate (508).

Although the reagent has not yet been applied to carbohydrate derivatives, iodinium nitrate was found to oxidise alcohols to carbonyl compounds.[594]

Manganese dioxide was found to be a convenient, selective oxidant for certain allylic alcohol groups (*e.g.* 315; R = CH_2OH).[385] It is notable that the β-anomer of the pentose analogue of (315; R = H) was resistant to oxidation by manganese dioxide.

Reduction

Optimum conditions for the reduction of D-xylose to xylitol using Raney nickel have been reported,[595] and the hydrogenation of 5-keto-D-gluconic acid and its sodium salt with a variety of metal catalysts has been studied.[596] Nickel and palladium catalysts were found to give more L-idonic acid than D-gluconic acid particularly if the catalysts were pre-treated with sodium borohydride. The mechanism of these hydrogenations was discussed.

592 G. Y. Wu and J. M. Sugihara, *Carbohydrate Res.*, 1970, **13**, 89.
593 S. J. Angyal and K. James, *Carbohydrate Res.*, 1970, **12**, 147.
594 U. E. Diner, *J. Chem. Soc.* (*C*), 1970, 676.
595 A. Tuskui, W. Tanimura, and K. Suminoe, *J. Agric. Chem. Soc. Japan*, 1970, **44**, 96.
596 C.-Y. Chen, H. Yamamoto, and T. Kwan, *Chem. and Pharm. Bull.* (*Japan*), 1970, **18**, 1305.

23

N.M.R. Spectroscopy and Conformational Features of Carbohydrates

As in previous volumes, n.m.r. spectroscopy and stereochemical aspects of carbohydrate chemistry will be treated together. This year the potential importance of ^{13}C and ^{19}F studies in stereochemical analysis has become even more obvious.

The use of europium tris(dipivaloylmethane) as a reagent for simplifying otherwise complex n.m.r. spectra has been demonstrated.[3] Unequivocal assignments were made for all the protons in 1,2:5,6-di-*O*-isopropylidene-α-D-glucofuranose using the europium complex, whereas such assignments are not possible with the normal 100 MHz spectrum.

A new convention for designating sugar conformations[10] was referred to in Chapter 2.

Pyranoid Systems

General Observations on Model Compounds.—A number of reports on conformational features of tetrahydropyran derivatives have appeared. The favoured conformation of the anomeric alkoxy-group in substituted 2-alkoxytetrahydropyrans has been estimated by dipolmetric measurements.[597] The conclusions reached were similar to those obtained by n.m.r. methods, which used coupling constants and chemical shift data to assign conformations and configurations.[598] For compounds with chloro substituents, the characteristic C–Cl i.r. stretching frequencies were used to confirm the n.m.r. assignments. Conformational studies of dialkoxy-2,6-tetrahydropyrans and 2,6-fused, bicyclic tetrahydropyrans have also been reported.[599] Results from n.m.r. and dipole moment measurements were in good agreement, and the latter measurements also permitted estimates of the preferred rotamer states of the 2-methoxy-groups to be given.

The equilibrium between (509) and (510) in acetonitrile and carbon tetrachloride solutions has been studied by n.m.r. spectroscopy.[600] For

[597] M. Gelin, Y. Bahurel, and G. Descotes, *Bull. Soc. chim. France*, 1970, 3723.
[598] G. Descotes, D. Sinou, and J.-C. Martin, *Bull. Soc. chim. France*, 1970, 3730.
[599] Y. Bahurel, M. Lissac-Cahu, G. Descotes, M. Gelin, J. Delmau, and J. Duplan, *Bull. Soc. chim. France*, 1970, 4006.
[600] N. S. Zefirov and N. M. Shekhtman, *Zhur. org. Khim.*, 1970, **6**, 863; *J. Org. Chem. (U.S.S.R.)*, 1970, **6**, 863.

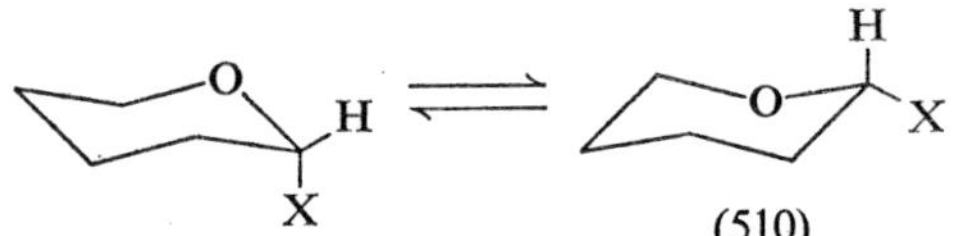

(509) (510)

X = N_3, NCO, or OAc, the conformer (509) preponderated, whereas (510) was the major conformer for X = NHAc.

Many of the factors which affect the conformations adopted by 1,3-dioxans, 1,3-dithians, and other saturated heterocyclic molecules have been summarised,[601] and another detailed paper has described the n.m.r. spectra and conformations of 1,3-dioxan derivatives.[602] Particular attention was focused on the use of CCl_4–C_6D_6 solvent shifts for assigning signals. A 220 MHz n.m.r. spectral analysis of some 2-substituted 1,3-dioxans (511)

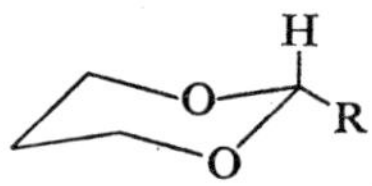

(511) R = Me, Pr^i, Bu^t, Ph, and *p*-ClC_6H_4

showed that there was no significant difference between the vicinal coupling constants of the C-4, C-5, and C-6 protons for the various R groups investigated, thereby showing that the ring-geometry was little affected by the nature of the substituent.[603]

Various isomers and derivatives of quinic acid have been prepared and their stereochemistry determined by n.m.r. methods.[471]

General Observations on Pyranoid Compounds.—A survey [604] of nearly 300 methyl glycosides has shown that in 209 compounds, where the methoxy-group was either axial or pseudoaxial in $CDCl_3$, 93% of the signals were in the range τ6·52—6·70, whereas in 74 compounds where the methoxy-group was equatorial or pseudoequatorial, 76% of the signals were in the range τ6·36—6·50. It has been suggested [605] that in 6-deoxypyranose derivatives, the 6-methyl doublet will resonate at higher field in the α-anomers than in the β-anomers as a consequence of electronic effects associated with the anomeric centre. Thus, configurational assignments at the anomeric centre can be made by comparison of the chemical shifts of the 6-methyl groups in mixtures of anomers.

The n.m.r. spectra of a series of sulphated monosaccharide derivatives have been measured.[606] It was found that H-1 *eq.* protons were deshielded by 0·27 p.p.m. by a *cis-eq.* 2-sulphate group and by 0·6 p.p.m. by an *eq.* 3-sulphate group; H-1 *ax.* protons were deshielded by 0·18 p.p.m. by an *eq.*

601 E. L. Eliel, *Bull. Soc. chim. France*, 1970, 517.
602 K. Pihlaja and P. Äyräs, *Acta Chem. Scand.*, 1970, **24**, 531.
603 H. R. Buys and E. L. Eliel, *Tetrahedron Letters*, 1970, 2779.
604 A. Konowal and A. Zamojski, *Roczniki Chem.*, 1970, **44**, 1607.
605 H. B. Sinclair and R. T. Sleeter, *Tetrahedron Letters*, 1970, 833.
606 M. J. Harris and J. R. Turvey, *Carbohydrate Res.*, 1970, **15**, 57.

2-sulphate group. Attention has been drawn [607] to the pronounced shielding effects on C-2 acetyl groups of 1-*O*-arylacetyl and 1-*O*-indolylacetyl groups in acetylated D-glucopyranoses, where C-2 acetyl groups appear at unusually high field (τ8·18—8·43).

Detailed n.m.r. analyses of a number of acetylated carbohydrate derivatives bearing 1,3-*O*-ethylidene or nitro substituents have been reported.[608] Possible reasons for benzene-induced chemical shifts in these compounds have been discussed, and a structure for a benzene–2-methyl-1,3-dioxan collision complex has been postulated.

N.m.r. spectroscopy and other methods have been used to study the solid-state transitions of levoglucosan and related 1,6-anhydrohexopyranoses.[609]

Application of the INDOR technique in the determination of coupling constants in complex n.m.r. spectra has been described. Examples of the use of the method included the spectra of sucrose octa-acetate and 2-deoxy-D-*arabino*-hexopyranose.[610]

N.m.r. spectra of fully trimethylsilylated oligosaccharides have been studied, and permitted determination of the configuration of glycosidic linkages.[611]

Specific Pyranoid Compounds.—Identification of the positions of substitution in partially and fully methylated derivatives of D-galactopyranose was shown to be feasible.[612] The methylated D-galactose derivatives were converted into the corresponding methyl glycosides and then perdeuteriomethylated, whereafter the n.m.r. spectra of the products in benzene were compared with the assigned spectra of methyl 2,3,4,6-tetra-*O*-methyl-α- and -β-D-galactopyranosides.

The n.m.r. parameters of a series of α- and β-D-glucopyranose and α-D-mannopyranose derivatives of general formula (512) have been examined.[613] Stereochemical and electronic factors were shown to be

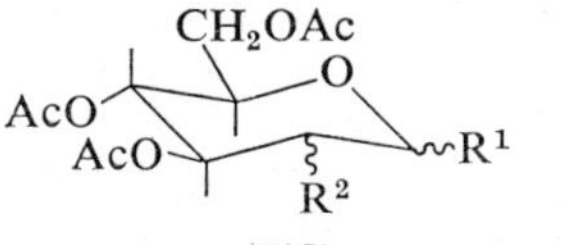

(512)

(a) $R^1 = R^2 = H$
(b) $R^1 = OH, R^2 = Cl$
(c) $R^1 = OAc, R^2 = Cl$
(d) $R^1 = R^2 = OAc$
(e) $R^1 = OMe, R^2 = OAc$
(f) $R^1 = Br, R^2 = OAc$
(g) $R^1 = OAc, R^2 = H$
(k) $R^1 = R^2 = Cl$
(i) $R^1 = OMe, R^2 = Cl$

607 N. Pravdić and D. Keglević, *Carbohydrate Res.*, 1970, **12**, 193.
608 K. D. Carlson, C. R. Smith jun., and I. A. Wolff, *Carbohydrate Res.*, 1970, **13**, 403.
609 F. Shafizadeh, G. D. McGinnis, R. A. Susott, and C. W. Philpot, *Carbohydrate Res.*, 1970, **15**, 165.
610 R. Burton, L. D. Hall, and P. R. Steiner, *Canad. J. Chem.*, 1970, **48**, 2679.
611 J. P. Kamerling, D. Rosenberg, and J. F. G. Vliegenthart, *Biochem. Biophys. Res. Comm.*, 1970, **38**, 794.
612 E. B. Rathbone and A. M. Stephen, *Tetrahedron Letters*, 1970, 1339.
613 G. Descotes, F. Chizat, and J.-C. Martin, *Bull. Soc. chim. France*, 1970, 2304.

important in determining chemical shifts and the magnitude of both geminal and vicinal coupling constants.

A comprehensive n.m.r. study has been made of a wide range of aldopentopyranosyl derivatives of general formula (513), in which the

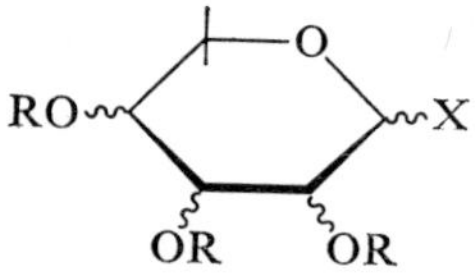

X = OAc, OBz, OMe, Cl, Br, or SAc
R = OAc or Bz

(513)

conformational equilibria between the two possible chair forms were considered.[614] The results obtained were discussed in detail with respect to the effect of steric and electronic factors, and with particular regard to the effect of the C-1 substituents. N.m.r. spectral investigations of peracetylated *N*-aryl-β-D-glucopyranosylamine, -β-D-xylosylamine, and -β-D-galactosylamine showed that they exist in the *C*1 conformation in $CDCl_3$ solution.[615]

N.m.r. studies have revealed that 1,2:4,6-di-*O*-benzylidene-α-D-glucopyranose derivatives have conformations in which the *m*-dioxan and pyranoid rings adopt chair and flattened-chair forms, respectively. It was suggested that the 4,6-*O*-benzylidene substituent increases the rigidity of the pyranoid ring since there was evidently far less distortion of the pyranoid ring in 1,2:4,6-di-*O*-benzylidene acetals than in 1,2-*O*-benzylidene acetals.[616] A detailed n.m.r. analysis of 3-*O*-benzoyl-1,2,4-*O*-benzylidyne-α-D-ribopyranose (514) has been used to test the applicability of equations

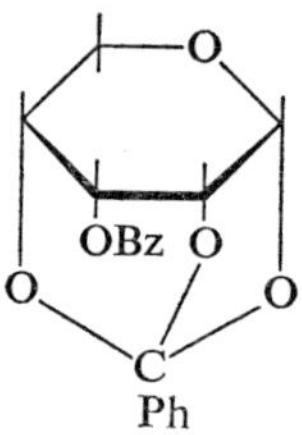

(514)

relating dihedral angles with geminal, vicinal, and long-range coupling constants.[617] Deviations were noted between approximate proton–proton dihedral angles calculated from a Karplus equation for vicinal couplings and those measured from a molecular model of the locked skew conformation. Moreover, no entirely satisfactory agreement was obtained between the results of equations relating the sign and magnitude of long-

[614] P. L. Durette and D. Horton, *Chem. Comm.*, 1970, 1608.
[615] Z. Smiatacz and J. Sokolowski, *Roczniki Chem.*, 1970, **44**, 1417.
[616] B. Coxon, *Carbohydrate Res.*, 1970, **14**, 9.
[617] B. Coxon, *Carbohydrate Res.*, 1970, **13**, 321.

range coupling constants with molecular geometry and those derived from measurements on molecular models.[617]

Configurational assignments to the anomers of 2-acetamido-5-ethoxyphenyl and 2-amino-5-ethoxyphenyl D-glucopyranosiduronic acid were only possible when the n.m.r. spectra were measured at 220 MHz.[618]

N.m.r. studies of tosyl derivatives of methyl 4,6-*O*-benzylidene-α- and -β-D-glucopyranosides have shown that the 2-tosyloxy-group in the β-D-anomer adopts a conformation such that H-1 is at right-angles to the aromatic ring, and that the 3-tosyloxy-group in both anomers is orientated so that the methine proton is at right-angles to the aromatic ring of the ester group.[198]

Furanoid Systems

The conformations of many cyclopentene and cyclopentenone derivatives have been studied by n.m.r. spectroscopy; the former assume envelope conformations whereas the latter adopt a *quasi*-coplanar shape.[619] The conformations of 1,3-dioxolans have been considered in detail with the conclusion that highly flexible rings are involved. Compounds with very bulky substituents showed signs of steric interactions.[620]

A number of 2,2,4-trisubstituted 1,3-dioxolans have been prepared and examined. Where one of the substituents was phenyl, the shifts of the other ring protons induced by the aromatic ring were used to assess the preferred conformation of the dioxolan ring. *trans*-4-Methoxycarbonyl-2-methyl-1,3-dioxolan is more stable than the corresponding *cis*-isomer, whereas the reverse holds for 2,4-dimethyl-1,3-dioxolan [621] (see Vol. 3, p. 175).

An excellent survey of furanoid ring conformations was provided in the introduction to a paper on the favoured ring forms of some pentofuranosyl fluoride derivatives.[622] The cycle of pseudorotation (CYCLOPS) for furanoid rings was depicted (using twist and envelope conformations only), and the conclusion was drawn that furanoid compounds should not be considered to adopt one conformation but rather that combined conformations of one section of CYCLOPS should be postulated.

The n.m.r. spectra of the 3-*O*-benzyl and 3-*O*-*p*-nitrobenzyl ethers of 5-deoxy-iodo-1,2-*O*-isopropylidene-α-D-xylofuranose have been measured at 220 MHz.[623] The diastereoisomers of 1,2-*O*-isopropylidene-3,5-*O*-(methoxymethylidene)-6-*O*-*p*-tolylsulphonyl-α-D-glucofuranose, (515) and (516), have been prepared, and their configurations were assigned on the basis of thermodynamic data and on evidence from vicinal coupling

618 J. Kiss and F. Burkhardt, *Carbohydrate Res.*, 1970, **12**, 115.
619 F. G. Cocu, G. Wolczunowicz, L. Bors, and T. Posternak, *Helv. Chim. Acta*, 1970, **53**, 739.
620 W. E. Willy, G. Binsch, and E. L. Eliel, *J. Amer. Chem. Soc.*, 1970, **92**, 5394.
621 T. D. Inch and N. Williams, *J. Chem. Soc.* (*C*), 1970, 263.
622 L. D. Hall, P. R. Steiner, and C. Pedersen, *Canad. J. Chem.*, 1970, **48**, 1155.
623 R. C. Young, P. W. Kent, and R. A. Dwek, *Tetrahedron*, 1970, **26**, 3983.

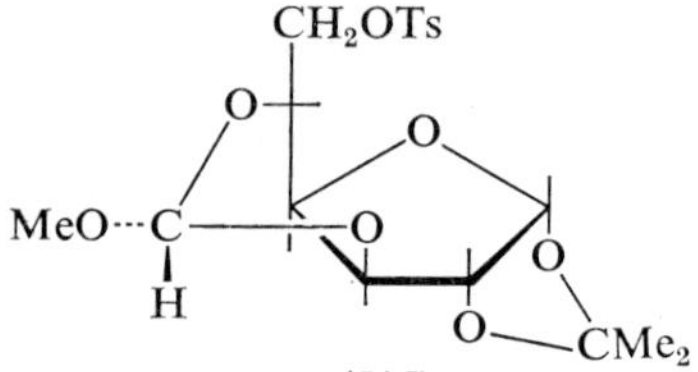

(515)

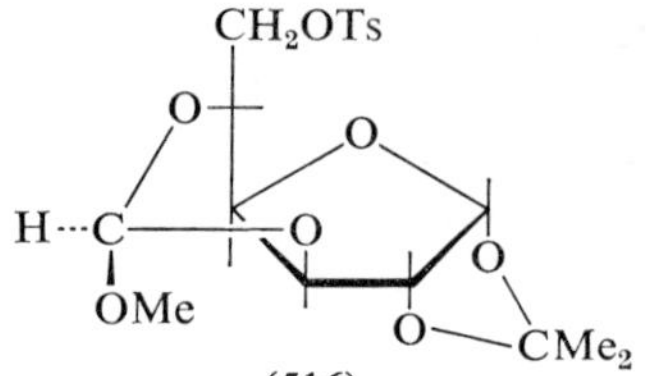

(516)

constants obtained by a computed spectral analysis.[624] Additionally, the nuclear Overhauser effect was used (for the first time in carbohydrate chemistry) to give a 25% enhancement of the intensity of the methylidene proton in (515) when the neighbouring *cis*-methoxy-group was irradiated.

An extensive study has been made of compounds possessing the structural unit Me–C–O–CH–O–. When this fragment forms part of a *cis*-2-methyl-1-oxa-5-*O*-substituted cyclopentane [*e.g.* (517)—(519)], ^5J couplings of

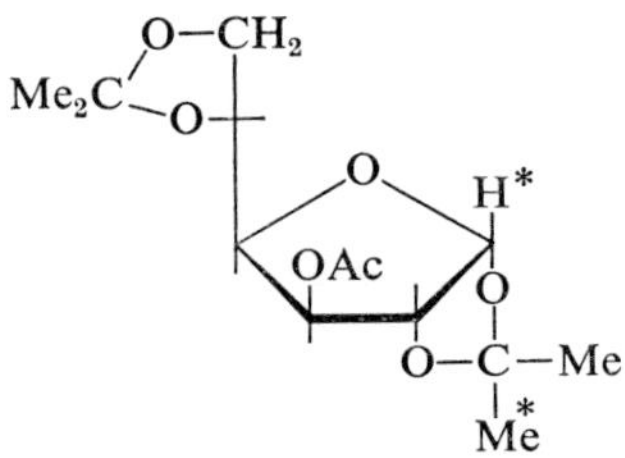

(517)

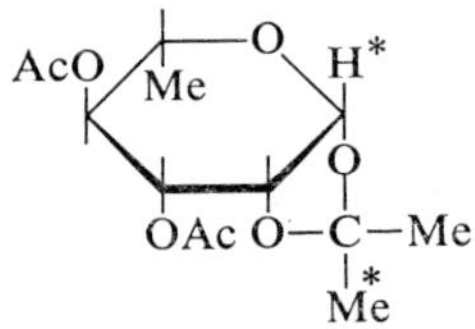

(518)

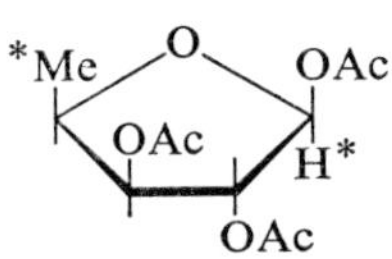

(519)

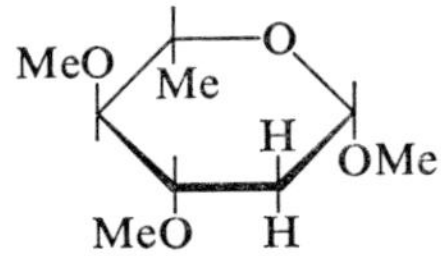

(520)

0·35—0·4 Hz are observable between the methyl protons and a ring proton located five bonds away.[625] Six-membered analogues [*e.g.* (520)] do not exhibit such couplings.

It has been pointed out that changes in the environment of nucleosides in crystalline form bring about changes in conformation.[626] The 220 MHz spectra of uridine and β-pseudouridine in aqueous solution have been determined.[627]

A 60 MHz study has been made of the hydroxy-protons of the D-ribose and 2-deoxy-D-ribose moieties of common nucleosides in mixtures of

[624] B. Coxon, *Carbohydrate Res.*, 1970, **12**, 313.
[625] J. C. Jochims and G. Taigel, *Chem. Ber.*, 1970, **103**, 448.
[626] H. R. Wilson, *Nature*, 1970, **225**, 545.
[627] B. J. Blackburn, A. A. Grey, and I. C. P. Smith, *Canad. J. Chem.*, 1970, **48**, 2866.

[$^2H_6$]DMSO and C_6H_6.[628] The observed coupling constants indicated that the O–(H5′) group rotates freely about the C(5′)–O(5′) bond. The O–H(3′) bond favours a *gauche* conformation relative to the C(3′)–H(3′) bond, but no such preference was indicated for O–H(3′) relative to C(2′)–H(2′).

Acyclic Systems

The conformations in solution of the tetra-acetates of *aldehydo*-D-ribose, -D-arabinose, -D-xylose, -D-lyxose, and -6-deoxy-D-galactose, and the diphenyl dithioacetals of tetra-*O*-acetyl-D-lyxose and tetra-*O*-acetyl-6-deoxy-L-mannose have been shown by n.m.r. spectroscopy to be as expected[629] (Vol. 3, p. 175). Similar studies have indicated the conformations in solution of the diphenyl dithioacetals of D-ribose, D-xylose, and D- and L-arabinose.[344]

X-Ray studies of nine alditols have shown that the carbon chain adopts the extended planar zig-zag conformation when the configurations at alternate centres are different, but is bent and non-planar when they are the same. The results indicated the importance of 1,3-parallel interactions in controlling the preferred conformations of acyclic molecules, and were compatible with results obtained for acetylated alditol derivatives in solution.[1] The n.m.r. spectra of the methyl 2,3,4,6-tetra-*O*-acetyl-5-hexulosonates also indicated that the shape of the molecules is governed by the presence of 2,4-interactions.[630]

Heteronuclear N.M.R. Studies

^{13}C N.M.R. Spectroscopy.—Roberts and his co-workers have carried out a series of ^{13}C magnetic resonance studies of particular significance to carbohydrate chemistry. An extensive investigation of the ^{13}C chemical shifts in acyclic and alicyclic alcohols[631] was extended to the inositols and related methylated derivatives[632] and, finally, to six-membered oxygen-containing rings including the aldopyranoses.[633] The stereochemical factors influencing chemical shifts were elucidated and, in consequence, an empirical method was developed for calculating the expected chemical shift of any carbon atom in a pyranoid ring.[633] Controversy over the assignment of the 3-^{13}C resonance in α- and β-D-glucopyranose was resolved by the synthesis of D-glucose-[3-^{2}H] as illustrated in Scheme 87.[634] A comparison of ^{13}C and ^{1}H chemical shifts for α- and β-D-glucose and of the corresponding hydroxy-proton resonances showed that there is widespread difference between the anomers in the polarisation of the various bonds. Thus, inversion of the anomeric hydroxy-group from an equatorial to an axial

[628] D. B. Davies and S. S. Danyluk, *Canad. J. Chem.*, 1970, **48**, 3112.
[629] D. Horton and J. D. Wander, *Carbohydrate Res.*, 1970, **15**, 271.
[630] S. J. Angyal and K. James, *Austral. J. Chem.*, 1970, **23**, 1223.
[631] J. D. Roberts, F. J. Weigert, J. I. Kroschwitz, and H. J. Reich, *J. Amer. Chem. Soc.*, 1970, **92**, 1338.
[632] D. E. Dorman, S. J. Angyal, and J. D. Roberts, *J. Amer. Chem. Soc.*, 1970, **92**, 1351.
[633] D. E. Dorman and J. D. Roberts, *J. Amer. Chem. Soc.*, 1970, **92**, 1355.
[634] H. J. Koch and A. S. Perlin, *Carbohydrate Res.*, 1970, **15**, 403.

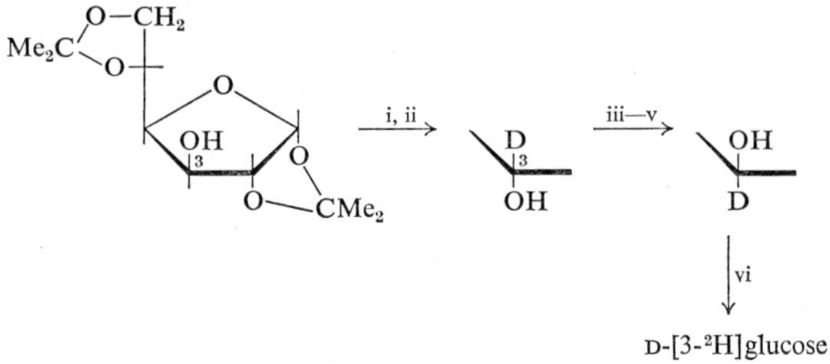

Reagents: i, Ac_2O, DMSO; ii, $NaBD_4$; iii, TsCl, py; iv, NaOBz, DMF; v, NaOMe; vi, H_3O^+

Scheme 87

orientation is associated with a uniform increase in shielding of ^{13}C, a decrease in shielding of the appended protons, and an increase in shielding of the hydroxy-protons at all positions except 4 and 6, which are virtually unaffected.[634] The effects of the configuration of adjacent carbon atoms on ^{13}C chemical shifts have been considered in more detail following a study of methyl glycosides.[634a] N.m.r. studies on some substituted cyclohexanes by Perlin and Koch have shown that ^{13}C shieldings increase as non-bonding interactions increase.[635] Protons are affected in the opposite sense. It was concluded that steric repulsive interactions alter the polarisation of C–H bonds making the carbon atoms more electronegative. Similar effects were found in a continuation of this work with pyranoid systems.

Natural abundance ^{13}C n.m.r. spectra of a number of pyrimidine nucleosides have been determined. The assignments made to the various carbon atoms of the sugar ring have enabled ribosides, deoxyribosides, arabinosides, and anhydro-derivatives to be characterised.[636]

It was evident from a ^{13}C n.m.r. study of 29 nucleosides that the ^{13}C resonances are conveniently divided into two groups — those arising from the sugars and those arising from the bases.[637]

^{13}C N.m.r. spectroscopy has been used in the elucidation of structure of the antibiotic hygromycin B.[638]

A preliminary report of the first ^{13}C Fourier-transform n.m.r. spectra of polyols has appeared.[639]

[634a] A. S. Perlin, B. Casu, and H. J. Koch, *Canad. J. Chem.*, 1970, **48**, 2596.

[635] A. S. Perlin and H. J. Koch, *Canad. J. Chem.*, 1970, **48**, 2639.

[636] A. J. Jones, D. M. Grant, M. W. Winkley, and R. K. Robins, *J. Phys. Chem.*, 1970, **74**, 2684.

[637] A. J. Jones, D. M. Grant, M. W. Winkley, and R. K. Robins, *J. Amer. Chem. Soc.*, 1970, **92**, 4079.

[638] N. Neuss, K. F. Koch, B. B. Molloy, W. Day, L. L. Huckstep, D. E. Dorman, and J. D. Roberts, *Helv. Chim. Acta*, 1970, **53**, 2314.

[639] W. Voelter, E. Breitmaier, G. Jung, T. Keller, and D. Hiss, *Angew. Chem. Internat. Edn.*, 1970, **9**, 803.

^{19}F N.M.R.—A number of joint papers by Hall's and Foster's groups have illustrated important aspects of ^{19}F n.m.r. behaviour. Thus, the angular dependence of ^{19}F–^{19}F geminal coupling constants has been investigated in 2-deoxy-2-fluoroglycosyl fluorides [640] and related compounds.[226] A full analysis of the 100 MHz spectra (^{1}H and ^{19}F) of (521) has been reported

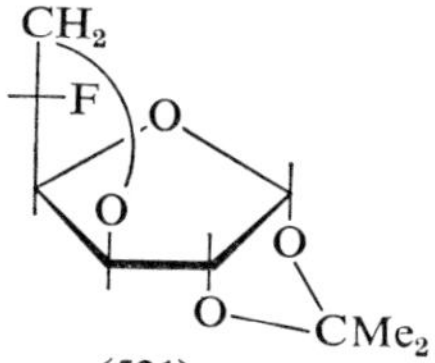

(521)

as a preliminary to the application of ^{19}F spectroscopy to the conformational analysis of furanoid rings.[641] The fluorine nucleus was found to couple with all the ring protons except H-2 ($J_{1,F}$ = 1·5; $J_{3,F}$ = 1·5; $J_{4,F}$ = 7·5; $J_{5,F}$ = 50·4; $J_{6,F}$ = 38·3; $J_{6',F}$ = 26·1 Hz). ^{1}H and ^{19}F studies of other furanoid derivatives [622] and, in particular, of 3,6-anhydrofuranosyl fluorides have been reported.[642] Long-range couplings between F-1 and H-4 of 4·8—6·3 Hz were detected in the latter compounds, but only very small F(1)–H(3) couplings were observed, although the geometrical relationships between F(1)–H(4) and F(1)–H(3) are similar. A ^{5}J coupling of 3—4 Hz was observed between F-4 and H-1 in methyl 4-deoxy-4-fluoro-2,3-di-*O*-methyl-α-D-glucopyranoside.[233]

The ^{19}F-n.m.r. spectra of anomeric pairs of trifluoroacetylated methyl glycosides have been compared. Assignments of the various CF_3 signals were made, and these measurements provided a simple and definitive method for identification of the glycosides.[643]

[640] L. D. Hall, R. N. Johnson, J. Adamson, and A. B. Foster, *Chem. Comm.*, 1970, 463.
[641] L. D. Hall and P. R. Steiner, *Canad. J. Chem.*, 1970, **48**, 451.
[642] L. D. Hall and P. R. Steiner, *Canad. J. Chem.*, 1970, **48**, 2439.
[643] J. Günther, W. Voelter, E. Breitmaier, and E. Bayer, *Annalen*, 1970, **734**, 136.

24
Other Physical Methods

Infrared Spectroscopy

The far-i.r. spectra of D-glucose and sucrose between 500 and 50 cm^{-1} have been published.[644]

The anomeric configuration of steroidal glucopyranosiduronic methyl esters were determined by i.r. methods; a band in the range 1146—1140 cm^{-1} was displayed only by the α-isomers.[74] Mitchell has published further spectra (of cellulose oligosaccharides, methyl β-D-glucopyranoside, and some polysaccharides) measured at room temperature and also at − 180 °C, and has concluded that the markedly better resolution obtained at low temperatures occurs for highly-ordered compounds with hydroxy-groups involved in strong intermolecular hydrogen bonding.[645]

Studies of the i.r. spectra of a series of sulphated monosaccharide derivatives have been reported. It was demonstrated that the precise position of the C–O–S bond (810—860 cm^{-1}) depends on a variety of factors and cannot be used, as was suggested previously, to distinguish between equatorial and axial sulphate groups with certainty.[646]

The solid-state transitions of levoglucosan and other 1,6-anhydrohexopyranoses have been studied by differential thermal analysis, and i.r. and n.m.r. spectroscopy.[609]

Mass Spectrometry

Mass spectrometry has proved to be a valuable method for locating the positions of the double bonds in various unsaturated compounds. For 2,3- and 3,4-unsaturated glycopyranosides, an important fragmentation pathway involves retrodienic cleavage whereby C-5 and O–R are lost for the first class, whereas O–R and C-1 are lost for the second class (Scheme 88). Recognition of the ions so formed provides means for locating the position of the unsaturated linkage. In some cases, however, (*e.g.* Scheme 89) thermal allylic rearrangement precedes fragmentation so that care has to be taken in interpreting the spectra.[647]

644 M. Hineno and H. Yoshinaga, *Bull. Chem. Soc. Japan*, 1970, **43**, 3308.
645 A. J. Mitchell, *Austral. J. Chem.*, 1970, **23**, 833.
646 M. J. Harris and J. R. Turvey, *Carbohydrate Res.*, 1970, **15**, 51.
647 R. J. Ferrier, N. Vethaviyasar, O. S. Chizov, V. I. Kadentsev, and B. M. Zolotarev, *Carbohydrate Res.*, 1970, **13**, 269.

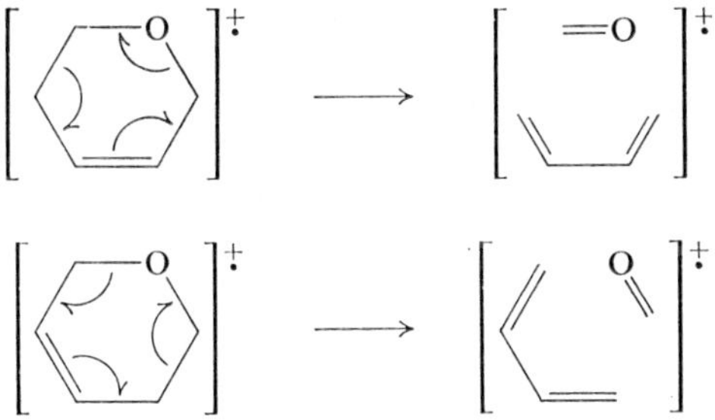

Scheme 88

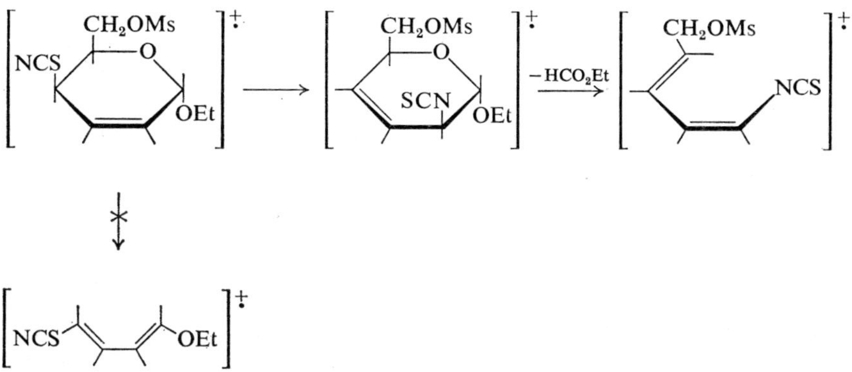

Scheme 89

A long paper has appeared providing the fullest details available on the mass spectra of adenine nucleosides. The effects of substitution and configurational changes were examined; anomers could be readily distinguished but the configurations at other centres were less readily determinable. However, the intensity of the $M - 30$ ion is dependent on the relative orientations of the base and C-5′.[573]

The spectra of the trimethylsilyl ethers of D-fructose, L-sorbose, methyl D-fructopyranoside, and methyl L-sorbofuranoside have been studied. Whereas the 2-ketohexose derivatives give characteristic peaks at *m/e* 437 and for the corresponding methyl glycosides at *m/e* 379, aldohexoses and methyl aldosides give such peaks at *m/e* 435 and 377, respectively. Thus, aldoses and ketoses may be easily differentiated. Also, peaks at *m/e* 437 and 204 are characteristic of ketopyranosides, whereas peaks at *m/e* 437 and 217 are given by the corresponding furanosides.[647a] Structural isomers of aldonic and deoxyaldonic acids can be distinguished by mass spectrometry of their TMS ethers, but the spectra of diastereoisomers are similar.[648]

[647a] S. Karady and S. H. Pines, *Tetrahedron*, 1970, **26**, 4527.
[648] G. Petersonn, *Tetrahedron*, 1970, **26**, 3413.

As a basis for the characterisation of the terminal amino units in naturally occurring glycopeptides, the high resolution mass spectra of the *N*-acyl derivatives of 2-acetamido-3,4,6-tri-*O*-acetyl-2-deoxy-β-D-glucopyranosylamines (522) have been studied.[649] Other compounds which have

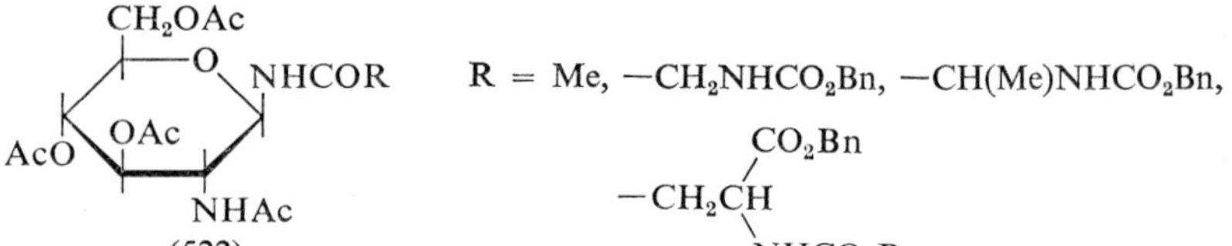

(522)

been examined as their peracetates are rutinose[650] and various naturally occurring *C*-glycosides.[651, 652] The fragmentations of methanesulphonyl and *p*-tolylsulphonyl esters of some methyl pentofuranosides have been investigated. Attention was drawn to differences in the fragmentation pathways of the two types of ester.[653] Detailed fragmentation mechanisms have been postulated for a number of partially methylated alditol acetates on electron impact. Use was made of specifically deuteriated alditols to obtain unequivocal assignments of the nature and origin of many of the fragments.[654]

Another report has appeared on the application of field ionisation mass spectrometry to aryl glucosides, which has enabled Hammett ρ values to be determined for the cleavage of the aglycone and for rearrangement.[655]

X-Ray Crystallography

The rate of publication of solved crystal structures continues to increase rapidly, and the situation has now been reached where generalisations can be made regarding structures and shapes of compounds in the solid phase. This is well illustrated by a paper in which the conformations of nine alditols were compared. As had been observed previously, the carbon chains adopt extended planar zig-zag conformations when the configurations at alternate asymmetric centres are different, but they are bent and non-planar when such configurations are alike.[1] These conclusions parallel those drawn for acyclic compounds in solution. A related paper has dealt with the D-glucitol–pyridine complex and also with the complex formed between α-D-glucose and urea; the primary binding forces of both complexes

[649] T. Komori, Y. Ida, Y. Inatsu, M. Kiyozumi, K. Kato, and T. Kawasaki, *Annalen*, 1970, **741**, 33.
[650] J. H. Looker, M. Sozmen, S. A. Kagal, and S. Meyerson, *Carbohydrate Res.*, 1970, **13**, 179.
[651] M. Aritomi, T. Komori, and T. Kawasaki, *Annalen*, 1970, **734**, 91.
[652] A. Prox, *Annalen*, 1970, **732**, 199.
[653] V. Kováčik, P. Kováč, and R. L. Whistler, *Carbohydrate Res.*, 1970, **14**, 133.
[654] H. Björndal, B. Lindberg, A. Pilotti, and S. Svensson, *Carbohydrate Res.*, 1970, **15**, 339.
[655] G. O. Phillips, W. G. Filby, and W. L. Mead, *Chem. Comm.*, 1970, 1269.

are exceptionally stable hydrogen-bonded arrangements, which overcome appreciable non-bonding interactions apparent in the crystal structures.[656]

Among the free sugars and their derivatives to have been examined by *X*-ray crystallography are the naturally occurring 3-heptulose coriose [which exists in the crystal in the α-furanose form (523)],[26] raffinose,[657] and

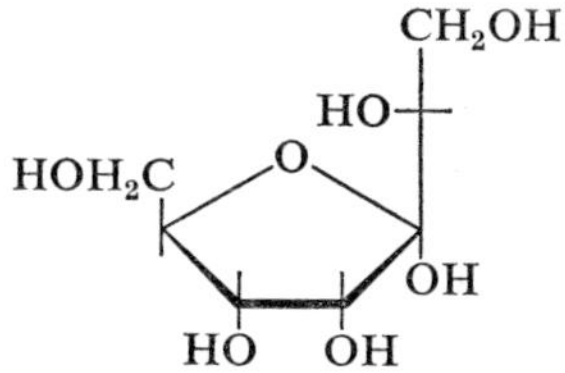

(523)

1,2:3,4-di-*O*-isopropylidene-5-*O*-chloroacetyl-α-D-glucoseptanose (Vol. 3, p. 38); the seven-membered ring in the latter compound adopts a conformation between the chair and the twist chair.[150]

Methyl α-D-mannopyranoside assumes the *C*1 chair conformation[658] but, surprisingly, methyl 4,6-dideoxy-4-(*NN*-dimethylamino)-α-D-talopyranoside methiodide occurs in the inverted chair form (524). This was

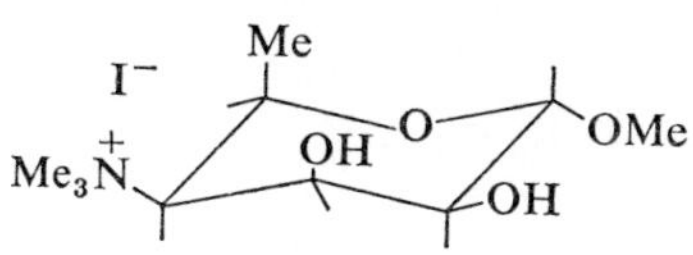

(524)

ascribed to the strong energetic preference of the bulky quaternary ammonium group for the equatorial orientation.[659] Other compounds to have been studied include the methyl β-cellobioside–methanol complex (an interesting feature here is a bifurcated, intramolecular hydrogen bond between the two D-glucose rings),[660] methyl 4,6-*O*-benzylidene-2-*O*-*p*-bromobenzenesulphonyl-3-cyano-3-deoxy-α-D-altropyranoside,[661] and the iodine azide adduct of 5,6-dideoxy-1,2-*O*-isopropylidene-α-D-*xylo*-hex-5-enofuranose (525) (Vol. 2, p. 100).[662] The crystal structures of compounds mentioned elsewhere have been established.[464, 465]

The structure of a new trianhydrosucrose was established by *X*-ray diffraction analysis.[134] Naturally-occurring compounds to have been so examined are gynecardin[87] and sinigrin.[663] Another report on 'α-D-isosaccharinic acid' has appeared.[433]

656 R. L. Snyder, R. D. Rosenstein, H. S. Kim, and G. A. Jeffrey, *Carbohydrate Res.*, 1970, **12**, 153.
657 H. M. Berman, *Acta Cryst.*, 1970, **26B**, 290.
658 B. M. Gatehouse and B. J. Poppleton, *Acta Cryst.*, 1970, **26B**, 1761.
659 R. E. Cook and M. D. Glick, *Acta Cryst.*, 1970, **26B**, 1741.
660 J. T. Ham and D. G. Williams, *Acta Cryst.*, 1970, **26B**, 1373.
661 B. E. Davison and A. T. McPhail, *J. Chem. Soc.* (*B*), 1970, 660.
662 J. S. Brimacombe, J. G. H. Bryan, and T. A. Hamor, *J. Chem. Soc.* (*B*), 1970, 514.
663 R. E. Marsh and J. Waser, *Acta Cryst.*, 1970, **26B**, 1030.

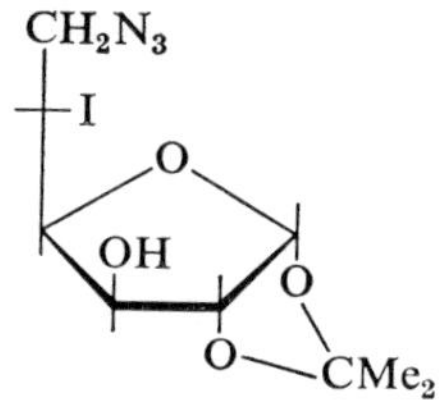

(525)

Most attention has, however, been devoted to nucleosides and related compounds, and a large wealth of information has been collected on this class of compound. The compounds which have been examined in 1970 include 3′-*O*-acetyladenosine,[664, 665] 2′-amino-2′-deoxy-α-D-adenosine,[666] showdomycin,[504] 8,5′-anhydro-2′,3′-isopropylidene-8-mercaptoadenosine,[667] ATP disodium salt,[668] 5-iodouridine[669, 670] (two molecules per asymmetric unit; each furanoid ring has a different envelope shape), α-pseudouridine,[670a] and triethylammonium uridine 2′,3′-cyclophosphorothioate (526). The

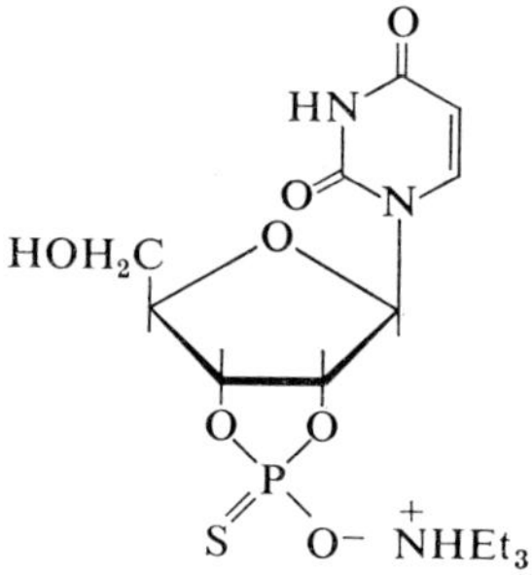

(526)

latter compound is a substrate for pancreatic ribonuclease, and the sugar adopts the unusual envelope conformation with the heteroatom exo-planar.[671]

2′-Deoxycytidine hydrochloride has a furanoid ring in the skew conformation, and the absolute configuration of the sugar was confirmed.[672]

664 S. T. Rao, M. Sundaralingam, and S. K. Arora, *Biochem. Biophys. Res. Comm.*, 1970, **38**, 496.
665 S. T. Rao and M. Sundaralingam, *J. Amer. Chem. Soc.*, 1970, **92**, 4963.
666 D. C. Rohrer and M. Sundaralingam, *J. Amer. Chem. Soc.*, 1970, **92**, 4956.
667 K. Tomita, T. Nishida, T. Fujiwara, and M. Ikehara, *Biochem. Biophys. Res. Comm.*, 1970, **41**, 1043.
668 O. Kennard, N. W. Isaacs, J. C. Coppola, A. J. Kirby, S. Warren, W. D. S. Motherwell, D. G. Watson, D. L. Wampler, D. H. Chenery, A. C. Larson, K. A. Kerr, and L. R. Di Sanseverino, *Nature*, 1970, **225**, 333.
669 A. Rahman and H. R. Wilson, *Nature*, 1970, **225**, 64.
670 A. Rahman and H. R. Wilson, *Acta Cryst.*, 1970, **26B**, 1765.
670a D. C. Rohrer and M. Sundaralingam, *J. Amer. Chem. Soc.*, 1970, **92**, 4950.
671 W. Saenger and F. Eckstein, *J. Amer. Chem. Soc.*, 1970, **92**, 4712.
672 E. Subramanian and D. J. Hunt, *Acta Cryst.*, 1970, **26B**, 303.

The dihydrates of guanosine and inosine are almost alike crystallographically; two nucleoside molecules are contained in each unit cell and both have different ring-conformations.[673] Other workers have examined anhydrous inosine and find only one molecule in the unit cell.[674]

Catalytic reduction of thymidine gave only the isomer (527), the configuration at the new asymmetric carbon atom being determined by *X*-ray crystallography.[675]

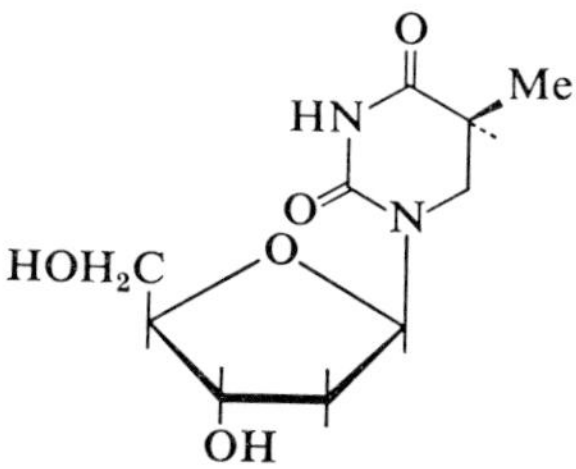

(527)

[673] U. Thewalt, C. E. Bugg, and R. E. Marsh, *Acta Cryst.*, 1970, **26B**, 1089.
[674] A. R. I. Munns and P. Tollin, *Acta Cryst.*, 1970, **26B**, 1101.
[675] J. Konnert, I. L. Karle, and J. Karle, *Acta Cryst.*, 1970, **26B**, 770.

25
Polarimetry

Developments in this area have been marked by the increased use of circular dichroism (c.d.), and there is a growing awareness that this technique has advantages over o.r.d. in many instances.

A notable development in the use of optical rotatory data has been reported by Lemieux and Martin. They followed Whiffen's pair-wise interaction approach but were able to reduce the parameters necessary for calculating the molecular rotation of a pyranoid compound to four, which they quantified by reference to suitable model compounds. It was conceded that the approach based on pair-wise interactions is oversimplified, but it still provided a useful method for calculating optical rotations of various structures and, hence, for determining the conformations of conformationally mobile species.[4] An essentially similar approach has been used to calculate the preferred conformations of acetyl α- and β-methylcholine in solution.[676]

Also, in the area of monochromatic polarimetry, a correlation has been found between the optical rotation and the carbon–halogen bond refraction of acylated glycosyl halides, amongst several other classes of compounds.[677] Yamana (*J. Org. Chem.*, 1966, **31**, 3698; *Tetrahedron*, 1968, **24**, 1559) has previously made similar observations regarding the relationship between optical activity and atomic refraction for these compounds.

It has been demonstrated that the molecular rotations of compounds containing 1,6-anhydro-β-D-hexopyranose structures can be calculated by summation of empirical rotatory contributions from various conformational elements of symmetry. The distortions in the bicyclo[3,2,1]octane skeleton present in such compounds were considered to invalidate calculations based on the simple approach described by Lemieux and Martin (see above). An approach necessitating the use of seven parameters for the 1,6-anhydro-β-D-hexopyranoses was developed, and an analogous approach requiring seven other parameters was applied to the 2,7-anhydro-β-D-heptulopyranoses.[678]

Rules for correlating the sign of the optical rotation at the sodium D line and the absolute configuration of 1,1-bis(acylamido)-1-deoxyalditols have

[676] T. D. Inch, R. Chittenden, and C. Dean, *J. Pharm. Pharmacol.*, 1970, **22**, 954.
[677] D. D. Davis and F. R. Jensen, *J. Org. Chem.*, 1970, **35**, 3410.
[678] D. Horton and J. D. Wander, *Carbohydrate Res.*, 1970, **14**, 83.

been reported.[679] The rules are similar to those derived for benzimidazole, phenylosotriazole, amide, and phenylhydrazide derivatives.

Polarimetry by the method of continuous variations has indicated that several glucopyranosides complex with zinc chloride.[680]

Circular dichroism of sugar–molybdate complexes can be useful in conformational studies. For example, characteristic and complex Cotton effects were observed between 220 and 350 nm with pyranoses in which the hydroxy-groups at C-1—C-3 have an axial : equatorial : axial relationship.[681]

A preliminary report on the c.d. spectra of cuprammonium complexes of furanosides has shown that the method is potentially useful for determining the conformations of such compounds, and the results were compatible with those obtained previously by n.m.r. methods.[682] Circular dichroism was also used in a stereochemical study of the cuprammonium complexes of pyranosides containing 1,2-diol and 1,2-amino-alcohol groups.[682a]

Conformational analysis by the benzoate-chirality method has been extended to the sugar series by use of *p*-chlorobenzoates. Diesters of methyl 2,6-dideoxyhexosides and triesters of pentosides and 6-deoxyhexosides were investigated.[683]

O.r.d. and c.d. of a number of steroidal 2-acetamido-2-deoxy-D-glucopyranosides have been reported, and solvent effects on the c.d. of some simple glycosides were described.[684] Conformational aspects of muramic acid have been discussed on the basis of c.d. measurements; it appears that O-3 of the 2-amino-2-deoxy-D-glucose residue and the double bond of the carboxylic ester group are eclipsed.[685]

The c.d. spectra of several sugar formazans and their peracetates have been recorded and, as expected, the signs of the Cotton effect curves correlate with the configuration at C-3.[686]

Several studies on nucleosides have been reported. A rule previously formulated to relate the o.r.d. spectra of pyrimidine D-ribonucleosides with their molecular features has been refined: the sign of the long-wavelength Cotton effects will be positive if (i) the preferred conformation is such that the sugar–base torsion angle is within the range $-75°$ to $+105°$, and (ii) a line from C-4 through C-2 of the base passes from above to below the plane of the furanoid ring.[576] The same group also discussed the relationship between the sign and the magnitude of the long-wavelength Cotton

[679] A. S. Cerezo, *Carbohydrate Res.*, 1970, **15**, 315.
[680] N. J. Richards and D. G. Williams, *Carbohydrate Res.*, 1970, **12**, 409.
[681] W. Voelter, G. Kuhfittig, G. Schneider, and E. Bayer, *Annalen*, 1970, **734**, 126.
[682] S. T. K. Bukhari and R. D. Guthrie, *Carbohydrate Res.*, 1970, **12**, 469.
[682a] S. T. K. Bukhari, R. D. Guthrie, A. I. Scott, and A. D. Wrixon, *Tetrahedron*, 1970, **26**, 3653.
[683] N. Harada, H. Sato, and K. Nakanishi, *Chem. Comm.*, 1970, 1691.
[684] D. K. Fukushima and M. Matsui, *Steroids*, 1969, **14**, 649.
[685] I. Listowsky, G. Avigad, and S. Englard, *Biochemistry*, 1970, **9**, 2186.
[686] L. Mester, G. Vass, and M. Mester, *Z. Chem.*, 1970, **10**, 395.

effect of purine nucleosides and molecular conformations. A correlation diagram was provided.[687]

For the study of the o.r.d. and c.d. of 'double-headed nucleosides', 5′-(adenin-9-yl)-5′-deoxythymidine (528) and 5′-(adenin-9-yl)-2,5′-dideoxy-

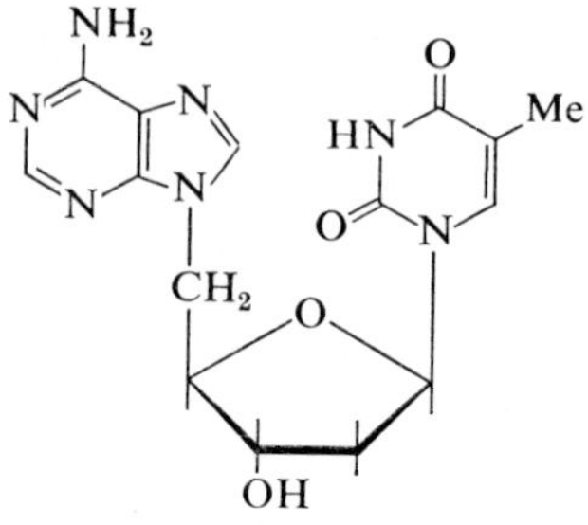

(528)

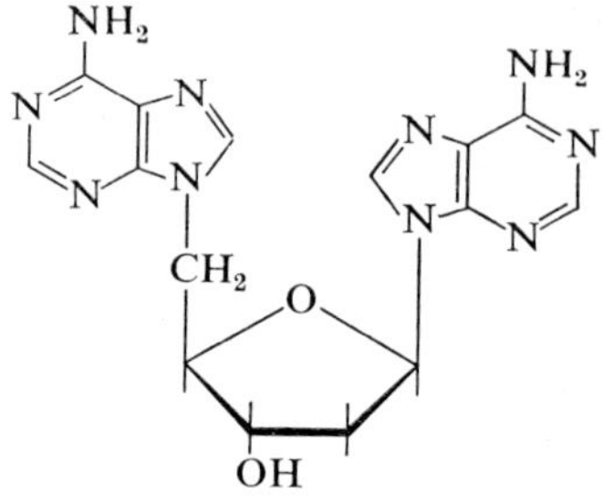

(529)

adenosine (529) have been synthesised from 5′-tosyl precursors by displacement of the tosyloxy-group with the sodium salt of adenine in DMF. The c.d. spectra of (528) and (529) show significant changes from the spectra of thymidine and adenosine.[688]

[687] G. T. Rogers and T. L. V. Ulbricht, *Biochem. Biophys. Res. Comm.*, 1970, **39**, 419.
[688] R. Fecher, K. H. Boswell, J. J. Wittick, and T. Y. Shen, *Carbohydrate Res.*, 1970, **13**, 105.

26
Separatory and Analytical Methods

Chromatographic Methods

Gas–Liquid Chromatography.—Reduction of oligo- and poly-saccharides followed by hydrolysis with acid gives a mixture of free sugars and polyols in a ratio which depends upon the molecular weight of the polymer. A model study has now been carried out on the determination of these ratios by g.l.c. of the TMS derivatives; the method appears to be of value as a means of determining the degree of polymerisation. It was noted that ethers of free sugars gave slightly lower responses from a flame ionisation detector than did ethers of polyols.[689] A further publication, giving an extensive survey of the relevant literature, has appeared on the analysis of free sugar mixtures (of the type obtained by hydrolysis of glycoproteins) by g.l.c. of the TMS ethers. Free sugars were anomerised before silylation, and molar response factors (flame ionisation detectors) were determined prior to evaluation of the results.[690] Hydrolysates of biological substances (*e.g.* saliva, bile, urine) have also been examined for carbohydrate constituents by g.l.c. of their TMS ethers,[691] and a simple, rapid, isothermal g.l.c. method for the analysis of mixtures of the non-acidic carbohydrate components of glycoproteins has been reported.[692] A further report on the g.l.c. of trimethylsilylated sugars has appeared.[693]

Gas chromatography and mass spectrometry have been used in combination for the analysis of disaccharides containing simple sugars and amino-sugars. The disaccharides were reduced to the alditols with sodium borodeuteride and were separated by g.l.c. after methylation. The position of the glycosidic linkage was then determined by mass spectrometry. Differentiation between 1 → 4- and 1 → 3-linked glycosylalditols was not possible without deuterium labelling, but 1 → 6-linked compounds could be distinguished by the presence of a fragment containing carbon atoms 1—4 of the alditol moiety. *N*-Methylation of *N*-acetamido-sugars was

[689] G. G. S. Dutton, P. E. Reid, J. J. M. Rowe, and K. L. Rowe, *J. Chromatog.*, 1970, **47**, 195.
[690] P. E. Reid, B. Donaldson, D. W. Secret, and B. Bradford, *J. Chromatog.*, 1970, **47**, 199.
[691] T. Gheorghiu and K. Oette, *J. Chromatog.*, 1970, **48**, 430.
[692] P. E. Reid, B. Donaldson, D. W. Secret, and B. Bradford, *J. Chromatog.*, 1970, **47**, 199.
[693] B. Arreguin and J. Taboada, *J. Chromatog. Sci.*, 1970, **8**, 187.

achieved by the use of methyl iodide in the presence of the methylsulphinyl carbanion.[694]

Gas chromatography–mass spectrometry was also used in a detailed study of the TMS derivatives of several sugar phosphates. Such ethers of aldohexose 6-phosphates, ketohexose 1- and 6-phosphates, aldopentose 5-phosphates, and hexonic acid 6-phosphates give characteristic mass spectra which can be used to identify these types of compound.[184] The enzymic condensation of glyceraldehyde and dihydroxyacetone phosphate was monitored by g.l.c. of the TMS derivatives of the sugar phosphates.[185]

Monomethyl ethers of D-glucose on reduction to the alditols followed by conversion to the trifluoroacetyl esters gave products which were readily separable. Trifluoroacetylation is sufficiently fast for the procedure to be used for routine analysis.[695] 2-Amino-2-deoxy-D-glucose and -D-galactose can be separated as their methyl 2-deoxy-3,4,6-tri-*O*-(trifluoroacetyl)-2-trifluoroacetamido-glycosides.[696]

Thermal fragmentation products of specifically labelled [^{14}C]sorboses and [6-^{3}H]sorbose have been separated and identified by gas radiochromatography.[697]

Column and Ion-exchange Chromatography.—In separations of monosaccharides in aqueous ethanol on ion-exchange resins in their lithium and sulphate forms, sugars were generally eluted in the order of increasing molecular weight. Most monosaccharides were conveniently separated in 86—90% ethanol on an anion-exchange resin in the sulphate form. With oligosaccharides there exists a linear relationship between the logarithm of the distribution coefficient and the number of monomeric units.[698] Sugar phosphates complexed with borate can be eluted from ion-exchange resins using gradient elution with ammonium chloride; phosphorylated glycolysis intermediates can be separated and recovered efficiently by this method.[699]

Separation of seven 2-amino-2-deoxy-D-hexoses[274] and all the methyl ethers of 2-amino-2-deoxy-D-glucose[700] has been achieved by ion-exchange chromatography. Adenosine, guanosine, uridine, and cytidine have also been separated quantitatively.[701] A procedure involving the use of charcoal columns and paper and ion-exchange chromatography has been reported for the isolation of pseudouridine from human urine.[702]

694 J. Kärkkäinen, *Carbohydrate Res.*, 1970, **14**, 27.
695 D. Anderle and P. Kováč, *J. Chromatog.*, 1970, **49**, 419.
696 D. J. Casagrande, *J. Chromatog.*, 1970, **49**, 537.
697 K. Heyns and R. Hauber, *Annalen*, 1970, **733**, 159.
698 E. Martinsson and O. Samuelson, *J. Chromatog.*, 1970, **50**, 429.
699 G. Bedetti, G. D'Agnolo, and F. Pocchiari, *J. Chromatog.*, 1970, **49**, 53.
700 G. A. Adams, M. Yaguchi, and M. B. Perry, *Carbohydrate Res.*, 1970, **12**, 267.
701 R. C. Van Den Bos, G. J. Van Kamp, and R. J. Planta, *Analyt. Biochem.*, 1970, **35**, 32.
702 R. T. Markiw, *J. Chromatog.*, 1970, **47**, 111.

Paper Chromatography and Electrophoresis.—Various aryl β-D-glucosides have been separated using different solvent systems, and components were detected with a reagent containing aniline and phosphoric acid.[703]

The R_F values of sucrose, glucose, and fructose in many solvent systems have been tabulated.[704]

A simple and rapid electrophoretic method for separating hexuronic acids in electrolytes of the acetates of various metals has been described. The best results were obtained with barium and zinc.[705]

Thin-layer Chromatography.—A densitometric method for determining monosaccharides on t.l.c. plates involves detection with aniline–diphenylamine–phosphoric acid and photography on Polaroid film.[706] Another general procedure is applicable to the quantitative recovery of sugars from silica-gel plates.[707] Perchloric acid is a useful reagent for the detection of sugars and their derivatives on cellulose t.l.c. plates.[708] The rapid separation of arabinose, ribose, and xylose from all other sugars (except fucose) which are normally found in biological fluids has been described.[709] A double elution method was recommended for the separation of sorbose from other common monosaccharides and for use in its quantitative determination.[710]

Amongst the glycosidic compounds to have been examined by t.l.c. are the glycosides of *Strophanthus*,[711] various maltoside derivatives,[712] mustard oil glucosides,[713] steroidal tri-*O*-acetyl-D-glucopyranosiduronic methyl esters and related glucoside derivatives (α-compounds are more mobile),[714] and aryl glucosides.[715]

Adenosine 3′,5′-cyclic-phosphate can be readily separated from other nucleoside monophosphates since it does not complex with borate ions and therefore has a high mobility on plates impregnated with tetraborate.[716]

Other Analytical Methods

Methods based on enzymic procedures continue to be developed: new techniques for the assay of ribitol,[717] lactose,[718] and D-glucose, D-galactose,

703 J. L. Garraway and S. E. Cook, *J. Chromatog.*, 1970, **46**, 134.
704 F. Said, M. M. Amer, A. Abou El-Khier, *Bull. Fac. Pharm. Cairo Univ.*, 1967, **6**, 217 (*Chem. Abs.*, 1970, **73**, 88090v).
705 M. J. St. Cyr, *J. Chromatog.*, 1970, **47**, 284.
706 P. C. Kelleher, *J. Chromatog.*, 1970, **52**, 437.
707 R. W. Scott, *J. Chromatog.*, 1970, **49**, 473.
708 K. Nagasawa, A. Ogamo, H. Harada, and K. Kumagai, *Analyt. Chem.*, 1970, **42**, 1436.
709 D. J. Bell and M. Q.-K. Talukder, *J. Chromatog.*, 1970, **49**, 469.
710 M. Iglóy and A. Mizsei, *J. Chromatog.*, 1970, **46**, 207.
711 T. Kartnig and R. Danhofer, *J. Chromatog.*, 1970, **52**, 313.
712 R. T. Sleeter and H. B. Sinclair, *J. Chromatog.*, 1970, **49**, 543.
713 M. Matsuo, *J. Chromatog.*, 1970, **49**, 323.
714 J. J. Schneider and D. K. Fukushima, *J. Chromatog.*, 1970, **48**, 509.
715 J. L. Garraway and S. E. Cook, *J. Chromatog.*, 1970, **46**, 134.
716 J. D. Upton, *J. Chromatog.*, 1970, **52**, 171.
717 D. R. D. Shaw and D. Mirelman, *Analyt. Biochem.*, 1970, **38**, 299.
718 R. G. Coffey and F. J. Reithel, *Analyt. Biochem.*, 1969, **32**, 229.

D-mannose, and L-fucose [719] have been reported. The last method was used for the determination of these sugars in glycoprotein hydrolysates.

Colorimetric analyses also continue to be developed and refined. A new medium—glycollic acid in benzyl alcohol and HMPT—has been used in place of the *o*-toluidine method for determining aldohexoses.[720] Attention has been drawn to the interference by dextrans in the determination of D-glucose by the *o*-toluidine method.[721] The cysteine–sulphuric acid method for the determination of D-galactose has been improved.[722]

A procedure for the determination of 2-amino-2-deoxyhexoses as their 2-(2,4-dinitroanilino)-derivatives has been described,[723] and D-glucitol and D-mannitol have been assayed by oxidation with periodic acid and determination of the formaldehyde liberated with 3-methylbenzothiazolin-2-one hydrazone.[724]

The reaction of ketoses with zirconium oxyfluoride gives a fluorescent complex which can be used in microanalysis,[725] and a fluorescence procedure has also been used to detect trioses in complex mixtures of monosaccharides.[726]

[719] P. R. Finch, R. Yuen, H. Schachter, and M. A. Moscarello, *Analyt. Biochem.*, 1969, **31**, 296.
[720] J. Bierens De Haan and M. Roth, *Z. klin. Chem. u. klin. Biochem.*, 1969, **7**, 624.
[721] C. S. Frings, *Clinical Chem.*, 1970, **16**, 618.
[722] J. A. Singer and H. Lyons, *Analyt. Biochem.*, 1970, **33**, 47.
[723] H. J. Haas and A. Weigerding, *Carbohydrate Res.*, 1970, **12**, 211.
[724] M. Pays and M. Beljean, *Ann. pharm. franç.*, 1970, **28**, 241.
[725] H. Trapmann and V. S. Sethi, *Z. analyt. Chem.*, 1969, **248**, 314.
[726] T. Nakai, M. Koyama, and H. Demura, *J. Chromatog.*, 1970, **50**, 338.

27
Alditols

The chemistry of xylitol and its derivatives has been reviewed.[727]

Volemitol (D-*glycero*-D-*manno*-heptitol) (530) has been isolated from avocado seeds, where it occurs together with perseitol (D-*glycero*-D-*galacto*-heptitol) (531), and D-*manno*-heptulose; other alditols isolated from this

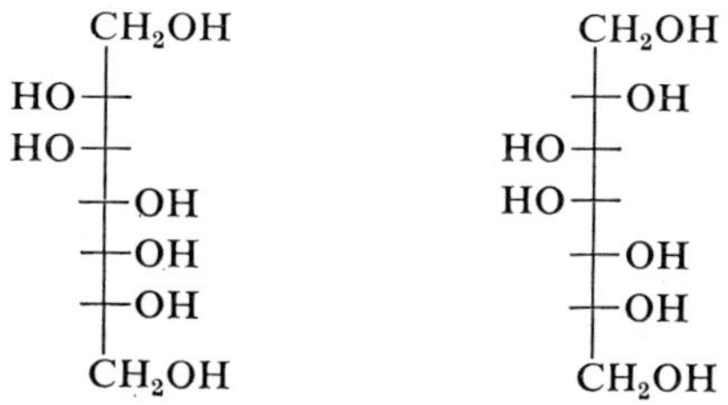

(530) (531)

source were glycerol, D-arabinitol, galactitol, and D-*erythro*-D-*galacto*-octitol.[728] Perseitol and volemitol also occur in *Sedum* species,[729] and the former heptitol was found among the alditols extracted from Pichi tops (the dried herbage of *Fibiana imbricata*).[730] Clusianose, a sugar extracted from the leaves of *Primula clusiana*, has been identified as 1-(*O*-α-D-galactopyranosyl)-hamamelitol.[731]

D-Mannitol has been used as starting material for a synthesis of 3-palmitoylglycerol-1-phosphorylcholine,[732] and hexitol phosphinates of the general formula (532) have been prepared from D-mannitol and D-glucitol.[733] The dehydration of D-glucitol in the presence of acidic resins has been examined.[131]

$$RP(O)H[OC_6H_8(OH)_5]; \quad (532) \quad R = Bu^i \text{ or } C_6H_{11}$$

[727] G. M. Zarubinskii, A. N. Anikeeva, L. G. Revelskya, and S. N. Danilov, 'Synthesis, Structure, and Polymer Properties,' Nauka, 1970, p. 307.
[728] N. K. Richtmyer, *Carbohydrate Res.*, 1970, **12**, 135.
[729] N. K. Richtmyer, *Carbohydrate Res.*, 1970, **12**, 139.
[730] N. K. Richtmyer, *Carbohydrate Res.*, 1970, **12**, 233.
[731] E. Beck, *Z. Pflanzenphysiol.*, 1969, **61**, 360 (*Chem. Abs.*, 1970, **72**, 101 031c).
[732] H. Eibl and O. Westphal, *Annalen*, 1970, **738**, 161.
[733] E. E. Nifant'ev, T. G. Shestakova, E. A. Kirichenko, and M. A. Yakunenkova, *Doklady Akad. Nauk S.S.S.R.*, 1969, **189**, 96.

N.m.r. spectroscopy has been used to assign structures to the three possible 1,2:5,6-bis-*O*-(trifluoromethyl)ethylidene-D-mannitols resulting from acetalation of D-mannitol with 1,1,1-trifluoroacetone,[141] and further work has been carried out on the stereochemistry of 1,2-*O*- and 1,2:5,6-di-*O*-bromoethylidene-D-mannitol.[142] 1,3:4,6-Di-*O*-benzylidene-D-mannitol was formed when D-mannitol was treated with benzaldehyde in acidified DMSO.[143] 2,3,4,5-Tetra-*O*-acetyl-1,6-dibromo-1,6-dideoxy-D-mannitol has been prepared by the action of phosphorus pentabromide on the corresponding hexa-acetate.[734]

The pyrrolidine derivatives (534)[735] and (536),[356] respectively, were obtained on reductive cyclisation of the azides (533) and (535) over Adam's

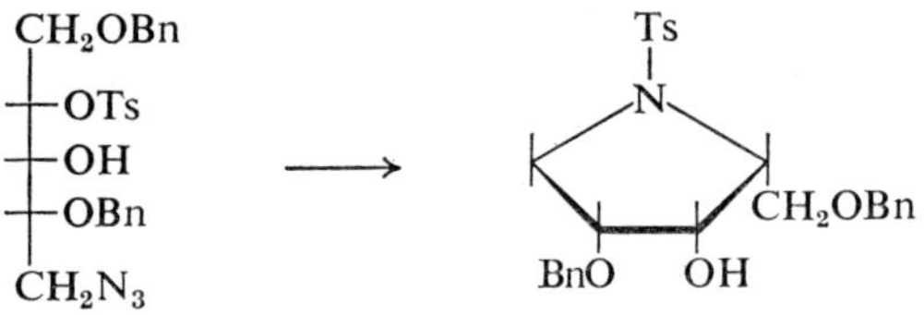

(533) (534)

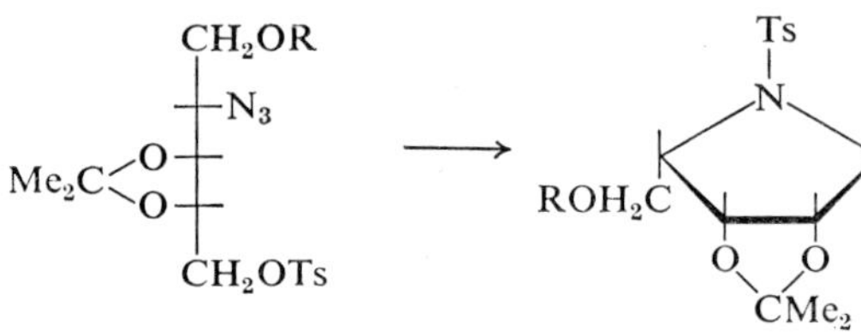

(535) (536)

R = tetrahydropyranyl

catalyst followed by *p*-tolylsulphonylation. Azide-displacement reactions involving the 5,6-ditosyl ester of 1,3:2,4-di-*O*-ethylidene-(or -benzylidene-) D-glucitol[325] and 3-*O*-*p*-bromobenzenesulphonyl-1,2:5,6-di-*O*-isopropylidene-D-mannitol[326] have been discussed already (see p. 74).

Empirical correlations have been made between the sign of the D-line optical rotation and the absolute configuration at C-2 of 1,1-bis(acylamido)-1-deoxyalditols (537).[679] The carbon-chain conformation of alditols in the crystalline state can be rationalised by the following rule: 'The carbon-chain adopts the extended, planar zig-zag when the *configurations* at alternative carbon centres are different, and is bent and non-planar when they are the same.' Thus, DL-arabinitol (LDD), D-mannitol (LLDD), and galactitol (DLLD) were observed to have extended carbon-chains, whereas

[734] L. Maijs, A. Berzate, and G. Fisers, *Latv. P.S.R. Zinat. Akad. Vestis, Kim. Ser.*, 1970, 119 (*Chem. Abs.*, 1970, **73**, 4128c).

[735] A. M. Sepulchre, A. Gateau, A. Gaudemer, and S. D. Gero, *Chem. Comm.*, 1970, 759.

HC(NHCOR)$_2$
|
(CHOH)$_n$
|
CH$_2$OH

(537)

R = Me, Et, or Ph

ribitol (DDD), xylitol (DLD), D-glucitol (DLDD), and D-iditol (DLLD) were found to have bent, non-planar carbon-chains. The rule predicts the most stable conformation for other unsubstituted alditols in the crystalline state, and it also provides a means of predicting the most probable rotameric states existing in solution.[1]

Mass spectral data for glycosylalditols[736] and partially methylated alditols have been reported.[654]

[736] O. S. Chizov, N. N. Malysheva, V. I. Kadentzev, and N. K. Kochetkov, *Doklady Akad. Nauk S.S.S.R.*, 1970, **194**, 836.

Part II

MACROMOLECULES

By

J. F. Kennedy

1
Introduction

The primary objectives of this section have followed the lines of previous Reports in this series. In view of the increasing number of relevant papers published in medical journals and in non-European languages, the journal coverage has been widened.

Chapter and section titles have been modified in keeping with developments in the various fields. Since glycosaminoglycans generally occur as glycoproteins, they have been included in the appropriate chapter. Although there is doubt in some cases whether phytohaemagglutinins contain carbohydrate, references to these molecules have been included under plant glycoproteins in view of their collective relevance to carbohydrate chemistry.

The chapter on enzymes has been considerably enlarged to include information published on carbohydrases, carbohydrate oxidases, and other enzymes which are glycoproteins. In view of possible problems of nomenclature, a sub-index incorporating the E.C. numbers is provided at the end of the chapter on enzymes. Although the enzymic activity of α-lactalbumin has not been proved beyond doubt, references to it are included in the chapter on enzymes since its structure is similar to that of lysozyme.

While not all enzymes are glycoproteins, references to all enzyme derivatives have been included in the section on modified glycoproteins in the expanded chapter on synthesis and modifications. The reader is thereby presented with all the new information on solid-phase derivatives and modes of coupling to insoluble phases.

The chapter on glycolipids and gangliosides has been re-introduced.

It is hoped that these arrangements will cater for a greater number of readers who are interested in macromolecules containing carbohydrate. Notification of any omissions will be welcomed.

2

Glycoproteins, Glycopeptides, and Animal Polysaccharides

Glycoproteins have been the subject of a recent review.[1] Sections are devoted to definition and distribution, the primary structure of the carbohydrate moieties and glycopeptide linkages, classification on the basis of structure, biosynthesis, catabolism, and correlation of the carbohydrate content with structure. A review on the quaternary structures of macromolecules included data on the subunit formation by some glycoproteins.[2]

The results of circular dichroism studies of a number of glycoproteins indicated the occurrence of significant amounts of the α-helix in some.[3] A short review of the heterogeneity and structure of antibody-combining sites included information on the carbohydrate structures which act as the antigenic determinants of various molecules.[4]

Glycoproteins in which the carbohydrate moiety is bound to the asparaginyl residue of sequences of the type asparagine-X-serine or -threonine have been listed.[5]

A method for the detection and measurement of glycoproteins present on the surface or in the interior of cells involved use of an insolubilised, cross-linked concanavalin A in combination with enzymes.[6] An automated spectrofluorimetric method for the determination of glycoproteins, glycopeptides, and certain amino-acids was based on oxidation with periodate.[7] Methods based on histochemical processes have been described for the localisation of glycoproteins in disc electrophoresis gels, three different techniques being used to identify tyrosine, tryptophan, and carbohydrate.[8]

A study has been made of procedures used for the acid hydrolysis and methanolysis of carbohydrate moieties in glycoproteins with a view to establishing the best methods for quantitative determination of the indi-

[1] R. G. Spiro, *Ann. Rev. Biochem.*, 1970, **39**, 599.
[2] I. M. Klotz, N. R. Langerman, and D. W. Darnall, *Ann. Rev Biochem.*, 1970, **39**, 25.
[3] B. Jirgensons, *Biochim. Biophys. Acta*, 1970, **200**, 9.
[4] E. A. Kabat, *Ann. New York Acad. Sci.*, 1970, **169**, 43.
[5] L. T. Hunt and M. O. Dayhoff, *Biochem. Biophys. Res. Comm.*, 1970, **39**, 757.
[6] S. Avramers, *Compt. rend.*, 1970, **270** *D*, 2205.
[7] H. Cho Tun, J. F. Kennedy, M. Stacey, and R. R. Woodbury, *Carbohydrate Res.*, 1969, **11**, 225.
[8] K. Felgenhauer, A. Weis, and G. G. Glenner, *J. Chromatog.*, 1970, **46**, 116.

vidual monosaccharides.[9] Neutral and amino-sugars in the products of hydrolysis and methanolysis were examined by colorimetric methods and g.l.c. Quantitative *N*-acetylation of 2-amino-2-deoxyhexoses was carried out using a modified method. Methanolysis was found to be preferable for the release of neutral sugars from glycoproteins, whereas hydrolysis with acid was advantageous for the release of amino-sugars. Other work showed that losses of monosaccharides occurred following hydrolysis of glycoproteins with hydrochloric acid due to transglycosylation (Maillard) reactions occurring during removal of the acid by evaporation.[10] These undesirable side-reactions were detected by the addition of small amounts of D-[^{14}C]glucose to the hydrolysates.

The use of an automatic amino-acid analyser for the simultaneous determination of amino-acids (including hydroxyproline), 2-amino-2-deoxy-D-glucose, and 2-amino-2-deoxy-D-galactose has been described.[11] An automatic analytical system for the simultaneous determination of amino-acids and 2-amino-2-deoxyhexoses by the ninhydrin method, and the latter also by the Elson–Morgan method, was suitable for column monitoring.[12]

Although the differential determination of 2-amino-2-deoxy-D-glucose and 2-amino-2-deoxy-D-galactose in mixtures by the methods of Gardell (*Acta Chem. Scand.*, 1953, **7**, 207) and Radhakrishnamurthy and Berenson (*Clin. Chim. Acta*, 1964, **10**, 562) gave identical results, the presence of 2-amino-2-deoxy-D-mannose gave rise to conflicting results.[13] This was due to the failure of the former method to separate 2-amino-2-deoxy-D-glucose from 2-amino-2-deoxy-D-mannose, and to the identical responses of 2-amino-2-deoxy-D-galactose and 2-amino-2-deoxy-D-mannose in the Radhakrishnamurthy assay. In glycoproteins where 2-amino-2-deoxy-D-mannose is a constituent monosaccharide it was necessary to use the two methods in conjunction.

The optimum conditions for the automated phenol–sulphuric acid procedure for the non-specific, continuous analysis of glycoproteins and glycopeptides have been reported.[14] The separation and detection of *O*-methyl derivatives of 2-amino-2-deoxy-D-glucose using an automated amino-acid analyser have been described.[15]

The reaction of alkaline sulphite with glycoproteins was potentially useful in differentiating between substitution at C-4 and at C-3 and C-6 of an *O*-(2-amino-2-deoxyhexopyranosyl) residue of a glycopeptide linkage.[16]

[9] G. A. Levvy, A. J. Hay, J. Conchie, and I. Strachan, *Biochim. Biophys. Acta*, 1970, **222**, 333.
[10] G. Spik, G. Strecker, and J. Montreuil, *Bull. Soc. Chim. biol.*, 1969, **51**, 1287.
[11] T. A. Mashburn and P. Hoffman, *Analyt. Biochem.*, 1970, **36**, 213.
[12] M. Monsigny, *Bull. Soc. Chim. biol.*, 1969, **51**, 1263.
[13] E. G. Brunngraber, A. Aro, and B. D. Brown, *Clin. Chim. Acta*, 1970, **29**, 333.
[14] M. C. Brummel, H. E. M. Ayer, and R. Montgomery, *Analyt. Biochem.*, 1970, **33**, 16.
[15] L. M. Likhosterstov, G. S. Kikot, V. A. Derevitskaia, N. P. Arbatskii, V. I. Fedorova, and N. K. Kochetkov, *Doklady Akad. Nauk S.S.S.R.*, 1970, **192**, 1046.
[16] P. Weber and R. J. Winzler, *Arch. Biochem. Biophys.*, 1970, **137**, 421.

It was concluded, from a review of the biosynthesis of glycoproteins, that transglycosylation was initiated principally, but not exclusively, at the ribosomes, where the glycopeptide linkage was formed.[17] The endoplasmic membrane region was responsible for addition of the intermediate monosaccharide components, and a peripheral site in the cytoplasm was responsible for termination of the biosynthesis of the carbohydrate moiety by addition of sialic acid or L-fucose.

Microbial Glycoproteins

A series of extracellular glycoproteins isolated from cultures of *Candida bogoriensis* contained different proportions of glucose, galactose, mannose, fucose, rhamnose, and hexuronic acid.[18] A glycoprotein extracted from the cells by treatment with alkali had a similar composition. β-Elimination studies suggested that both threonine and serine were involved in the glycopeptide linkages.

The source of contaminant 2-amino-2-deoxyglucose in acid hydrolysates of deoxyribonucleic acid isolated from various strains of *Escherichia coli* was shown to be a glycoprotein, which was not associated with the deoxyribonucleic acid.[19]

Physicochemical studies have been carried out on glycoproteins from the cell walls of various strains of yeast.[20]

Plant Glycoproteins

The plant (lectins) agglutinins have been reviewed recently.[21]

The topography of the saccharide-binding region of concanavalin A and the forces involved in the stabilisation of the haemagglutinin–saccharide complex were explored using the hapten inhibition technique with a variety of deoxy, *O*-alkyl, halogeno, thio, and acetamido derivatives of D-glucose and D-mannose.[22] It was concluded that the protein moiety of concanavalin A possessed specific binding loci capable of interacting with the oxygen atoms of the C-1, C-2, and C-3 hydroxy-groups as well as with the C-4 hydroxy-group of α-D-mannopyranosyl residues. It also appeared that the protein moiety was in close proximity to the β-glycosidic oxygen atom and the C-2 hydroxy-group of bound β-D-glucopyranosyl residues, but did not bind to the carbohydrate at these points, and that the saccharide probably became bound when in a *C*1 conformation. The polar saccharide binding site of the concanavalin A molecule was illustrated by a diagrammatic representation (1). Studies employing alkyl β-D-glucopyranosides as inhibitors lead to the theory that the region of the concanavalin A

[17] P. Louisot and R. Got, *Bull. Soc. Chim. biol.*, 1970, **52**, 455.
[18] R. N. Mullinger and L. Do Carmo-sousa, *Biochim. Biophys. Acta*, 1970, **222**, 348.
[19] Y. F. Drygin, I. A. Kozlov, and A. A. Bogdanov, *Biokhimiya*, 1970, **35**, 1014.
[20] T. P. Lyons and J. S. Hough, *Biochem. J.*, 1970, **117**, 44P.
[21] W. C. Boyd, *Ann. New York Acad. Sci.*, 1970, **169**, 168.
[22] R. D. Porter and I. J. Goldstein, *Biochemistry*, 1970, **9**, 2890.

(1)

molecule in juxtaposition to the aglycone of the bound glycoside was able to accommodate branching at the β- but not at the α-carbon atom of the alkyl aglycone. U.v. spectral studies on the interaction of concanavalin A with a series of glycosides demonstrated that the molecule specifically bound low molecular weight carbohydrates at much lower pH than was previously believed.[23] Although polysaccharides were also bound at low pH, they were not precipitated. Molecular weight studies in acidic media indicated that concanavalin A did not dissociate, and it was suggested that electrostatic repulsion of the molecules, due to their high net positive charge, prevented concanavalin A–polysaccharide lattice formation and hence precipitation of the complex. Concanavalin A, insolubilised by cross-linking[6, 24] and by covalent attachment to agarose,[24, 25] has been used for the detection, isolation, and separation of various glycoproteins.

A non-specific phytohaemagglutinin from lentil (*Lens esculenta*) seeds contained glucose and traces of xylose (1·5% total), which were considered to be genuine components of the molecule.[25a] High aspartic acid and threonine values were noteworthy aspects of the amino-acid composition, which also showed the absence of sulphur-containing amino-acids.

Two haemagglutinins isolated from pea (*Pisum sativum*) seeds apparently contained neutral sugar (0·3%) which might be due to the ability of the molecules to complex with carbohydrates.[26]

[23] G. S. Hassing and I. J. Goldstein, *European J. Biochem.*, 1970, **16**, 549.
[24] K. O. Lloyd, *Arch. Biochem. Biophys.*, 1970, **137**, 460.
[25] K. Aspberg and J. Porath, *Acta Chem. Scand.*, 1970, **24**, 1839.
[25a] M. Tichà, G. Entlicher, J. V. Koštiř, and J. Kocourek, *Biochim. Biophys. Acta*, 1970, **221**, 282.
[26] G. Entlicher, J. V. Koštiř, and J. Kocourek, *Biochim. Biophys. Acta*, 1970, **221**, 272.

A phytohaemagglutinin, which showed activity against erythrocytes from several species, has been isolated from the lichen *Parmelia michauxiana*.[27] The agglutinin, which is possibly a glycoprotein, contained hexose and hexitol (80% total) comprising D-glucose, D-galactose, and mannitol (1·00 : 1·76 : 1·75).

The addition of kidney bean (*Phaseolus vulgaris*) phytohaemagglutinin to cultures of human peripheral blood lymphocytes doubled the rate of incorporation of 2-amino-2-deoxy-D-[1-^{14}C]glucose.[28] Glycoprotein II isolated from *P. vulgaris* seeds possessed a molecular weight of $1{\cdot}4 \times 10^5$, and contained mannose (3·20%), xylose (0·25%), 2-amino-2-deoxyglucose (0·99%), and traces of galactose and arabinose.[29] Amino-acid analyses revealed a deficiency in the sulphur-containing amino-acids.

The 12*S* globulins isolated from two species of rape seed, *Brassica campestris* and *B. nupus*, had similar amino-acid compositions, *N*-terminal amino-acids, and carbohydrate contents, *viz*. glucose and arabinose (1·0–1·5% total) and 2-amino-2-deoxygalactose (0·15—0·20%).[30] One major subunit, the 2·7*S* globulin, present in both of the 12*S* aggregates was purified by gel filtration, and its carbohydrate moiety was found to contain glucose, arabinose, and 2-amino-2-deoxyglucose.

Rice embryo γ_1-globulin contained hexose (18 residues), pentose (3), and 2-amino-2-deoxyhexose (6) as well as 1751 amino-acid residues;[31] it had a molecular weight of $2{\cdot}0 \times 10^5$ and was considerably asymmetric. *N*-Terminal amino-acid analyses suggested that the glycoprotein was composed of ten subunits.

It was concluded from an investigation of the biosynthesis of glycoproteins in carrot by radioactive tracer and other methods that the biosynthesis of the hydroxyproline-rich cell-wall glycoproteins of plants involved several discrete cytoplasmic steps, *viz*. the assembly of the polypeptide chain, the modification of this polypeptide through the hydroxylation of proline residues, and the glycosylation of most of the hydroxyproline residues with arabinose.[32] Finally, the glycoprotein was transported to the cell wall.

Blood-group Substances

An extensive review of the molecular aspects of human blood-group specificity included a comprehensive section on the structures of the carbohydrate determinants of the blood-group glycoproteins;[33] the methods used to elucidate the structures are described and discussed. Descriptions of the proposed structures of the reactive groups of blood-group A, B, and

[27] M. L. Howe and J. T. Barrett, *Biochim. Biophys. Acta*, 1970, **215**, 97.
[28] G. A. Hayden, G. M. Crowley, and G. A. Jamieson, *J. Biol. Chem.*, 1970, **245**, 5827.
[29] A. Pusztai and W. B. Watt, *Biochim. Biophys. Acta*, 1970, **207**, 413.
[30] L. A. Goding, R. S. Bhatty, and A. J. Finlayson, *Canad. J. Biochem.*, 1970, **48**, 1096.
[31] H. Sawai and Y. Morita, *Agric. and Biol. Chem. (Japan)*, 1970, **34**, 61.
[32] M. J. Chrispeels, *Biochem. Biophys. Res. Comm.*, 1970, **39**, 732.
[33] W. T. J. Morgan, *Ann. New York Acad. Sci.*, 1970, **169**, 118.

H substances are incorporated in a review on agglutinins.[21] A shorter review on the heterogeneity and structure of antibody-combining sites included information on the carbohydrate structures which act as antigenic determinants of the blood-group substances.[4]

A general scheme for the structure of the carbohydrate chains of blood-group substances (2) was achieved by combining previously published data with new data obtained from periodate oxidation and methylation studies.[34, 35]

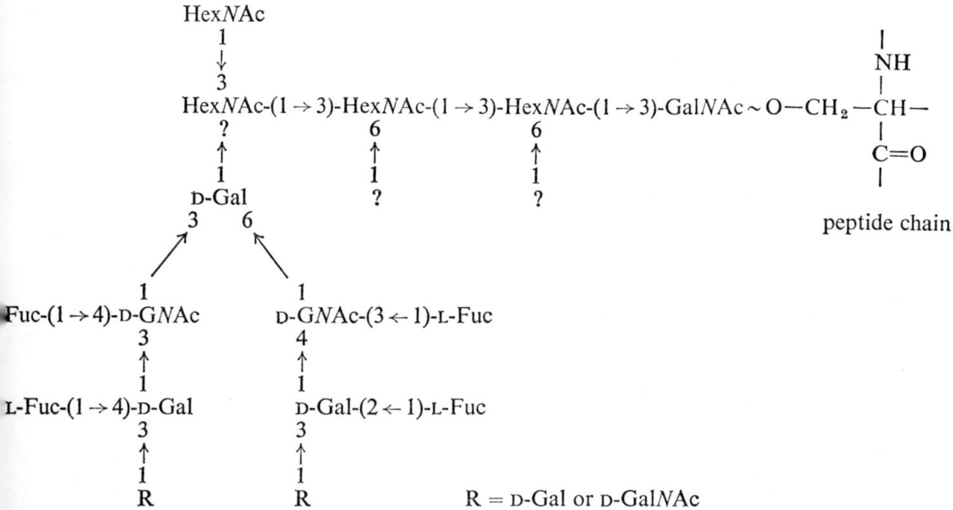

(2)

The structure of the carbohydrate chains of porcine blood-group (A + H) substance in the region adjacent to the peptide backbone has been investigated by acid hydrolysis, periodate oxidation, and degradation with extracellular enzymes from cultures of *Clostridium welchii*.[36] In preparation for studies involving hydrolysis with alkali, the alkaline degradation of substituted D-galactoses was studied with model compounds such as 2,3-di-*O*-methyl-D-galactose (Scheme 1).[37] Alkaline degradation of the blood-group substance was followed by measurement of the

[34] N. K. Kochetkov, V. A. Derevitskaya, L. M. Likhosherstov, M. D. Martynova, S. N. Senchenkova, G. S. Kikot, and L. S. Bogdashova, *Biochem. Biophys. Res. Comm.*, 1970, **39**, 583.

[35] N. K. Kochetkov, V. A. Derevitskaya, L. M. Likhosherstov, M. D. Martynova, S. N. Senchenkova, G. S. Kikot, and L. S. Bogdashova, *Doklady Akad. Nauk S.S.S.R.*, 1970, **193**, 1181.

[36] N. K. Kochetkov, V. A. Derevitskaya, L. M. Likhosherstov, M. D. Martynova, and S. N. Senchenkova, *Carbohydrate Res.*, 1970, **12**, 437.

[37] V. A. Derevitskaya, L. M. Likhosherstov, S. N. Senchenkova, and N. K. Kochetkov, *Izvest. Akad. Nauk S.S.S.R., Ser. Khim.*, 1970, 2067.

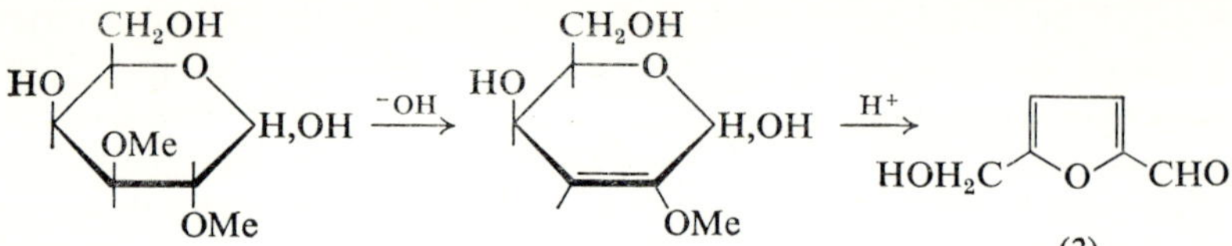

(3)

Scheme 1

degradation products 5-hydroxymethyl-2-furaldehyde (3) and metasaccharinic acid (from D-galactose) and 3-acetamido-5-dihydroxyethylfuran (from 2-acetamido-2-deoxyhexoses).[38] The results showed that the glycopeptide linkage contained a 2-acetamido-2-deoxy-D-galactose residue, and that adjacent to this region was a chain of several 2-acetamido-2-deoxyhexose residues bound by (1 → 3)-linkages, some of these residues having C-6 branch points. Porcine α-L-fucosidase was inactive against intact porcine gastric mucosal (A + H) substance,[38, 39] but the enzyme was active against oligosaccharides obtained from the substance after degradation with alkali.[39] From these studies, partial structures (4), (5), and (6) for the carbohydrate moieties were deduced, and the role of α-L-fucosidase in the metabolism of blood-group substances was discussed. Removal of

```
                    α-L-Fuc             α-L-Fuc
                       1                   1
                       ↓                   ↓
                       2                   4
α-D-GalNAc-(1 → 3)-β-D-Gal-(1 → 3)-β-D-GNAc-(1 → 3)-β-D-Gal-(1 → 3)-D-GalNAc
```

A substance, type 1 chain

(4)

```
                    α-L-Fuc             α-L-Fuc
                       1                   1
                       ↓                   ↓
                       2                   3
α-D-GalNAc-(1 → 3)-β-D-Gal-(1 → 4)-β-D-GNAc-(1 → 3)-β-D-Gal-(1 → 3)-D-GalNAc
```

A substance, type 2 chain

(5)

```
α-L-Fuc         α-L-Fuc
   1               1
   ↓               ↓
   2               3
β-D-Gal-(1 → 4)-β-D-GNAc-(1 → 3)-β-D-Gal-(1 → 3)-D-GalNAc
```

H substance

(6)

[38] G. Y. Wiederschain and E. L. Rosenfeld, *Bull. Soc. Chim. biol.*, 1969, **51**, 1075.
[39] G. Y. Wiederschain, L. M. Likhosherstov, and S. N. Senchenkova, *Biokhimiya*, 1970, **35**, 831.

the *N*-acetyl groups from the terminal, non-reducing 2-acetamido-2-deoxy-D-galactose residues of hog gastric mucin blood-group (A + H) substance with the A-degrading enzyme of *Clostridium tertium*, followed by acid hydrolysis and *N*-acetylation, resulted in an increased yield of the active disaccharide, *O*-(2-acetamido-2-deoxy-α-D-galactosyl)-(1 → 3)-D-galactose, compared with that obtained by direct acid hydrolysis.[40] The structures of other oligosaccharides which survived acid hydrolysis were *O*-(2-acetamido-2-deoxy-α-D-glucosyl)-(1 → 4)-D-galactose, *O*-(2-acetamido-2-deoxy-β-D-glucosyl)-(1 → 6)-D-galactose, *O*-(2-acetamido-2-deoxy-α-D-glucosyl)-(1 → 4)-*O*-β-D-galactosyl-(1 → 4)-2-acetamido-2-deoxy-D-glucose, and *O*-(2-acetamido-2-deoxy-α-D-glucopyranosyl)-(1 → 4)-*O*-β-D-galactopyranosyl-(1 → 3)-2-acetamido-2-deoxy-D-galactose. These structures were involved in a new antigenic determinant of the (A + H) substance unrelated to ABH activity.

Treatment of human blood-group H substance[41] and blood-group H specific porcine submaxillary glycoprotein[41a] with α-fucosidases resulted in loss of activity.

The structures of two blood-group specific oligosaccharides isolated from the urine of human O(H)- and B-secretors were determined by methylation of the original and partially hydrolysed oligosaccharides.[42] The oligosaccharide characteristic of O(H)-secretors was identical with lactodifucotetraose (7), and the B-secretor oligosaccharide (8) had D-galactose linked to the D-galactose residue of lactodifucotetraose.

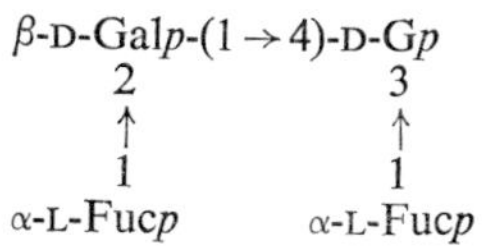

(7)

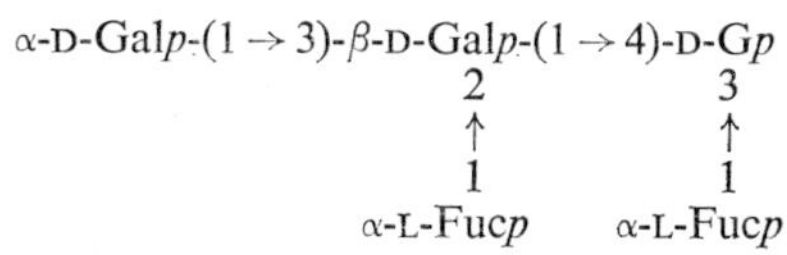

(8)

A new compound was found in the hydrolysates of several alkaline sulphite-treated blood-group M,N-active glycopeptides from human erythrocytes and bovine and canine submaxillary mucins. It was identified as a 2-amino-2-deoxyhexose sulphonic acid in which the sulphonic acid residue appeared to be attached to C-3.[16]

Alkaline degradation of an 'inactive' blood-group OG substance from human ovarian cyst fluid with sodium deuteroxide–sodium borodeuteride gave a number of compounds which were shown to be identical to compounds previously isolated and characterised from Le^{a}, A, B, and H substances.[43] These included three oligosaccharides: *O*-β-D-galacto-

[40] M. E. Etzler, B. Anderson, S. Beychok, F. Gruezo, K. O. Lloyd, N. G. Richardson, and E. A. Kabat, *Arch. Biochem. Biophys.*, 1970, **141**, 588.

[41] O. P. Bahl, *J. Biol. Chem.*, 1970, **245**, 299.

[41a] D. Aminoff and K. Furukawa, *J. Biol. Chem.*, 1970, **245**, 1659.

[42] H. Bjorndal and A. Lundblad, *Biochim. Biophys. Acta*, 1970, **201**, 434.

[43] G. Vicari and E. A. Kabat, *Biochemistry*, 1970, **9**, 3414.

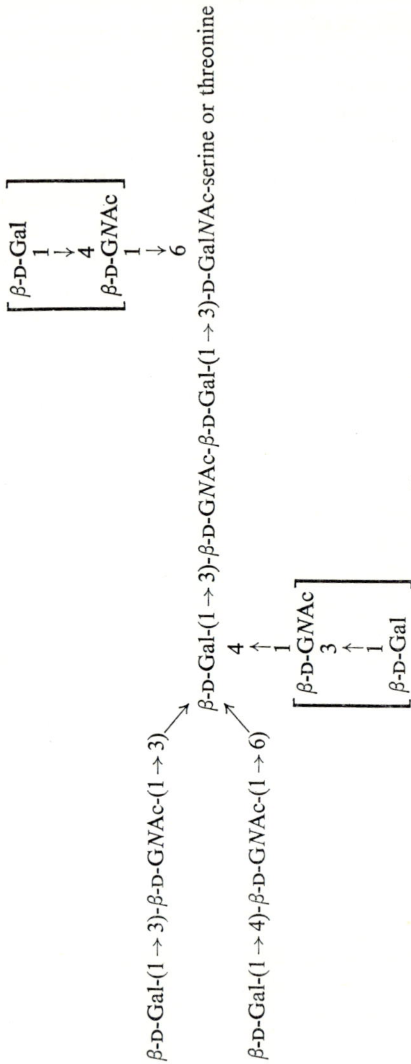

(9)

pyranosyl-(1 → 4)-*O*-(2-acetamido-2-deoxy-β-D-glucopyranosyl)-(1 → 6?)-3-hexenetetrol(s), *O*-β-D-galactopyranosyl-(1 → 4)-*O*-(2-acetamido-2-deoxy-β-D-glucopyranosyl)-(1 → 6)-1,2,4,5,6-hexanepentol(s), and *O*-β-D-galactopyranosyl-(1 → 4)-*O*-(2-acetamido-2-deoxy-β-D-glucopyranosyl)-(1 → 6)-[*O*-β-D-galactopyranosyl-(1 → 3)]-2-acetamido-2-deoxy-D-galactitol. The data obtained for OG substance were consistent with the overall structure (9); the substituent bracketed onto C-4 of the 3,4,6-linked D-galactose residue could be D-galactose or a sequence capable of giving galactitol on alkaline 'peeling'. The data also indicated that the OG substance could be considered as a precursor of the blood-group A,B,H,Le^a, and Le^b substances as well as a new determinant in which an L-fucosyl residue was linked to the 2-acetamido-2-deoxy-D-glucose residue of an *O*-β-D-galactopyranosyl-(1 → 4)-*O*-(2-acetamido-2-deoxy-β-D-glucopyranosyl) chain.

Sedimentation equilibrium techniques in a caesium chloride gradient have been applied to the preparation of blood-group specific glycoproteins from human ovarian cyst fluids, and the physicochemical characteristics of the glycoproteins isolated in this way have been defined.[44]

Amphibian blood-group substances obtained from the gastric mucosa of *Xenopus laevis* and bull-frogs (*Rana catesbiana*) were glycoprotein in nature.[45] The carbohydrate moieties consisted of galactose, glucose, mannose, fucose, 2-amino-2-deoxygalactose, and 2-amino-2-deoxyglucose.

A particle-bound enzyme from submaxillary glands and stomach mucosal tissue of human subjects catalysed the transfer of 2-acetamido-2-deoxy-D-galactose from UDP-2-acetamido-2-deoxy-D-galactose to a peptide acceptor isolated from human blood-group substance.[46] The monosaccharide was transferred in α-glycosidic linkage, and alkaline borohydride treatment liberated 2-acetamido-2-deoxy-D-galactitol indicating that the carbohydrate was bound to the hydroxyamino-acid residues in the peptide. The enzyme was distinguishable from the 2-acetamido-2-deoxy-α-D-galactosyltransferase associated with blood-group A character by its acceptor specificity.

Collagens

The structures involved in collagen have been briefly reviewed.[47]

Glycopeptides from human and guinea-pig skin, and carp swim bladder collagens were prepared by sequential collagenase and trypsin digestions.[48] The glycohexapeptide, glycyl-methionyl-(galactosyl-glucosyl)-hydroxylysyl-glycyl-histidyl-arginine, was found in significant quantities in all three

[44] J. M. Creeth and M. A. Denborough, *Biochem. J.*, 1970, **117**, 879.
[45] S. Yamamoto and S. Iseki, *Proc. Japan. Acad.*, 1970, **46**, 191.
[46] V. M. Hearn, S. D. Goodwin, and W. M. Watkins, *Biochem. Biophys. Res. Comm.*, 1970, **41**, 1279.
[47] The Connective Tissue, in: Nineteenth Rheumatism Review, *Arthritis and Rheumatism*, 1970, **13**, 603.
[48] P. H. Morgan, H. G. Jacobs, J. P. Segrest, and L. W. Cunningham, *J. Biol. Chem.*, 1970, **245**, 5042.

collagens, but the carp and human collagens each contained an additional major glycopeptide identified as glycyl-isoleucyl-(galactosyl-glucosyl)-hydroxylysyl-glycyl-histidyl-arginine and glycyl-phenylalanyl-(galactosyl-glucosyl)-hydroxylysyl-glycyl-isoleucyl-arginine, respectively. The structure of the carbohydrate moiety of the former was shown to be *O*-α-D-glucosyl-(1 → 2)-*O*-β-D-galactosyl-hydroxylysine. A possible role of the carbohydrate chains in the organisation of the collagen fibrils was discussed.

Collagen peptides were found tightly bound to glycosaminoglycan proteins from human costal cartilage.[49] Although the correlation between stiffness and the chemical constituents of cartilage on the human femoral head was significant for the total glycosaminoglycan content, the correlation was not significant with respect to the collagen content.[50]

Two intermediate intermolecular cross-links have been isolated and characterised from collagen obtained from bovine achilles tendon;[51] one structure was relatively stable and the other was a labile aldimine.

An insoluble, collagenase-digestible, glycoprotein was isolated from bovine nasal septum by repeated homogenisation in magnesium chloride solution.[52] The glycopeptide released by collagenase was purified; its carbohydrate moiety differed from that present in cartilage proteoglycan and from the hydroxylysine-linked disaccharide of collagen fibres.

Alkaline hydrolysis of low molecular weight collagen fractions, obtained from calf corneal stroma and pig aorta, yielded galactosyl-hydroxylysine and glucosyl-galactosyl-hydroxylysine.[53] A generally applicable method based on paper electrophoresis was devised for the quantitative analysis of lysine, hydroxylysine, galactosyl-hydroxylysine, and glucosyl-galactosyl-hydroxylysine. Studies have been made of the collagen contents of different topographical zones of rat, dog, and pig kidney.[54]

The salt- and acid-extracted collagens from earthworm cuticle differed little in their compositions, and contained galactose (14%), mannose, fucose, xylose, 2-amino-2-deoxygalactose, and 2-amino-2-deoxyglucose (2% or less of each).[55] The collagens all yielded the same ratio of galactobiitol to galactotriitol on alkaline borohydride treatment, and the oligosaccharide chains appear to be linked directly to threonine and serine residues.

Connective tissue of the invertebrate *Hippospongia gossypina* yielded two distinct collagen fibres, both of which contained a range of monosaccharide units.[56] A homogeneous glycopeptide isolated from the collagen

[49] K. T. Kao, J. G. Leslie, W. E. Hitt, and T. H. McGavack, *Biochim. Biophys. Acta*, 1970, **207**, 367.
[50] G. E. Kempson, H. Muir, S. A. V. Swanson, and M. A. R. Freeman, *Biochim. Biophys. Acta*, 1970, **215**, 70.
[51] A. J. Bailey, C. M. Peach, and L. J. Fowler, *Biochem. J.*, 1970, **117**, 819.
[52] K. I. Bruerton and W. H. Murphy, *Proc. Austral. Biochem. Soc.*, 1970, **3**, 78.
[53] E. Moczar and M. Moczar, *J. Chromatog.*, 1970, **51**, 277.
[54] V. H. Kresse and A. Grossmann, *Z. Klin. Chem. Klin. Biochem.*, 1970, **8**, 420.
[55] L. Muir and Y. C. Lee, *J. Biol. Chem.*, 1970, **245**, 502.
[56] R. L. Katzman, E. Lisowska, and R. W. Jeanloz, *Biochem. J.*, 1970, **119**, 17.

of the sea anemone, *Metridium dianthus*, contained 3- and 4-hydroxyproline (8%) and carbohydrate (30%).[57] The latter consisted of mannose, fucose, xylose, 2-amino-2-deoxygalactose, and 2-amino-2-deoxyglucose (11·7 : 2·3 : 1·0 : 1·7 : 14·3).

The trace amount of a protease present in crude bacterial α-amylase was capable of cleaving the β-chains of tropocollagen to α-chains.[58] However, the protease could be specifically inhibited with phenylmethylsulphonyl fluoride, thereby rendering the α-amylase entirely satisfactory for the preparation of polymerised collagen fibrils. Tropocollagen was shown to be depolymerised by L-ascorbic acid.[59]

Glycogens

A previously published method (E. Y. C. Lee and W. J. Whelan, *Arch. Biochem. Biophys.*, 1966, **116**, 162) for the microdetermination of glycogen and its unit chain-length based on α-amylase and pullulanase digestion has been critically assessed and shown to be extensively inaccurate on occasions.[60] An alternative procedure (Vol. 2, ref. 333) based on the same principles was found to be reliably accurate.

The sedimentation characteristics of liver glycogens from human glycogen storage diseases have been studied.[61] The molecular weight spectra of glycogens obtained from the livers of patients with types I, II, and VI glycogen storage disease were characteristic of the specific type of glycogenosis. The spectrum obtained for the glycogen of type III disease resembled either that from type I or VI. Analysis of the fine structure of the glycogen from type IV glycogen storage disease, which was superficially similar to plant amylopectin, showed marked differences between the profiles of the chain units that made up the macromolecules.[62] The type IV glycogen contained a significant proportion of short branches, an aspect consistent with the enzyme deletion which characterised the glycogenosis. Glycogen with a lower than normal molecular weight has been isolated from human liver in a case of generalised glycogen storage disease by two methods devised to avoid destruction during isolation.[63]

In cases of diabetes mellitus, the glycogen concentration of muscle quadriceps femoris was found to be decreased below the normal level.[64] Human blood platelet glycogen was studied ultracentrifugally under various conditions, and an isolation procedure devised.[65]

[57] R. L. Katzman and R. W. Jeanloz, *Biochem. Biophys. Res. Comm.*, 1970, **40**, 628.
[58] F. S. Steven, M. E. Grant, S. Ayad, J. B. Weiss, and S. J. Leibovich, *Biochim. Biophys. Acta*, 1970, **214**, 564.
[59] T. Miyata, S. Kawai, A. L. Rubin, and K. H. Stenzel, *Biochim. Biophys. Acta*, 1970, **200**, 576.
[60] W. Banks and C. T. Greenwood, *Arch. Biochem. Biophys.*, 1970, **136**, 320.
[61] E. Bueding, J. Sidbury, and S. A. Orrell, *Biochem. Med.*, 1970, **3**, 355.
[62] C. Mercier and W. J. Whelan, *European J. Biochem.*, 1970, **16**, 579.
[63] R. D. Edstrom, *Arch. Biochem. Biophys.*, 1970, **137**, 293.
[64] A. E. Roch-Norlund, J. Bergström, H. Castenfors, and E. Hultman, *Acta Med. Scand.*, 1970, **187**, 445.
[65] R. B. Scott and W. J. S. Still, *Blood*, 1970, **35**, 517.

Diethylaminoethyl (DEAE-) cellulose chromatography has been used to fractionate rabbit liver glycogen,[66] and distinct fractions were obtained using lithium chloride as eluant. The glycogen content of rabbit skeletal muscle under various conditions has been determined by enzymic assay and the results were compared with those obtained from acid hydrolysis and anthrone assay.[67] A series of reagents for the extraction of rat foetal and mother liver glycogen has been evaluated, and the amounts and characteristics of the glycogen obtained in each case compared.[68-70]

Glycogen was digested and cellular glycogen levels were increased during encystment in salt medium of amoeba (*Acanthamoeba castellanii*) cells.[71] It was proposed that the glycogen was utilised in the formation of the cellulose-containing cyst wall.

The order of *in vivo* incorporation of D-[U-^{14}C]glucose into starved rat livers was UDP-D-glucose > D-glucosyl oligosaccharides > glycogen, whereas in hydrocortisone-induced glycogenosis the incorporation into the latter two increased and that into UDP-D-glucose decreased.[72] The results signified that glycogen and D-glucosyl oligosaccharides were products evolving simultaneously from the same transfer reaction catalysed by UDP-glucose-glucosyl transferase. Cyclic AMP and its dibutyryl derivative had opposite effects on glycogen levels in HeLa S3 cells.[73] Recent advances in glycogen metabolism and its control have been reviewed.[74]

Glycosaminoglycuronans, Glycosaminoglycans, and their Protein and Peptide Derivatives

Occurrence, Isolation, Measurement, and Structure.—The structures of the glycosaminoglycan-proteins have been briefly reviewed.[47]

A method has been described (see p. 119) for the selective cleavage of glycuronosidic linkages in polysaccharides which involves Hofmann degradation of hexopyranuronamide residues followed by mild acid hydrolysis of the resulting 5-aminopentopyranose.[75] The pentodialdoses so produced from several model hexopyranosiduronic acids were characterised as their corresponding pentitol acetates following reduction with borohydride and acetylation.

Pulse radiolysis has been used to examine the effects of ionising radiations on glycosaminoglycans.[76] The results indicate that the presence of the

[66] L. N. Bobrava and B. N. Stepanenko, *Doklady Akad. Nauk S.S.S.R.*, 1970, **191**, 468.
[67] J. A. Johnson and R. M. Fusaro, *Analyt. Biochem.*, 1970, **37**, 298.
[68] R. Vaillant, *Bull. Soc. Chim. biol.*, 1970, **52**, 269.
[69] R. Vaillant, *Bull. Soc. Chim. biol.*, 1970, **52**, 751.
[70] R. Vaillant, *Bull. Soc. Chim. biol.*, 1970, **52**, 291.
[71] R. A. Weisman, R. S. Spiegel, and J. G. McCauley, *Biochim. Biophys. Acta*, 1970, **201**, 45.
[72] H. Sie, I. Das, and W. H. Fishman, *Arch. Biochem. Biophys.*, 1970, **138**, 679.
[73] H. Hilz and W. Tarnowski, *Biochem. Biophys. Res. Comm.*, 1970, **40**, 973.
[74] C. Villar-Palasi and J. Larner, *Ann. Rev. Biochem.*, 1970, **39**, 639.
[75] N. K. Kochetkov, O. S. Chizhov, and A. F. Sviridov, *Carbohydrate Res.*, 1970, **14**, 277.
[76] J. S. Moore, G. O. Phillips, J. V. Davies, and K. S. Dodgson, *Carbohydrate Res.*, 1970, **12**, 253.

2-amino-2-deoxyhexose moiety was responsible for the enhanced reactivity of the polymers towards hydrated electrons.

A method for the determination of the sulphate contents of glycosaminoglycans on a microgram scale involved the conversion of the macromolecule into its free acid form, and reaction of the *O*-sulphate and *N*-sulphate groups with n-butylamine.[77] After volatilisation of the excess of reagent, the bound n-butylamine was released by treatment with alkali and was determined by g.l.c.

Deamination of *N*-deacetylated 2-acetamido-2-deoxyhexoses and their glycosides with nitrous acid yielded a mixture of aldehydes or ketones, which was quantitatively assayed by radiochromatography.[78] The technique was applied to the depolymerisation of hyaluronic acid, chondroitin 4-sulphate, chondroitin 6-sulphate, dermatan sulphate, heparin, and heparan sulphate. The glycosaminoglycuronans were then hydrolysed and the products subjected either to borotritide reduction or deamination and borotritide reduction, followed by radiochromatography. The data showed that recoveries of 2-amino-2-deoxyhexose and uronic acid before deamination were very close to theoretical values, whereas losses could be considerable in conventional hydrolyses. The radiochromatographic profiles before and after deamination were unique for each polymer, and served to distinguish between them with the exception of chondroitin 4-sulphate and chondroitin 6-sulphate. The data also permitted an approximation to be made of the percentage of the total 2-amino-2-deoxyhexopyranosyl and hexopyranosyluronic acid bonds cleaved by hydrolysis. The percentage of 2-amino-2-deoxyhexopyranosyl bond cleavage was hyaluronic acid (100%), chondroitin 4-sulphate (77%), chondroitin 6-sulphate (93%), dermatan sulphate (82%), heparin ($< 21\%$), and heparan sulphate ($< 45\%$), respectively, whereas the corresponding figures for the hexopyranosyluronic acid bonds were 49, 49, 40, > 90, 85, and 63%, respectively. These values reflect the relative stabilities of the glycosidic bonds in the different polymers. It also appeared that heparin and heparan sulphate might contain an additional uronic acid different from D-glucuronic, D-galacturonic, and L-iduronic acids.

Water-sorption isotherms obtained for various hyaluronic acid samples indicated that the physical state (such as degree of crystallinity) greatly influenced the water-sorption characteristics.[79] Possible mechanisms of water sorption were discussed.

A ^{59}Fe-labelled iron-chondroitin sulphate colloid has been used in an investigation of the iron distribution in mice.[80]

[77] S. R. Srinivasan, B. Radhakrishnamurthy, E. R. Dalferes, and G. S. Berenson, *Analyt. Biochem.*, 1970, **35**, 398.
[78] J. E. Shively and H. E. Conrad, *Biochemistry*, 1970, **9**, 33.
[79] A. Block and F. A. Bettelheim, *Biochim. Biophys. Acta*, 1970, **201**, 69.
[80] Y. Nakanishi and M. Kishi, *Yakagaku Zasshi*, 1970, **90**, 170.

The o.r.d. and c.d. characteristics of dermatan sulphate were different from those of heparan sulphate thus providing a physicochemical method of distinguishing between these glycosaminoglycuronans.[81]

L-Iduronic acid has been shown to be a major constituent of heparin.[82] Heparin was degraded by a sequence involving methanolysis, deamination, reduction, and hydrolysis with acid to give predominantly 1,6-anhydro-L-idopyranose and 2,5-anhydro-D-mannitol, which were derived from L-iduronosyl and 2-amino-2-deoxy-D-glucopyranosyl residues, respectively. Products such as methyl α-D-glucopyranoside, derivable from D-glucopyranosyluronic acid residues, were relatively minor in occurrence, and the ratio of 1,6-anhydro-L-idopyranose : 2,5-anhydro-D-mannitol : methyl α-D-glucopyranoside was 5 : 5 : 1. Periodate oxidation data in the form of a threitol : erythritol ratio of 5 : 1 confirmed these results. The latter data also showed that the uronic acid residues were substituted at C-4, while other evidence indicated a 2-amino-2-deoxyhexosyl residue to be the substituent. Tritium-labelling experiments involving reduction of the methyl ester of the heparin carboxy-groups indicated that little, if any, of the 1,6-anhydro-L-idopyranose was derived by isomerisation of D-glucosyluronic acid residues. The monosaccharides obtained from heparin by treatment with heparin-induced *Flavobacterium* enzymes did not contain D-glucuronic acid.[83]

A method involving stepwise elution from an anion exchange resin has been developed whereby tritium-labelled heparin could be stored and purified from the products of radiolytic degradation.[84] The sulphate content and molecular weight of the chromatographic fractions were inversely related to their specific activities, but the biological activity of the fractions of low specific activity was superior to that of the crude material.

The hydrolysis of a single *N*-sulphate residue from heparin and related hexa- and tetra-saccharides, produced by enzymic degradation, was catalysed by a sulphamidase isolated from lymphoid tissues of man, dogs, and rats.[85] Heparin and heparan sulphate were degraded to products containing α-L-*threo*-hex-4-enepyranosyluronic acid residues by a heparinase and a heparitinase, respectively, obtained from heparin-induced *Flavobacteria*.[86]

Keratan sulphate was degraded to D-galactose, 2-acetamido-2-deoxy-D-glucose, and sulphate by a multienzyme system found in an extract from a marine gastropod, *Charonia lampas*.[87] D-Galactose was liberated from highly purified keratan sulphate by an enzyme present in human and rat tissues.[88]

[81] A. L. Stone, G. Constantopoulos, S. M. Sotsky, and A. Dekaban, *Biochim. Biophys. Acta*, 1970, **222**, 79.
[82] A. S. Perlin and G. R. Sanderson, *Carbohydrate Res.*, 1970, **12**, 183.
[83] J. C. Karapally and C. P. Dietrich, *Canad. J. Biochem.*, 1970, **48**, 164.
[84] N. DiFerrante and E. A. Popenoe, *Carbohydrate Res.*, 1970, **13**, 306.
[85] C. P. Dietrich, *Canad. J. Biochem.*, 1970, **48**, 725.
[86] P. Hovingh and A. Linker, *J. Biol. Chem.*, 1970, **245**, 6170.
[87] M. Nishida-Fukuda and F. Egami, *Biochem. J.*, 1970, **119**, 39.
[88] P. A. Öckerman, *Carbohydrate Res.*, 1970, **12**, 429.

A method for the microanalysis and characterisation of glycosaminoglycuronans from human skin, involving zone electrophoresis on cellulose acetate, avoided the use of ion-exchange columns;[89] alcian blue was used to detect the glycosaminoglycuronans. The amount of dye bound to each was dependent on the number of moles of disaccharide repeating unit and the number of dissociated carboxylic and sulphate groups per repeating unit. Combination of the technique with chemical analyses permitted the identification of several glycosaminoglycans. Comparison of the glycosaminoglycans in human skin during foetal development and adult life showed that early foetal skin contained only hyaluronic acid, chondroitin 4-sulphate, and chondroitin 6-sulphate, but at twenty times the adult level.[90] During gestation this level dropped primarily due to a fall in hyaluronic acid content, and at term, when dermatan sulphate was present, was only twice the adult level. Further changes in the composition and concentration of glycosaminoglycans occurred during childhood, but very little change was found during adult life.

Heparan sulphate has been isolated from human umbilical cords,[91] and showed similarities to the material obtained from human aortic tissue in having low total sulphate and *N*-sulphate contents and relatively low electrophoretic mobility. Reaction with nitrous acid gave products which indicated a structural similarity to bovine lung heparan sulphate, although it contained fewer sulphate ester groups on the sections containing 2-acetamido-2-deoxy-D-glucose and a larger proportion of sections containing multiple *O*-hexosyluronic acid-2-acetamido-2-deoxy-D-glucose repeating units.

Since the cetyl pyridinium chloride–cellulose method does not permit an accurate determination of keratan sulphate in tissues and cartilages, a modification of the method was developed.[92] Cetyl pyridinium chloride was removed from an aqueous solution of keratan sulphate and glycopeptides by extraction with isoamyl alcohol, and the aqueous phase was then fractionated on ECTEOLA-cellulose to yield one keratan sulphate and one glycopeptide fraction. The method was applied to the estimation of keratan sulphate in artificial mixtures as well as in human knee-joint cartilage and rabbit cornea.

Dermatan sulphate samples have been analysed by the cetyl pyridinium chloride-cellulose microcolumn technique before and after treatment with testicular hyaluronidase.[93] 'Neutral magnesium chloride elution profiles'

89 M. Breen, H. G. Weinstein, M. Anderson, and A. Veis, *Analyt. Biochem.*, 1970, **35**, 146.

90 M. Breen, H. G. Weinstein, R. L. Johnson, A. Veis, and R. T. Marshall, *Biochim. Biophys. Acta*, 1970, **201**, 54.

91 J. A. Cifonelli and J. King, *Biochim. Biophys. Acta*, 1970, **215**, 273.

92 A. Anseth, C. A. Antonopoulos, A. Bjelle, and L.-Å. Fransson, *Biochim. Biophys. Acta*, 1970, **215**, 522.

93 L.-Å. Fransson, A. Anseth, C. A. Antonopoulos, and S. Gardell, *Carbohydrate Res.*, 1970, **15**, 73.

reflected the molecular size polydispersity of the material obtained after hyaluronidase digestion, whilst 'acid magnesium chloride elution profiles' were not influenced by molecular weight, but were governed by the proportions of D-glucuronic acid, L-iduronic acid, and sulphate. Barium acetate–ethanol cellulose elution profiles separated dermatan sulphate samples according to their uronic acid compositions. By comparing elution profiles of unknown samples with those of well-characterised dermatan sulphate polymers, data permitting the formulation of the general hybrid properties of human knee-joint capsule, and bovine aorta and sclera dermatan sulphate were obtained.

Automated amino-acid and 2-amino-2-deoxyhexose analyses of hydrolysates of glycosaminoglycan proteins extracted from human and bovine cartilage have revealed inherently widespread limits of composition. There was little glycosaminoglycan in human articular cartilages.[11] However, based on the assumption that 2-amino-2-deoxy-D-glucose represented keratan sulphate and that 2-amino-2-deoxy-D-galactose represented chondroitin sulphate, bovine nasal septum contained the highest chondroitin sulphate level, which decreased only slightly with age, whilst the relatively low keratan sulphate level remained constant. Bovine articular cartilage provided a striking contrast, the chondroitin sulphate level decreasing drastically with age and the keratan sulphate level increasing slightly. The automated analysis was proposed as a screening method for investigations of the microheterogeneity of cartilage. Changes with age of the glycosaminoglycans of human femoral condylar articular cartilage have been investigated quantitatively.[94] Significant correlations were found between the contents of glycosaminoglycan, chondroitin sulphate, and 'keratan sulphate' with stiffness of cartilage on the human femoral head.[50]

The protein polysaccharide light (PPL) fraction from human costal cartilage was separated into three fractions, and it was estimated that 40—60% of the glycosaminoglycans present were keratan sulphates.[49] Although the carbohydrate contents of the three fractions differed, the amino-acid compositions were similar to one another and to that of bovine nasal cartilage PPL. Two of the fractions appeared to contain tightly bound collagen peptides whereas one was collagen free.

In an investigation of the glycosaminoglycan proteins of mature human tracheal cartilage, separation into PPL and protein polysaccharide heavy (PPH) fractions was achieved.[95] Compositional analyses were made of both PPL and PPH, and it was found that the keratan sulphate content of the cartilage attained maximum values by the beginning of the fourth decade and thereafter remained constant. The glycosaminoglycans of human tracheobronchial cartilages were liberated by proteolysis and were then separated by ion-exchange chromatography and precipitation with cetylpyridinium chloride.[96] The total chondroitin sulphate content of the

[94] R. A. Stockwell, *Ann. Rheumat. Dis.*, 1970, **29**, 509.
[95] R. M. Mason, *Biochem. J.*, 1970, **119**, 599.
[96] R. M. Mason and F. S. Wusteman, *Biochem. J.*, 1970, **120**, 777.

cartilages decreased linearly with increasing age, and age-dependent changes in the chemical heterogeneity of chondroitin sulphate were observed.

Chromatographic studies have indicated that the glycosaminoglycan isolated from both cultured human bone-marrow cells and the culture medium was a polydisperse chondroitin sulphate.[97] The secreted polymer was of larger molecular size than that remaining in the cells.

The urinary glycosaminoglycan peptides have been reviewed.[98] The urinary excretion of glycosaminoglycuronans by normals has been measured,[99] and the excretion of the chondroitin sulphate isomers determined using chondroitinases and sulphatases.[100] Gel filtration of testicular hyaluronidase digests of urinary glycosaminoglycans demonstrated the presence of higher proportions of enzyme-resistant glycosaminoglycans in adult specimens than in those from children;[101, 102] this difference may reflect changes in the levels of enzymes involved in the degradative sequences. The amounts of glycosaminoglycans in urine, and the relation of the value to creatinine excretion have been critically examined.[103] A significant correlation between the excretion of glycosaminoglycan and 17-hydroxycorticosteroid was demonstrated in males undergoing lunar flight simulation.[104]

The structure of dermatan sulphate–chondroitin sulphate copolymers isolated from horse aorta has been examined.[105] A large proportion of the galactosaminoglycans was obtained as a discrete polysaccharide fraction, which had an L-iduronic acid to D-glucuronic acid molar ratio of 1 : 2. This and other data indicated that the polymer contained approximately equimolar proportions of the three repeating disaccharide units: 2-acetamido-2-deoxy-*O*-(β-D-glucopyranuronosyl)-D-galactose 4-sulphate (*A*), 2-acetamido-2-deoxy-*O*-(α-L-idopyranuronosyl)-D-galactose 4-sulphate (*B*), and 2-acetamido-2-deoxy-*O*-(β-D-glucopyranuronosyl)-D-galactose 6-sulphate (*C*). Degradation of the polymer with a chondroitinase yielded a large quantity of unsaturated disaccharides and a tetrasaccharide corresponding to an *AB* sequence in the original polymer. Degradation with testicular hyaluronidase, however, produced a tetrasaccharide of the sequence *CA*, in addition to higher oligosaccharides of the type *ABA* or *ABC*. A detailed examination of the fragments obtained by enzymic degradation revealed that the majority of the *B* units were located in clusters, particularly in the immediate vicinity of the glycopeptide linkage. The *A* units were predominantly in positions adjacent to *B* units, whereas

[97] P. Lau, G. G. Cornwell, and W. J. Williams, *J. Lab. Clin. Med.*, 1970, **76**, 739.
[98] E. H. F. McGale, *Advances Carbohydrate Chem. Biochem.*, 1969, **24**, 435.
[99] J. Clausen, H. V. Dyggve, and J. C. Melchior, *Clin. Chim. Acta*, 1970, **29**, 197.
[100] K. Murata, T. Ishikawa, and Y. Oshima, *Clin. Chim. Acta*, 1970, **28**, 213.
[101] D. Allalouf and A. Ber, *Biochim. Biophys. Acta*, 1970, **201**, 61.
[102] D. Allalouf and A. Ber, *Biochim. Biophys. Acta*, 1970, **208**, 141.
[103] N. DiFerrante and H. S. Lipscomb, *Clin. Chim. Acta*, 1970, **30**, 69.
[104] M. Kastelan, N. DiFerrante, and H. S. Lipscomb, *Experientia*, 1970, **26**, 937.
[105] L. Fransson and B. Havsmark, *J. Biol. Chem.*, 1970, **245**, 4770.

the *C* units occurred preferentially in the centre of sections of the chain containing D-glucuronic acid.

Aortic glycosaminoglycuronans have been studied in a number of species including the rhesus monkey, cow, dog, cat, rabbit, hamster, guinea-pig, rat, mouse, pheasant, chicken, turtle, frog, dog-fish, and dusky shark.[106] Dermatan sulphate was not observed in tissue from dog-fish and shark, the two most primitive species studied. High correlation between the inclusion of β-lipoproteins into the aortic wall of various animals and the glycosaminoglycan content was demonstrated *in vitro*.[107]

Of the total glycosaminoglycans of bovine aortic tissue, 24% was extractable with potassium chloride; gel filtration after digestion with hyaluronidase yielded two macromolecules, the more retarded of which corresponded to single polysaccharide chains.[108] Treatment of the higher molecular weight fraction with either alkali or papain produced a material resembling the other fraction, while analysis of the intact material showed approximately equal amounts of dermatan sulphate and heparan sulphate. It was concluded that these glycosaminoglycuronans occurred in the aortic wall as multichain proteoglycans. Calf aortic wall contained hyaluronic acid (30%), chondroitin sulphate (49%), dermatan sulphate (9·5%), and heparan sulphate (11·5% of the total glycosaminoglycuronans).[109]

Several glycosaminoglycan-proteins have been isolated from the soluble extracts of bovine heart valves by sedimentation equilibration in a density gradient.[110] Compositional and structural studies indicated that the polysaccharide is chondroitin sulphate. The molecular weights of the samples lay in the range 4·2—6·5 × 10^4, and differences were observed in the protein contents. The results were discussed in terms of a basic model of two polysaccharide chains linked by a protein of variable size. The monosaccharide compositions of the glycosaminoglycan fractions from bovine heart aortic and pulmonary valves have been compared.[111]

An examination of the fractions containing glycosaminoglycans from bovine heart, ovine heart, lung, and liver, and avian heart, liver, skin, and bone showed that most of the glycosaminoglycans had electrophoretic mobilites near to those of hyaluronic acid and chondroitin sulphate.[112] Heparin appeared in ovine liver and lung. On the basis of these results and those of an investigation of the carbohydrate moieties of other glycoproteins present in the samples, it was suggested that a gross analytical differentiation could be achieved by proteolysis of the tissues followed by fractional precipitation of the carbohydrate moieties with ethanol.

106 V. Stefanovich and K. Akiyama, *Comparative Biochem. Physiol.*, 1970, **34**, 125.
107 S. Virag, R. Denesh, and T. Pozhonji, *Biulleten Eksperimentalnoi Biologii i Meditsiny*, 1970, **70**, 37.
108 L. Jansson and U. Lindahl, *Biochem. J.*, 1970, **117**, 699.
109 H. Kresse and E. Buddecke, *Z. physiol. Chem.*, 1970, **351**, 151.
110 D. A. Lowther, B. N. Preston, and F. A. Meyer, *Biochem. J.*, 1970, **118**, 595.
111 Y. Kanke, R. I. Bashey, Y. Mori, and A. A. Angrist, *Life Sciences Part II*, 1970, **9**, 1081.
112 J. C. Hilborn and P. A. Anastassiadis, *Biochim. Biophys. Acta*, 1970, **215**, 57.

A wide range in the content of D-galactose, D-xylose, and 2-acetamido-2-deoxy-D-glucose in the region of the glycopeptide linkage was found in the heparins of bovine lung, liver, and mucosa, porcine and ovine mucosa, whale tissue, and other heparins.[113] Bovine lung heparin was separated into two components by electrophoresis.

An attempt was made to isolate heparin proteoglycan from bovine liver capsule by fractionation of the polysaccharide extract with cetyl pyridinium chloride followed by hyaluronidase digestion.[114] Gel chromatography indicated the presence of single heparin chains in addition to a dermatan sulphate proteoglycan. Analysis of the purified heparin preparation gave xylose (0·1), galactose (0·2), and serine (0·4 residues per polysaccharide chain, molecular weight $7{\cdot}4 \times 10^3$), and it was concluded that the material had been degraded by a polysaccharidase with endo-activity.

A modification of the method for the separation of glucosaminoglycuronans and galactosaminoglycuronans, based on stepwise elution of the cetyl pyridinium-glycosaminoglycan complexes from cellulose with specific eluants, permitted the fractionation of glycosaminoglycuronans from bovine corneal tissue.[115] A small amount of 2-amino-2-deoxy-D-glucose in the galactosaminoglycan fraction was explained by the existence in the cornea of certain forms of glycosaminoglycan inseparable by the cetyl pyridinium technique.

The glycosaminoglycans of bovine elastic cartilages comprised chondroitin 4- and 6-sulphates (90%), and a small amount of keratan sulphate.[116] This composition was similar to that of the glycosaminoglycans of hyaline cartilages, but could not be correlated with the presence of elastic fibres.

A rapid extraction procedure has been developed to provide high yields of proteoglycan from bovine nasal septum.[117] Alkaline cleavage and borotritide reduction of the proteoglycan (molecular weight, $1{\cdot}23 \times 10^6$) followed by ECTEOLA-cellulose chromatography gave three radioactive fractions. One of these fractions, containing all the hexuronic acid, was subfractionated after chondroitinase treatment to yield glycopeptide-linkage oligosaccharides of keratan sulphate and chondroitin sulphate, and an unlabelled disaccharide fraction. The radioactivity of the oligosaccharides was confined to xylitol.

Most of the proteoglycan of bovine nasal cartilage could be recovered in a purified, polydisperse proteoglycan subunit containing chondroitin sulphate (87%), keratan sulphate (6%), and protein (7%).[118] A method was described by which the macromolecules of the subunit could be partitioned into three major fractions with average molecular weights of 1·8, 2·5, and $3{\cdot}0 \times 10^6$. Analyses and physicochemical studies of these fractions suggested that the proteoglycan subunit contained only one type of

[113] J. A. Cifonelli and J. King, *Carbohydrate Res.*, 1970, **12**, 391.
[114] U. Lindahl, *Biochem. J.*, 1970, **116**, 27.
[115] R. Praus, *J. Chromatog.*, 1970, **48**, 535.
[116] G. C. Gillard and F. S. Wusteman, *Biochem. J.*, 1970, **118**, 25P.
[117] J. J. Hopwood and H. C. Robinson, *Proc. Austral. Biochem. Soc.*, 1970, **3**, 78.
[118] V. C. Hascall and S. W. Sajdera, *J. Biol. Chem.*, 1970, **245**, 4920.

protein, and that polydispersity was the result of differences in the number of chondroitin sulphate chains attached to the protein. Similar conclusions were reached from a comparison of four different methods of fractionation, one of which was capable of dealing with gram quantities and involved differential sedimentation in neutral salt solutions at high concentrations.[119] Electron microscopic studies have been conducted on the fractions isolated from bovine nasal cartilage.[120]

Hyaluronic acid, chondroitin sulphate, dermatan sulphate, and heparan sulphate have been identified in the kidneys of rat, dog, and pig; comparative studies have been made of the glycosaminoglycan content in different topographical zones.[54]

The glycosaminoglycans isolated from rabbit renal papillae have been shown to be hyaluronic acid (50%), chondroitin 4- and 6-sulphate (45%), dermatan sulphate (5% of total), and possibly heparan sulphate.[121] Various methods, including electrophoresis, precipitation, and chromatography on ECTEOLA-cellulose, have been used to demonstrate the presence of chondroitin 4- and 6-sulphates and a minor keratan sulphate component in rabbit bone.[122]

A proteoglycan isolated from guinea-pig Kurloff cells behaved as a multichain molecule on gel filtration.[123] Its molecular size was decreased by proteolysis, which yielded a material resembling chondroitin 4-sulphate. The chains were longer than those of cartilage chondroitin sulphate-protein, as judged by their 2-amino-2-deoxy-D-galactose to D-xylose molar ratios, and contained equimolar amounts of D-xylose, D-galactose, and L-serine, and underwent β-elimination.

One clonal line (B-6) among hybrids derived from the fusion of mouse and Chinese hamster cell lines synthesised hyaluronic acid and secreted it into the medium.[124]

Fractionation of the glycosaminoglycuronans of neonatal rat skin showed the presence of hyaluronic acid (56%), chondroitin 6-sulphate (9%), and dermatan sulphate (16% of total), together with some chondroitin 4-sulphate, heparin, heparan sulphate, and two further fractions of short chain length and low sulphate content resembling heparan sulphate and dermatan sulphate.[125] Hyaluronic acid, chondroitin sulphate, dermatan sulphate, and heparan sulphate were present in rat brain in the proportions 33 : 48 : 11 : 8; the absence of heparin and keratan sulphate was also confirmed.[126] Miniaturised methods were used to extract the glycosaminoglycans from rat cartilage.[127] Investigation of the fractions

119 L. Rosenberg, S. Pal, R. Beale, and M. Schubert, *J. Biol. Chem.*, 1970, **245**, 4112.
120 L. Rosenberg, W. Hellmann, and A. K. Kleinschmidt, *J. Biol. Chem.*, 1970, **245**, 4123.
121 S. J. Farber and D. Van Praag, *Biochim. Biophys. Acta*, 1970, **208**, 219.
122 J. Burckard and P. Degand, *Compt. rend. Soc. Biol.*, 1970, **164**, 432.
123 M. F. Dean and H. Muir, *Biochem. J.*, 1970, **118**, 783.
124 H. Koyama and T. Ono, *Biochim. Biophys. Acta*, 1970, **217**, 477.
125 T. E. Hardingham and C. F. Phelps, *Biochem. J.*, 1970, **117**, 813.
126 J. M. Goldberg and W. L. Cunningham, *Biochem. J.*, 1970, **120**, 15P.
127 J. C. Pita, L. A. Cuervo, J. E. Madruga, F. J. Muller, and D. S. Howell, *J. Clin. Invest.*, 1970, **49**, 2188.

provided evidence for a role of these macromolecules in the regulation of mineral phase separation in calcifying tissue.

Proteoglycan was easily extracted from adult chicken cartilage with edta–sodium chloride, and purification was effected by gel filtration.[128] The macromolecule contained mostly chondroitin 4-sulphate together with some chondroitin 6-sulphate and keratan sulphate. The matrix produced in chondrocyte cell culture contained a proteoglycan of which chondroitin 4-sulphate and keratan sulphate were components.

The keratan sulphate fraction from whale nasal cartilage was fractionated into at least four fractions having sulphate to 2-amino-2-deoxyhexose molar ratios of 0·75—1·80 : 1·00.[129] All fractions contained galactose, mannose, fucose, 2-amino-2-deoxy-D-glucose, 2-amino-2-deoxy-D-galactose, and sialic acid ($\leqslant$ 17%) in various amounts. The sialic acid was identified as *N*-acetylneuraminic acid, and enzymic studies revealed that these residues were attached to the non-reducing end of the carbohydrate moiety. Studies of the glycopeptide linkage of whale cartilage keratan sulphate demonstrated the involvement of an *O*-glycosidic bond between 2-acetamido-2-deoxy-D-glucose and the hydroxyamino-acids.[130]

β-Galactosidase released a small amount of D-galactose from shark cartilage keratan sulphate indicating that non-sulphated D-galactose residues occupy the terminal non-reducing positions in the carbohydrate moieties.[87]

Hyaluronic acid has been isolated from *Pasteurella multocida* cells by proteolysis and extraction with alkali, and the crude material was further purified.[131] On comparison with the hyaluronic acid from other bacterial sources, where alkaline extraction was not used, the material did not appear to be appreciably degraded. A purified pyrogenic exotoxin from *Streptococcus pyrogenes* filtrates has been characterised as a protein complexed with hyaluronic acid.[132] Evidence has been presented for the existence of a glycosaminoglycuronan resembling dermatan sulphate in a strain of *Clostridium welchii*.[133]

Clinical Conditions.—Heritable diseases of connective tissue involving carbohydrates have been reviewed.[47]

Several simple screening methods have been used to measure the excess of glycosaminoglycan excreted in urine, and the relative merits of the methods were discussed.[134]

[128] H. J. Shulman and K. Meyer, *Biochem. J.*, 1970, **120**, 689.
[129] N. Toda and N. Seno, *Biochim. Biophys. Acta*, 1970, **208**, 227.
[130] N. Seno and N. Toda, *Biochim. Biophys. Acta*, 1970, **215**, 544.
[131] J. A. Cifonelli, P. A. Rebers, and K. H. Heddleston, *Carbohydrate Res.*, 1970, **14**, 272.
[132] Y. B. Kim and D. W. Watson, *J. Extpl. Med.*, 1970, **131**, 611.
[133] G. K. Darby, A. S. Jones, J. F. Kennedy, and R. T. Walker, *J. Bacteriol.*, 1970, **103**, 159.
[134] C. A. Pennock, M. G. Mott, and G. F. Batstone, *Clin. Chim. Acta*, 1970, **27**, 93.

The measurement of urinary glycosaminoglycans and the relationship of the values to creatinine excretion have been critically examined.[103] The urinary excretion of the chondroitin sulphates from normals and patients with connective tissue diseases was determined enzymically using chondroitinases and sulphatases.[100] A moderately increased excretion was found in cases of dermatomyositis, polymyositis, aortic syndrome, and progressive systemic sclerosis; the method was also applied to cases of Hunter and Hurler syndromes.

A method has been described for the determination of heparan sulphate and dermatan sulphate in the urine and tissues of cases of Hunter and Hurler syndromes.[135] The microscale technique was generally applicable to the differential determination of the two polysaccharides and provided a chemical diagnosis of the two syndromes. Ion-exchange chromatography of heparan sulphate from cases of Hunter syndrome gave results significantly different from those obtained for similar analyses of the heparan sulphate from cases of Hurler syndrome.[136] The results indicated that the anomalous glycosaminoglycuronan degradation was different in the two diseases. Distinctive excretion patterns of glycosaminoglycans from patients with Hurler, Hunter, and Sanfilippo syndromes were also demonstrated by ion-exchange chromatography.[137] The dermatan sulphate and heparan sulphate varied in molecular size and degree of sulphation, and probably represented a series of degradation products of the native polysaccharides. Molecular weight distributions were obtained for the heparan sulphate excreted by cases of Sanfilippo syndrome, and for the heparan and dermatan sulphates excreted by cases of Hurler syndrome.[81] The average molecular weight was one-third of the normal value for the Sanfilippo heparan sulphate, which could be separated further into two fractions differing in molecular weight, sulphate content, and primary structure. Physicochemical and other studies showed that one of these fractions possessed a conformational structure similar to that found in normal tissues but had a low sulphate content; the other fraction had a heparin-like conformation and a high *N*-sulphate content. The o.r.d. and c.d. characteristics of dermatan sulphate were greatly different from those of heparan sulphate thereby providing an optical method of distinguishing between the syndromes. Analysis of Sanfilippo syndrome brain tissue indicated that the major glycosaminoglycan accumulated was chondroitin 4-sulphate.[138]

Data obtained for a series of related cases of Morquio–Ullrich disease revealed only a slight increase above normal in the daily glycosaminoglycan excretion, but significant changes were observed in the proportions of hyaluronic acid, chondroitin 4- and 6-sulphate, dermatan sulphate, and

135 J. F. Kennedy, C. H. Sinnette, and J. B. Familusi, *Clin. Chim. Acta*, 1970, **29**, 37.
136 P. Maroteaux, *Rev. Europ. Etudes Clin. Biol.*, 1970, **15**, 203.
137 B. A. Gordon and M. D. Haust, *Clin. Biochem.*, 1970, **3**, 203.
138 E. George and B. K. Bachhawat, *Clin. Chim. Acta*, 1970, **30**, 317.

heparan sulphate.[99] Particularly significant was the proportionally increased excretion of desulphated polysaccharides.

Determination of the excretion of glycosaminoglycans precipitable by aminoacridine in the urine of patients with rheumatoid arthritis and various connective tissue diseases revealed that only 20% of the rheumatoid patients and those with systemic lupus erythematosus showed raised levels.[139] Application of an automated method for analysis of cartilage glycosaminoglycans to arthritic cartilage showed the presence of little polysaccharide.[11] An automated spectrofluorimetric method for the determination of carbohydrates and proteins was applicable to monitoring the fractionation of synovial fluids.[7]

Assessment of the individual glycosaminoglycans in human skin in cases of lipoid proteinosis (hyalinosis cutis et mucosae), Ehlers Danlos syndrome, Dupuytren's contracture, and ichthiosiform erythroderma indicated that all but the last involved mucopolysaccharide disorders.[140]

The glycosaminoglycans, which accumulated in greatly increased amounts in the livers of cases of G_{M1}-gangliosidosis Type I, contained D-galactose and 2-amino-2-deoxy-D-glucose as the major carbohydrate constituents, but not in equimolar ratio.[141] These compounds contained an alkali-labile glycopeptide linkage, were polydisperse, and, apart from a low sulphate content, resembled human cartilage keratan sulphate. The keratan sulphate-like glycosaminoglycans found in the urine of a patient with G_{M1}-gangliosidosis Type II could be classified into two types.[142]

Biosynthesis and Metabolism.—The biosynthesis and metabolism of glycosaminoglycans has been briefly reviewed.[47] The preparation of L-[6-^{14}C]iduronic acid and 2-([1-^{14}C]acetamido)-2-deoxy-D-glucose and -D-galactose for use in biosynthetic studies has been reported.[143]

A sulphotransferase system that catalysed the sulphation of UDP-2-acetamido-2-deoxy-D-galactose in the presence of adenosine-3′-phosphate-5′-sulphatophosphate was obtained in particulate form from the isthmus of hen oviduct.[144] One of the products of sulphation was UDP-2-acetamido-2-deoxy-D-galactose 4-sulphate, and further investigation revealed the presence of four other sulphated nucleotides which were characterised by conversion into identifiable monosaccharide sulphates (Schemes 2–5).[145] The new nucleotides were UDP-2-acetamido-2-deoxy-D-galactose 4,6-disulphate (10) (Scheme 2), UDP-2-acetamido-2-deoxy-6-*O*-phospho-D-galactose 4-sulphate (11) (Scheme 3), UDP-(*O*-α-D-galactopyranosyl 2- or 4-

[139] T. Bitter, P. Siegenthaler, T. DePreux, and E. Martin, *Ann. Rheumat. Dis.*, 1970, **29**, 427.
[140] S. A. Barker and J. F. Kennedy, *Life Sciences Pt. II*, 1969, **8**, 989.
[141] J. W. Callahan and L. S. Wolfe, *Biochim. Biophys. Acta*, 1970, **215**, 527.
[142] L. S. Wolfe, J. Callahan, J. S. Fawcett, F. Andermann, and C. R. Scriver, *Neurology*, 1970, **20**, 23.
[143] J. F. Kennedy, *J. Labelled Compounds*, 1970, **6**, 201.
[144] M. Tsuji, S. Shimizu, Y. Nakanishi, and S. Suzuki, *J. Biol. Chem.*, 1970, **245**, 6039.
[145] Y. Nakanishi, H. Sonohara, and S. Suzuki, *J. Biol. Chem.*, 1970, **245**, 6046.

sulphate)-(1 → 6)-2-acetamido-2-deoxy-6-*O*-phospho-D-glucose (12) (Scheme 4), and UDP-D-galactose 4-sulphate (13) (Scheme 5).

Incubation of ^{35}S- or ^{14}C-labelled chondroitin sulphate with a supernatant fraction of chick embryo cartilage containing a glycosaminoglycan sulphating system indicated that no significant loss of previously incorporated

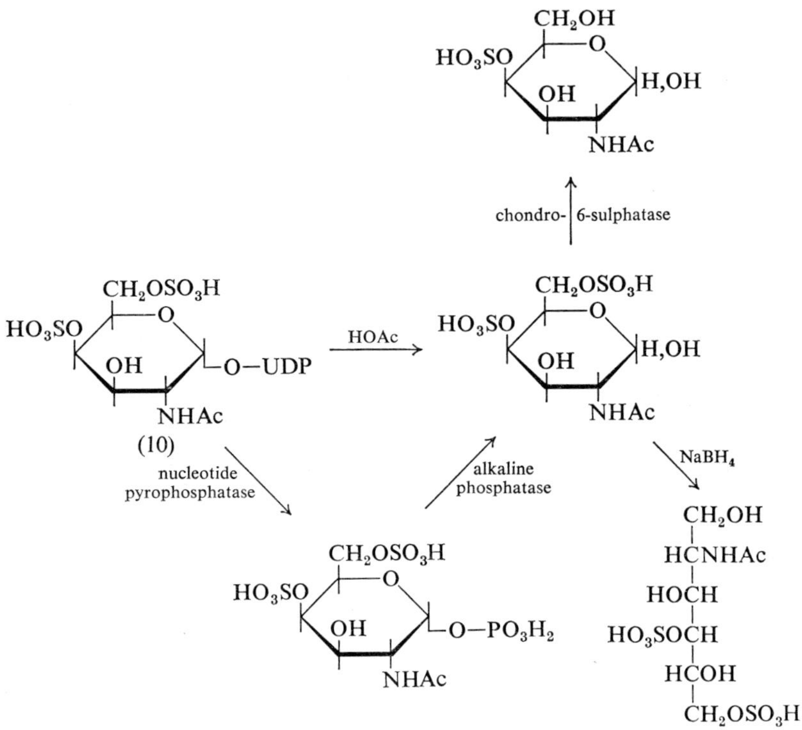

Scheme 2

sulphate groups occurred, but that the chondroitin sulphate was still capable of incorporating more sulphate.[146] It was concluded that chondroitin sulphate could not exchange sulphate groups and further incorporation of sulphate groups was part of the biosynthetic sulphation step.

The incorporation of [^{35}S]sulphate *in vivo* into the acid-soluble intermediates extracted from neonatal rat skin revealed three sulphated components containing 2-amino-2-deoxyhexose.[147] Their rate of synthesis and other information indicated that these components were not precursors of the sulphated 2-amino-2-deoxyhexoses present in the glycosaminoglycuronans. An investigation of the effect of temperature on the biosynthesis of porcine cartilage chondroitin 4-sulphate *in vitro* showed that [^{35}S]sul-

[146] J. P. Bently, *Carbohydrate Res.*, 1970, **12**, 281.
[147] T. E. Hardingham and C. F. Phelps, *Biochem. J.*, 1970, **119**, 885.

phate was incorporated preferentially into the high molecular weight molecules if the tissue was cultured directly.[148] However, the incorporation became symmetrical if the tissue was incubated before culture.

As determined by the incorporation of uniformly labelled D-[^{14}C]glucose, a decrease in the rate of biosynthesis of chondroitin and chondroitin sulphate and an increase in that of keratan sulphate were observed when

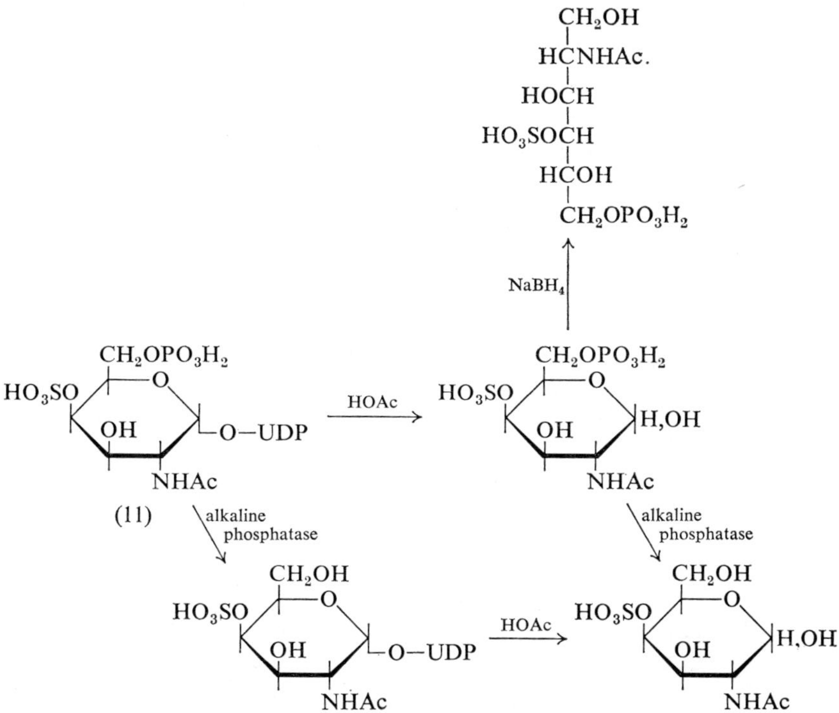

Scheme 3

samples of bovine cornea were incubated with UDP-D-xylose.[149] The UDP-D-glucuronic acid isolated after incubation had a lower specific activity than that isolated after incubation in the absence of UDP-D-xylose. This suggested that UDP-D-xylose inhibited UDP-D-glucose dehydrogenase causing an accumulation of UDP-D-glucose and, consequently, an increase in the formation of UDP-D-galactose and keratan sulphate. Determination of the specific activity of the 2-amino-2-deoxyhexoses isolated from the polysaccharides indicated that polysaccharide biosynthesis occurred *in vitro*, and that the regulatory effect of UDP-D-xylose was active at the

[148] T. E. Hardingham and H. Muir, *F.E.B.S. Letters*, 1970, **9**, 145.
[149] C. Balduini, A. Brovelli, and A. A. Castellani, *Biochem. J.*, 1970, **120**, 719.

monosaccharide level. The involvement of UDP derivatives in the transformation of D-[^{14}C]glucose into both uronic acid and 2-amino-2-deoxyhexose units of chondroitin sulphate by extracts of chick embryonic cartilage has been studied.[150]

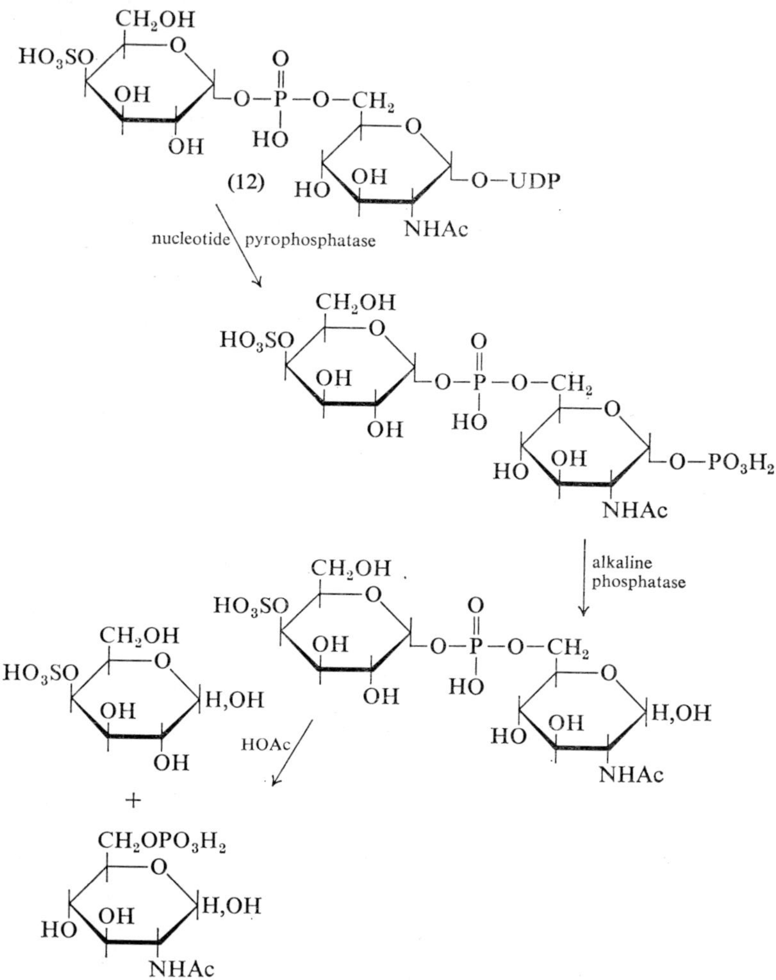

Scheme 4

Relative rates of synthesis of glycosaminoglycans of calf aortic wall were obtained using [^{35}S]sulphate and uniformly labelled D-[^{14}C]glucose.[109] The highest rate was found for ^{14}C-labelled hyaluronic acid, and the rates of incorporation for both labels decreased in the order dermatan sulphate,

[150] P. L. Jeffrey and K. G. Rienits, *Biochim. Biophys. Acta*, 1970, **208**, 147.

heparan sulphate, and chondroitin sulphate. Preliminary observations with bovine aortic and pulmonary heart valves showed that the carbohydrate moieties of other glycoproteins were synthesised more rapidly than those of glycosaminoglycan-proteins.[111]

[^{35}S]Sulphate was incorporated *in vivo* into chondroitin sulphates (71%), dermatan sulphate (14%), and heparan sulphate (14%) in rat brain; one of the chondroitin sulphate isomers incorporated the label at a higher rate

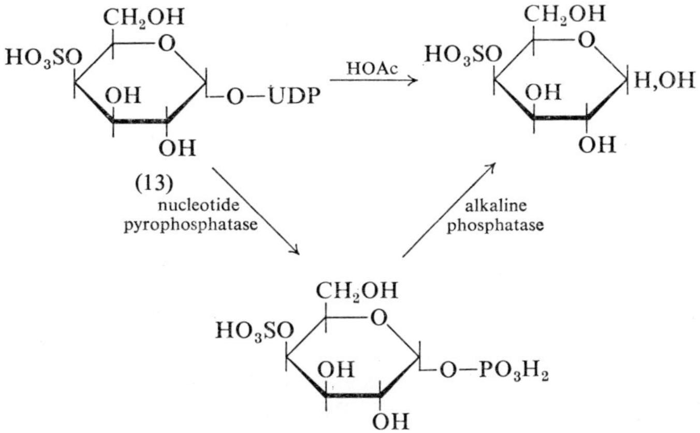

Scheme 5

than the other.[126] Radioactivity was not found in fractions corresponding to heparin and keratan sulphate confirming that these glycosaminoglycans were not present in the brain tissue.

The sulphation of glycosaminoglycans in rat embryonic tissue and the effect of cyclophosphamide have been investigated.[151] Treatment of guinea-pigs with oestrogen increased the number of Kurloff cells accumulating and their chondroitin 4-sulphate content.[123] Cortisol and cortisone treatment of chick embryo femur cultured *in vitro* increased the incorporation of [^{35}S]sulphate and D-[^{14}C]glucose, but the rate of decay of chondroitin sulphate remained unaltered.[152] The significance of these results, which differed from those published previously, was discussed.

Chondroitin 4-sulphate sulphotransferase was absent in Sanfilippo syndrome brain tissue, whereas dermatan sulphate and heparan sulphate sulphotransferase activities were normal.[138] A relative transfer of sulphate by the heparan sulphate sulphotransferase was doubled, indicating that *N*-sulphation was more predominant than *O*-sulphation in Sanfilippo syndrome.

151 K. Schimmelpfennig, *Arkiv. Pharmakol.*, 1970, **266**, 439.

152 M. Calcagno, H. Goyena, E. Arrambide, and C. A. deUrse, *Exptl. Cell. Res.*, 1970, **63**, 131.

The mode of aggregation of hyaluronic acid-protein on the surface of human articular cartilage has been studied.[153] A possible pathway for the complete desulphation of heparin *in vivo* was discussed on the basis of the properties of a lymphoid sulphamidase.[85]

Hormonal Glycoproteins

The chemistry of gonadotrophins in relation to their biological and antigenic properties has been reviewed.[154] The involvement of follicle stimulating and luteinizing hormones in the female reproductive cycle has been summarised.[155] Papers presented at a meeting on the chemistry of gonadotrophins have been published.[156]

Electrophoretic separation of the pituitary glycoprotein hormones, follicle stimulating hormone, luteinizing hormone, thyroid stimulating hormone, growth hormone, and prolactin from rats has been achieved using polyacrylamide.[157] A procedure for the further purification of human pituitary follicle stimulating hormone using gel filtration has been described.[158] The product, type CP150, had a potency of 1200 i.u. per mg, sedimented as a single component, and had a molecular weight of $3{\cdot}5 \times 10^4$. The amino-acid and carbohydrate compositions of CP150 were similar to preparations of follicle stimulating hormone prepared by other methods. 2-Amino-2-deoxy-D-galactose was shown to be a very small part of the 2-amino-2-deoxyhexose content of CP150,[7] and other structural information was deduced from a series of reactions including enzymic hydrolysis and oxidation, and hydrolysis with alkali and alkaline borohydride.[159] Testing of modified follicle stimulating hormones showed that the sites responsible for *in vivo* biological and immunological activities were not identical, certain carbohydrate components being essential for the former but not for the latter activity.[160] Although essential for *in vivo* biological activity, *N*-acetylneuraminic acid was not essential for *in vitro* activity of the hormone.[161]

A highly purified preparation of ovine pituitary follicle stimulating hormone contained 133 units of ovine NIH-FSH-S1 per mg, but showed only a very low luteinizing hormone activity;[162] the molecular weight of the material was $3{\cdot}3 \times 10^4$. C.d. studies indicated that 25—33% of the amino-acid residues of ovine follicle stimulating hormone were in an α-helical

[153] P. S. Walker, A. Unsworth, D. Dowson, J. Sikorski, and V. Wright, *Ann. Rheumat. Dis.*, 1970, **29**, 591.
[154] W. R. Butt, *Acta Endocr.*, 1969, **63**, Suppl. 142, 13.
[155] V. Petrow, *Chem. in Britain*, 1970, **6**, 167.
[156] 'Gonadotrophins and Ovarian Development,' eds W. R. Butt, A. C. Crooke, and M. Ryle, Livingstone, Edinburgh, 1970.
[157] D. R. Hodges and W. H. McShan, *Acta Endocr.*, 1970, **63**, 378.
[158] S. A. Barker, C. J. Gray, J. F. Kennedy, and W. R. Butt, *J. Endocr.*, 1969, **45**, 275.
[159] M. F. Chaplin, C. J. Gray, and J. F. Kennedy, ref. 156, p. 77.
[160] W. R. Butt, S. S. Lynch, M. F. Chaplin, C. J. Gray, and J. F. Kennedy, ref. 156, p. 171.
[161] M. Ryle, M. F. Chaplin, and J. F. Kennedy, ref. 156, p. 98.
[162] O. D. Sherwood, H. J. Grimek, and W. H. McShan, *J. Biol. Chem.*, 1970, **245**, 2328.

conformation.[163] An immunological study showed that rabbit anti-ovine follicle stimulating hormone cross-reacted with the hormone from a number of species.[164]

Ovine pituitary follicle stimulating hormone was dissociated into subunits on treatment with 1M-propionic acid.[165] The subunits, FSH-α and FSH-β, had low biological activities individually, but recombination restored 60% of the original activity. FSH-α and FSH-β contained hexose (2·8 and 13·8%), 2-amino-2-deoxyhexose (2·3 and 7·1%), and sialic acid (0·8 and 3·7%, respectively). Combination of a subunit (LH-CII) of luteinizing hormone with FSH-β generated luteinizing hormone activity while the other subunit (LH-CI) with FSH-β generated follicle stimulating hormone activity.

A method has been described for the purification of follicle stimulating hormone and luteinizing hormone from horse pituitaries.[166] The final follicle stimulating hormone preparation (90 units NIH-FSH-S1 per mg) was homogeneous, had a molecular weight of $3{\cdot}3 \times 10^4$, and an isoelectric point of 4·1. The luteinizing hormone (5·3 units NIH-LH-S1 per mg) had at least four active components with isoelectric points of 4·5, 5·9, 6·6, and 7·3.

Luteinizing hormone (8·9 units NIH-LH-S1 per mg) was isolated from human pituitaries by isoelectric focusing in a sucrose gradient; its homogeneity was demonstrated by disc electrophoresis and immunoelectrophoresis.[167] A molecular weight of $2{\cdot}675 \times 10^4$ was calculated from the Stokes radius, and analyses showed that the hormone contained sialic acid (1·8%), hexose (12·1%), 2-amino-2-deoxyhexose (5·0%), and a high proportion of proline; serine was detected at the *C*-terminus. The hormone did not cross-react with anti-human follicle stimulating hormone.

Bovine luteinizing hormone (1·5 units NIH-LH-S1 per mg, molecular weight $2{\cdot}8 \times 10^4$) was prepared free from follicle stimulating hormone activity.[168] The material, which contained hexose (5·3%), L-fucose (0·9%), and 2-amino-2-deoxyhexose (7·0%), closely resembled the ovine variety. Subunits, LH-CI and LH-CII, were formed by counter-current distribution in a solvent system containing dichloroacetic acid, and contained hexose (7·3 and 3·7%), L-fucose (1·6 and 0·9%), and 2-amino-2-deoxyhexose, (8·7 and 5·5%, respectively). Although the subunits had individually low biological activities, recombination with restoration of 50% of the original hormonal activity could be effected by incubation at room temperature and pH 6·0.

Hydroxyapatite chromatography of highly purified ovine pituitary luteinizing hormone (1·6—1·9 units NIH-LH-S1 per mg), containing

163 M. Ekblad, T. A. Bewley, and H. Papkoff, *Biochim. Biophys. Acta*, 1970, **221**, 142.
164 A. J. Rao and N. R. Moudgal, *Arch. Biochem. Biophys.*, 1970, **138**, 189.
165 H. Papkoff and M. Ekblad, *Biochem. Biophys. Res. Comm.*, 1970, **40**, 614.
166 W. E. Braselton and W. H. McShan, *Arch. Biochem. Biophys.*, 1970, **139**, 45.
167 P. Rathnam and B. B. Saxena, *J. Biol. Chem.*, 1970, **245**, 3725.
168 H. Papkoff and J. Gan, *Arch. Biochem. Biophys.*, 1970, **136**, 522.

0·025% follicle stimulating hormone, gave three major active fractions among which differences were noted by physicochemical techniques.[169] Two of these fractions had potencies of 2·0—2·6 units NIH-LH-S1 per mg and contained hexose, L-fucose, and 2-amino-2-deoxyhexose. Complete compositional analyses showed differences between the fractions, but dissimilarity in molecular size and immunological specificity was not detected. Further subfractionation was achieved by electrophoresis, and it was concluded that the hormone contained a broad spectrum of physicochemically similar components.

S-Sulphonated ovine luteinizing hormone, produced by oxidative sulphitolysis of the disulphide bonds, yielded two subunits (LH-A and LH-S) on treatment with 8M-urea and gel filtration.[170] The carbohydrate and amino-acid compositions of these subunits were compared with those of LH-I and LH-II separated by counter-current distribution.[171]

Bovine thyroid stimulating hormone (thyrotropin) was separated into two subunits, TSH-I and TSH-II, by urea treatment and chromatography; these subunits were of identical molecular weight.[172] TSH-I contained D-glucose, D-mannose, L-fucose, 2-acetamido-2-deoxy-D-glucose, and 2-acetamido-2-deoxy-D-galactose, and TSH-II contained D-glucose and D-galactose. In comparison, LH-CI had a similar overall composition to TSH-I and contained D-mannose, L-fucose, 2-acetamido-2-deoxy-D-glucose, and 2-acetamido-2-deoxy-D-galactose, and traces of D-glucose and D-galactose; LH-CII contained D-mannose, L-fucose, 2-acetamido-2-deoxy-D-glucose, and 2-acetamido-2-deoxy-D-galactose. Although TSH-I retained some activity, full activity depended on the integrity of the tertiary structure, and attempted recombination of TSH-I and TSH-II regenerated 80% of the original activity. The similarity between TSH-I and LH-CI was sufficient to allow association between LH-CI and TSH-II subunits with a resulting increase in thyroid stimulating activity, whereas association with LH-CI and LH-CII suppressed this activity.

Similar results have been obtained from a comparison of the bovine thyroid stimulating hormone subunits TSH-α and TSH-β, produced by urea treatment, and bovine LH-CI and LH-CII.[173] TSH-α and -β resembled one another more closely in amino-acid composition than did LH-CI and LH-CII, whereas TSH-α and LH-CI were closely related in their carbohydrate and amino-acid compositions. LH-CI could substitute fully for TSH-α in reconstitution of thyroid stimulating activity, and TSH-α and LH-CII combined to form a hybrid having luteinizing hormone activity. The amino-acid compositions of TSH-β and LH-CII differed markedly.

169 O. D. Sherwood, H. J. Grimek, and W. H. McShan, *Biochim. Biophys. Acta*, 1970, **221**, 87.
170 W. M. Lamkin, M. Fujino, J. D. Mayfield, G. N. Holcomb, and D. N. Ward, *Biochim. Biophys. Acta*, 1970, **214**, 290.
171 H. Papkoff and T. S. A. Samy, *Biochim. Biophys. Acta*, 1967, **147**, 175.
172 G. Hennen, G. Maghuin-Rogister, and G. Mamoir, *F.E.B.S. Letters*, 1970, **9**, 20.
173 T. Liao and J. G. Pierce, *J. Biol. Chem.*, 1970, **245**, 3275.

It was proposed that different hormonal activities arose from the combination of hormone specific subunits with a common subunit.

The origin of the multiple components of thyroid stimulating hormone has been investigated using the hormone from whale, since the size of the pituitary allowed the work to be carried out on one gland, thereby eliminating differences between individuals.[174] Isoelectric focusing showed at least four biologically-active components in the preparation. Human and simian thyroid stimulating hormone were considered to be similar on the basis of immunological cross-reactions.[175]

By a combination of gel filtration and ion-exchange chromatography, two highly purified human chorionic gonadotrophin fractions (20 000 i.u. per mg) were obtained containing galactose, mannose, 2-amino-2-deoxyhexose, and sialic acid.[176] One fraction had follicle stimulating hormone activity and the other luteinizing hormone-like activity. This and compositional data suggested that human chorionic gonadotrophin comprised two components, the ratio, and hence the biological and physicochemical properties, of which altered according to the stage of pregnancy. The gonadotrophin was dissociated into two non-identical subunits A and B by treatment with 8M-urea, and these were separated by DEAE-Sephadex chromatography.[177] The subunits A and B contained D-galactose (1·52 and 7·50%), D-mannose (5·40 and 4·80%), L-fucose (0·36 and 1·30%), 2-acetamido-2-deoxy-D-glucose (8·55 and 7·40%), 2-acetamido-2-deoxy-D-galactose (0·19 and 2·00%), and sialic acid (3·9 and 10·20%, respectively). The subunits re-associated on incubation.

Milk Glycoproteins

Human colostrum contained a non-specific inhibitor of influenza virus haemagglutination, whose activity was mediated by the *N*-acetylneuraminic acid units of a 14·5*S* glycoprotein;[178] the antiviral properties of this factor have been assessed.

The stabilising action of two carboxymethylated cellulose derivatives on α-, αs-, β-, and κ-caseins at an acid pH was studied, and selective precipitation was achieved in the case of mixtures.[179]

Improved chemical and chromatographic methods have been developed for the isolation of electrophoretically pure κ-casein.[180] The use of 6·6M-urea did not adversely affect the κ-casein, and the final product sedimented as a single peak. The amino-acid composition, sialic acid content (1·89%), and phosphorus content were comparable with values previously reported.

174 H. Tamura and N. Ui, *Biochim. Biophys. Acta*, 1970, **214**, 566.
175 B. R. Webster, J. C. Paice, and C. C. Gale, *Nature*, 1970, **227**, 712.
176 Y. Ashitaka, Y. Tokura, M. Tane, M. Mochizuki, and S. Tojo, *Endocrinology*, 1970, **87**, 233.
177 N. Swaminatham and O. P. Bahl, *Biochem. Biophys. Res. Comm.*, 1970, **40**, 422.
178 K. F. Shortridge, *Clin. Chim. Acta*, 1970, **29**, 233.
179 Y. Abano, *Agric. and Biol. Chem. (Japan)*, 1970, **34**, 102.
180 K. K. Tripathi and C. W. Gehrke, *J. Chromatog.*, 1970, **46**, 280.

The monosaccharide components of the glycopeptides released from κ-casein by the action of rennin on the milk of a single cow comprised D-galactose, 2-acetamido-2-deoxy-D-glucose, *N*-acetylneuraminic acid, and, in addition, D-mannose.[181] The glycopeptides present in the trichloroacetic acid filtrate from milk treated with rennin were separated into six fractions by ion-exchange chromatography.[182] The first and second fractions contained D-galactose and 2-acetamido-2-deoxy-D-galactose; the former also contained D-mannose, and the latter small amounts of *N*-acetylneuraminic acid. All other fractions contained D-galactose (2), 2-acetamido-2-deoxy-D-galactose (1), and *N*-acetylneuraminic acid (1 mole per mole glycopeptide).

The action of rennin on the acid-soluble fraction of casein gave two fragments, one of which had the same end-groups and amino-acid composition as a glycopeptide from κ-casein and contained sialic acid (7·8%);[183] it was proposed that the acid-soluble fraction was a subunit of κ-casein. Non-enzymic browning in a lactose–casein model system has been studied.[184]

Oligosaccharides containing sialic acid have been isolated from cow colostrum. In addition to two known isomers of monosialyl-lactose, partial acid hydrolysis and oxidation with periodate permitted the characterisation of a disialyl-lactose, *viz.* (*N*-acetylneuraminyl)-(2 → 8)-(*N*-acetylneuraminyl)-(2 → 3)-*O*-β-D-galactopyranosyl-(1 → 4)-D-glucopyranose.[185]

Streptococcal hydrolase degradation of 275 nm positive and negative cow colostrum whey glycoproteins showed that their carbohydrate moieties were identical in composition.[186] The purified carbohydrate moiety of the whey M-1 glycoprotein, obtained by β-elimination and column chromatography, had an approximate molecular weight of $1{\cdot}2 \times 10^3$ and contained D-galactose (2), 2-amino-2-deoxy-D-glucose (1), 2-amino-2-deoxy-D-galactose (1), and sialic acid (1—2 residues per carbohydrate chain). The detection of 2-amino-2-deoxy-D-glucitol in the isolated carbohydrate moiety and α-aminobutyric acid in the protein residues inferred the presence of an *O*-glycosidic linkage involving 2-amino-2-deoxy-D-galactose and threonine. It was proposed that one of the two whey glycoproteins possessed two carbohydrate chains, whereas the other had three or four. A component of the acid casein fraction of bovine colostrum possessed a carbohydrate moiety whose structure closely resembled that of the whey M-1 glycoprotein.

A variant (Dr) of β-lactoglobulin occurred in the milk of a few Droughmaster cattle,[187] together with equal amounts of either the A or B variant.

181 G. Sinkinson and J. V. Wheelock, *Biochim. Biophys. Acta*, 1970, **215**, 517.
182 J. V. Wheelock and G. Sinkinson, *Biochem. J.*, 1970, **119**, 13P.
183 R. Beeby, *Biochim. Biophys. Acta*, 1970, **214**, 364.
184 A. Ferretti, V. P. Flanagan, and J. M. Ruth, *J. Agric. Food Chem.*, 1970, **18**, 13.
185 R. Öhman and O. Hygstedt, *Analyt. Biochem.*, 1968, **23**, 395.
186 A. Bezkorovainy, D. Grohlich, and J. K. Pinter, *J. Agric. Food Chem.*, 1970, **18**, 624.
187 K. Bell, H. A. McKenzie, W. H. Murphy, and D. C. Shaw, *Biochim. Biophys. Acta*, 1970, **214**, 427.

In both cases, the Dr variant had the same amino-acid composition as the former, but also contained a carbohydrate moiety comprising D-galactose (0·8), D-mannose (1·9), 2-amino-2-deoxy-D-galactose (0·9), 2-amino-2-deoxy-D-glucose (3·4), and *N*-acetylneuraminic acid (1·0 residues per monomer of molecular weight $2{\cdot}0 \times 10^4$).

Milk from mice containing a mammary tumour agent contained a unique chemical determinant, which was isolated, purified, and shown to contain carbohydrate (38%) and protein (62%).[188] The antigenicity of the molecule in rabbits was not inactivated by treatment with neuraminidase, amylase, and lysozyme, but was inactivated by pronase.

Serum, Plasma, and Blood Cellular Element Glycoproteins, and Immunoglobulins

Serum and Plasma Glycoproteins.—The normal ranges of human serum glycoprotein levels have been determined.[189] Normal human serum α_1-acid glycoprotein contained D-galactose and D-mannose (14·3% combined), L-fucose (1·0%), 2-amino-2-deoxy-D-glucose (14·3%), and *N*-acetylneuraminic acid (8·8%).[190] The tryptic digest of *S*-carboxymethylated, human α_1-acid glycoprotein was separated by gel filtration into a glycopeptide and peptide fragments for which some amino-acid sequences were derived.[191]

α_1-Acid glycoprotein (molecular weight $3{\cdot}9$—$4{\cdot}0 \times 10^4$) has been isolated in homogeneous form from chimpanzee plasma.[192] The glycoprotein contained D-galactose (6·7%), D-mannose (4·8%), L-fucose (1·1%), 2-acetamido-2-deoxy-D-galactose (1·2%), 2-acetamido-2-deoxy-D-glucose (13·4%), and sialic acid (11·9%); a comparison of its composition and physicochemical properties with those of human α_1-acid glycoprotein showed them to be almost identical. The immunological differences between the two were attributed to sequence differences.

A method has been described for the preparation of rabbit serum α_1-acid glycoprotein using ammonium sulphate precipitation and chromatography on carboxymethylcellulose.[193] The glycoprotein had a molecular weight of $3{\cdot}3 \times 10^4$, and contained D-galactose (9·1%), D-mannose (9·1%), L-fucose (0·4%), 2-amino-2-deoxy-D-glucose (13·6%), and sialic acid (15·6%).

The partitioning of human serum albumin, bovine serum albumin, and ovalbumin in a dextran–poly(ethylene glycol)–water system has been investigated,[194] and the partition coefficient of bovine serum albumin with

[188] G. Miroff, H. M. Meade, M. Winetrout, and H. V. Lamberson, *Nature*, 1970, **227**, 1243.
[189] C. Peyer and H. P. Rieder, *Clin Chim. Acta*, 1970, **30**, 295.
[190] K. Hotta, M. Horikawa, and M. Kurokawa, *Analyt. Biochem.*, 1970, **33**, 463.
[191] R. Bourrillon, D. Meyer, and M. A. Parnaudeau, *Biochim. Biophys. Acta*, 1970, **200**, 378.
[192] Y. Li and S. Li, *J. Biol. Chem.*, 1970, **245**, 825.
[193] J. Marçais, C. Nicot, and J. Moretti, *Bull. Soc. Chim. biol.*, 1970, **52**, 741.
[194] G. Johansson, *Biochim. Biophys. Acta*, 1970, **222**, 381.

various grades of Sephadex has been measured.[195] Selective precipitation of bovine serum albumin from mixtures of closely associated glycoproteins was achieved using tungstophosphoric, tungstosilicic, and molybdosilicic acids.[196] Precipitates formed when the pH of the polyacid–glycoprotein solutions were close to the isoelectric point of the glycoprotein; the groups which combined with the polyacid were the guanido, imidazole, and ε-amino-groups, rather than the α-amino-groups. Data were obtained for potentiometric and spectrophotometric titrations of bovine serum albumin in dioxan–water mixtures,[197] and for changes in the macrostructure of the spin-labelled glycoprotein on γ-irradiation.[198] Dimeric bovine serum albumin, prepared by oxidation of mercaptoalbumin, reverted to the monomer on mild treatment with alkali.[199]

Liver-cell potassium ions play a role in the secretion of rat serum albumin.[200] Differences in the chemical and physicochemical characteristics of serum albumin from two species of frog have been observed.[201]

A procedure for the fractionation and isolation, in an electrophoretically pure form, of α_2-glycoprotein from rat inflammatory serum has been reported.[202] A 4*S* α_2-β_1-glycoprotein was isolated from Cohn fraction VI of normal human plasma by chromatography on diethylaminoethyl- and carboxymethyl-cellulose, DEAE-Sephadex, hydroxyapatite, and Sephadex G-100.[203] The glycoprotein had a molecular weight of $6{\cdot}0 \times 10^4$, and the carbohydrate moiety (30%) comprised D-galactose (14), D-mannose (14), L-fucose (2), 2-amino-2-deoxy-D-glucose (30), and sialic acid (14 residues, out of a total of 465). The terminal positions of the sialic acid residues were indicated by their liberation with neuraminidase.

There is apparently no significance in the measured hexose contents (1·0—4·6%) of foetal and adult fibrinogen.[204] The hexoses were principally D-galactose and D-mannose, and the presence of the 2-amino-2-deoxyhexose (0·5—1·0%) and sialic acid (0·8%) was also indicated. The action of plasmin on fibrinogen and fibrin resulted in the formation of almost identical high molecular weight products.[205]

Two glycopeptides of molecular weight 2015 and 2180 have been isolated from pronasic hydrolysates of human transferrin.[206] The glycopeptide

195 A. G. Ogston and P. Silpananta, *Biochem. J.*, 1970, **116**, 171.
196 M. Z. Sternberg, *Biotechnol. and Bioeng.*, 1970, **12**, 1.
197 S. F. Sun, *Biochim. Biophys. Acta*, 1970, **200**, 433.
198 V. P. Yuferov, G. L. Grigoryan, E. G. Rosantsev, and A. E. Kalmanson, *Biofizika*, 1970, **15**, 404.
199 L. Anderson, *Biochem. Biophys. Acta*, 1970, **200**, 363.
200 J. D. Judah and M. R. Nichols, *Biochem. J.*, 1970, **116**, 663.
201 V. N. Vitvitskii, *Doklady Akad. Nauk S.S.S.R.*, 1970, **194**, 950.
202 F. F. Menninger, H. J. Esber, and A. E. Bogden, *Clin. Chim. Acta*, 1970, **27**, 385.
203 T. Iwasaki and K. Schmid, *J. Biol. Chem.*, 1970, **245**, 1814.
204 I. Witt and H. Müller, *Biochim. Biophys. Acta*, 1970, **221**, 402.
205 G. A. Dudek, M. Kloczewiak, A. Z. Budzyński, Z. S. Latallo, and M. Kopeć, *Biochim. Biophys. Acta*, 1970, **214**, 44.
206 G. Spik and J. Montreuil, *Bull. Soc. Chim. biol.*, 1969, **51**, 1271.

linkage regions (14) and (15), respectively, involved 2-acetamido-1-{(*N*-β-L-aspartyl)amino}-2-deoxy-β-D-glucosylamine bonds.

Studies have been carried out on the enzymic fragmentation and molecular weight of a single component (6·5*S* human cryoglobulin) isolated from a case of idiopathic cryoglobulinemia.[207] Carbohydrate analyses showed the presence of D-galactose and D-mannose (10 combined), L-fucose (2), 2-amino-2-deoxy-D-glucose (6), and sialic acid (2 residues).

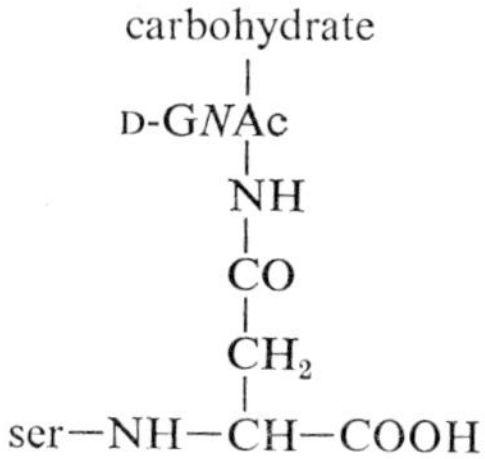

(14)

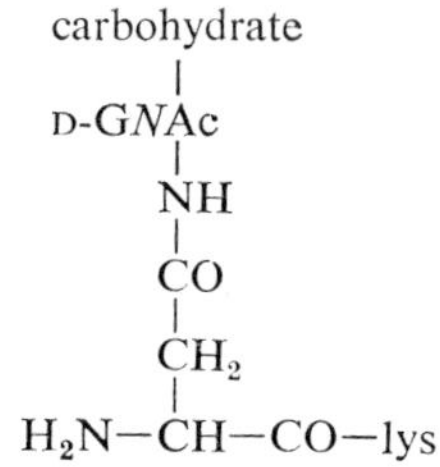

(15)

Lipid-free fractions, obtained from low- and high-density rat serum lipoprotein fractions, contained carbohydrate (5·2 and 2·2%, respectively) consisting of D-galactose, D-glucose, D-mannose, 2-amino-2-deoxy-D-glucose, and *N*-acetylneuraminic acid.[208]

Blood Cellular Element Glycoproteins.—Proteolysis of intact human platelets with trypsin, pronase, or papain released up to 60% of the total cell sialic acid in the form of three distinct size classes of glycopeptides (GP-1—3).[209] GP-1 ($M = 1{\cdot}2 \times 10^5$) was released by trypsin but was resistant to further proteolysis, GP-2 ($M = 2{\cdot}25 \times 10^4$) was released in small amounts by all procedures, and GP-3 ($M = 5{\cdot}0 \times 10^3$) was preferentially released by pronase or papain; GP-1 and GP-2 were apparently homogeneous whereas GP-3 could be subfractionated. All three classes contained D-galactose, D-glucose, L-fucose, 2-amino-2-deoxy-D-glucose, 2-amino-2-deoxy-D-galactose, and sialic acid; in the case of GP-1, 50% of the 2-amino-2-deoxyhexose units were involved in alkali-labile linkages.

Solubilised glycoproteins of human erythrocyte membrane have been separated by electrophoresis on acrylamide gel, and the pattern was compared with that obtained from the membrane of patients with hereditary spherocytosis.[210] A series of sialoglycoprotein fractions containing 2-amino-2-deoxy-D-galactose and 2-amino-2-deoxy-D-glucose, and sialic acid (4·7—31·4%) have been prepared from human erythrocyte membranes.[211] One of the fractions contained virus receptor activity. A

207 A. Saha, P. Chowdhury, S. Sambury, K. Smart, and B. Rose, *J. Biol. Chem.*, 1970, **245**, 2730.
208 J. B. Marsh and R. Fritz, *Proc. Soc. Exptl. Biol. Med.*, 1970, **133**, 9.
209 D. S. Pepper and G. A. Jamieson, *Biochemistry*, 1970, **9**, 3706.
210 G. K. Limber, R. F. Davis, and S. Bakerman, *Blood*, 1970, **36**, 111.
211 O. O. Blumenfeld, P. M. Callop, C. Howe, and L. T. Lee, *Biochim. Biophys. Acta*, 1970, **211**, 109.

study of the interaction of a sialoglycopeptide, released from human erythrocytes by trypsin treatment, with positively-charged substances (*e.g.* bovine albumin, toluidine blue, and sodium ions) suggested that the glycoprotein possessed ion-exchange properties and that the sialic acid residues could act as receptor sites for positively-charged molecules.[212]

Human peripheral blood lymphocytes in culture incorporated 2-amino-2-deoxy-D-[1-^{14}C]glucose at a constant rate, which doubled on the addition of a phytohaemagglutinin from *Phaseolus vulgaris*, but which was inhibited by puromycin. The radioactivity in an acid-soluble extract appeared mainly in UDP-2-acetamido-2-deoxy-D-glucose. Incorporated 2-amino-2-deoxy-D-glucose was released from the intact lymphocyte by proteolysis; papain digestion of a purified membrane fraction yielded three classes of glycopeptides having molecular weights of $2{\cdot}3 \times 10^5$, $1{\cdot}3 \times 10^5$, and $6{\cdot}0 \times 10^3$.[28] The *N*-acetylneuraminic acid contents of human normal and leukemic lymphocytes and granulocytes in relation to their electrophoretic mobilites have been investigated.[213]

Immunoglobulins.—The structure and heterogeneity of antibody-combining sites have been surveyed briefly,[4] and the structure of IgM antibodies and their function as rheumatoid factors have been briefly reviewed.[214]

Human IgG has been separated into fractions of different contents of *N*-acetylneuraminic acid by DEAE-cellulose chromatography.[215] The biological properties of these fractions before and after treatment with neuraminidase were compared.

The isolation of guinea-pig IgGI, IgGII, and IgM has been described together with methods for separating the subunits of each.[216] Approximate molecular weights were determined by sedimentation and gel filtration studies as IgM, $9{\cdot}5 \times 10^5$; IgGI, $1{\cdot}57 \times 10^5$; and IgGII, $1{\cdot}55 \times 10^5$. The compositions of the intact molecules and their subunits were determined, and each immunoglobulin contained hexose, L-fucose, 2-amino-2-deoxyhexose, and sialic acid (maximum proportions 5·2, 0·8, 2·9, and 1·6%, respectively).

The immunoglobulin IgM secreted by a mouse plasmacytoma reacted with a number of dextrans in a manner characteristic of antibody–antigen reactions, *viz.* precipitation of soluble dextran, agglutination of dextran-coated erythrocytes, and inhibition of dextran precipitation by small oligosaccharides.[217] The order of efficiency for inhibition of dextran–IgM reactions by α-linked D-glucose oligosaccharides was $1,3 \gg 1,6 > 1,2 > 1,4$; nigerotetraose and nigeropentaose were the best inhibitors of the nigerodextrin series.

212 S. Ohkuma, *Yakagaku Zasshi*, 1970, **90**, 647.
213 M. A. Lichtman and R. I. Weed, *Blood*, 1970, **35**, 12.
214 S. I. Chavin and E. C. Franklin, *Ann. New York Acad. Sci.*, 1969, **168**, 84.
215 F. Berger-du Bois, D. Des Gouttes, and H. Isliker, *Ann. Inst. Pasteur*, 1970, **119**, 8.
216 G. Q. Leslie and S. Cohen, *Biochem. J.*, 1970, **120**, 787.
217 M. A. Leon, N. M. Young, and K. R. McIntire, *Biochemistry*, 1970, **9**, 1023.

A glycine-rich glycoprotein isolated from normal human plasma by successive fractionations with 6,9-diamino-2-ethoxyacridine lactate, ammonium sulphate, DEAE-Sephadex, and carboxymethyl-Sephadex migrated as an IgG on agarose gel electrophoresis.[218] The glycoprotein, which contained hexose (8·5%) and *N*-acetylneuraminic acid (1·3%), was unrelated antigenically to a wide variety of known plasma macromolecules.

Insoluble derivatives of β-galactosidase,[219] insulin,[220] and *Escherichia coli* lipopolysaccharide[221] antibodies have been prepared.

Urinary Glycoproteins

The compositions of urinary glycoproteins and glycopeptides of high and low molecular weight have been reviewed.[98]

The carbohydrate and amino-acid compositions of human Tamm-Horsfall glycoprotein have been revised; the new values for the carbohydrate moiety are hexose (11·7%), L-fucose (0·8%), 2-acetamido-2-deoxyhexose (11·2%), and sialic acid (4·4%).[222] Treatment with alkali did not release carbohydrate, suggesting that the predominant glycopeptide linkages were not of the *O*-glycosidic type. *N*-Terminal amino-acid was not detected, and no significant differences in the amino-acid composition were observed for glycoprotein from normals and patients with cystic fibrosis. Subunit molecular weights of 7·6—8·2 × 10^4 were obtained for the native and alkylated glycoprotein by gel filtration in the presence of sodium dodecyl sulphate.[223] A further estimate of 7·9 × 10^4 was obtained for the molecular weight by disc electrophoresis in sodium dodecyl sulphate, whilst a minimum value of 7·9 × 10^4 was obtained from the number of *N*-terminal amino-acids released by treatment with cyanogen bromide. Similar values were obtained for the subunit molecular weight of the glycoprotein from patients with cystic fibrosis. Treatment of the glycoprotein with 6M-guanidine hydrochloride also produced a homogeneous subunit of molecular weight 1·0 × 10^5.[224] Component analyses of a glycoprotein isolated from urinary casts demonstrated the presence of a small amount of associated lipid.[225] The glycoprotein was indistinguishable from Tamm-Horsfall glycoprotein.

A method for the paper chromatographic isolation and characterisation of 2-acetamido-1-{(*N*-β-L-aspartyl)amino}-2-deoxy-β-D-glucosylamine from urine has been reported, and was recommended for the correct diagnosis of aspartylglycosaminuria.[226]

218 T. Boenisch and C. A. Alper, *Biochim. Biophys. Acta*, 1970, **214**, 135.
219 R. P. Erickson and E. Steers, *Arch. Biochem. Biophys.*, 1970, **137**, 399.
220 L. Kågedal and S. Åkerström, *Acta Chem. Scand.*, 1970, **24**, 1601.
221 M. Esknazy, *Nature*, 1970, **226**, 855.
222 A. P. Fletcher, A. Neuberger, and W. A. Ratcliffe, *Biochem. J.*, 1970, **120**, 417.
223 A. P. Fletcher, A. Neuberger, and W. A. Ratcliffe, *Biochem. J.*, 1970, **120**, 425.
224 F. K. Stevenson and P. W. Kent, *Biochem. J.*, 1970, **116**, 791.
225 A. P. Fletcher, J. E. McLaughlin, W. A. Ratcliffe, and D. A. Woods, *Biochim. Biophys. Acta*, 1970, **214**, 299.
226 J. Palo and K. Mattsson, *J. Chromatog.*, 1970, **50**, 534.

Two glycopeptides (SGP1 and SGP2), rich in sialic acid, were isolated from normal human urine, and contained D-galactose (20·7 and 18·1%), D-mannose (8·3 and 1·4%), L-fucose (3·5 and 1·4%), 2-acetamido-2-deoxy-D-galactose (5·2 and 11·3%), 2-acetamido-2-deoxy-D-glucose (20·6 and 5·1%), and sialic acid (24·2 and 52·1%, respectively).[227] Their compositions resembled those of carbohydrate-containing compounds from the brain and erythrocyte membrane. The glycopeptide linkage of SGP1 involved 2-amino-2-deoxy-D-galactose and threonine, but treatment of SGP2 with alkaline borohydride did not significantly decrease its serine and threonine contents. SGP2 contained a sialic acid derivative which was not identical with *N*-acetyl- or *N*-glycolyl-neuraminic acids.

Glucosyl-galactosyl-δ-hydroxylysine and galactosyl-δ-hydroxylysine were isolated in crystalline form for the first time in a study of urinary glycosylated hydroxylysine and *N*-methyl derivatives of lysine and arginine.[228]

Miscellaneous Glycoproteins

Enzymic digestion with pronase and collagenase of human kidney glomerular and tubular basement membranes yielded two glycopeptides from each source, which were separated by gel filtration and ion-exchange chromatography.[229] One of the glomerular glycopeptides was identical to one of the tubular pair, and contained a disaccharide composed of D-galactose and D-glucose, probably linked to hydroxylysine. The other glomerular and tubular glycopeptides had molecular weights of $3·6 \times 10^3$ and $3·4 \times 10^3$, respectively; both contained D-galactose, D-mannose, L-fucose, 2-amino-2-deoxyhexose, and sialic acid.

An electro-immunodiffusion technique was suitable for analysis of nanogram amounts of prealbumin, albumin, α_1-acid glycoprotein, α_1-antitrypsin, α_2-haptoglobin, α_2-macroglobulin, and transferrin in human cerebrospinal fluid.[230] An improved method has been reported for the isolation of pure α_1-acid glycoprotein from ascitic fluid; the preparation was homogeneous on immunoelectrophoresis.[190] Its carbohydrate composition, D-galactose and D-mannose (14·6% combined), L-fucose (2·1%), 2-amino-2-deoxy-D-glucose (14·1%), and *N*-acetylneuraminic acid (9·8%), was very similar to that of normal serum α_1-acid glycoprotein.

A single glycopeptide, isolated from a peptic digest of membrane protein bovine visual pigment$_{500}$ by ion-exchange chromatography, contained D-mannose (3·1) and 2-amino-2-deoxy-D-glucose (2·7 moles per mole glycopeptide).[231] The carbohydrate moiety was linked to aspartyl residue-2 of the established peptide sequence through an alkali-stable bond. Treatment with *β*-acetamidodeoxyglucosidase liberated two residues of 2-amino-2-deoxy-D-glucose, whereas α-mannosidase liberated 86% of the D-mannose

[227] Y. Goussault and R. Bourrillon, *Biochem. Biophys. Res. Comm.*, 1970, **40**, 1404.
[228] Y. Karimoto and S. Akazawa, *J. Biol. Chem.*, 1970, **245**, 5751.
[229] P. Mahieu and R. J. Winand, *European J. Biochem.*, 1970, **15**, 520.
[230] E. Schuller, L. Tömpe, M. Lefèvre, and P. Moreno, *Clin. Chim. Acta*, 1970, **30**, 73.
[231] J. Heller and M. A. Lawrence, *Biochemistry*, 1970, **9**, 864.

residues. It was concluded that the parent glycoprotein (molecular weight $2{\cdot}8 \times 10^4$) contained a single carbohydrate moiety in which two of the 2-amino-2-deoxy-D-glucose residues and all of the D-mannose residues were linked peripherally to the third 2-amino-2-deoxy-D-glucose residue, which was involved in a 2-acetamido-1-{(*N*-β-L-aspartyl)amino}-2-deoxy-β-D-glucosylamine glycopeptide linkage.

Sialoglycoproteins were determined in bovine brain by measurement of the sialic acid, and their distribution was assessed.[232] The carbohydrate compositions of sialoglycoproteins from bovine heart aortic and pulmonary valves have been determined.[111] The rate of biosynthesis of these glycoproteins was higher in the aortic than in the pulmonary valve, and, in both cases, was higher than that of the glycosaminoglycans. The carbohydrate compositions of fractions containing glycoproteins from bovine heart, ovine heart, lung, and liver, and avian heart, liver, skin, and bone have been determined and compared with a view to gross analytical differentiation.[112]

D-Galactose, D-mannose, 6-deoxyhexose, 2-amino-2-deoxy-D-galactose, 2-amino-2-deoxy-D-glucose, uronic acids, *N*-acetylneuraminic acid, and *N*-glycolylneuraminic acid were detected in the glycoproteins of bovine myocardium.[233] During hydrolysis with acid, these carbohydrates were largely liberated in the order: sialic acids, 6-deoxyhexoses, hexoses, uronic acids, and 2-amino-2-deoxyhexoses.

Treatment of bovine and ovine submaxillary mucins with alkaline borohydride converted the carbohydrate moieties into dialysable fragments;[234] β-elimination of these moieties from the serine and threonine residues of the protein was shown to be 95% of quantitative. However, 10 and 22% of the 2-amino-2-deoxyhexose residues of the ovine and bovine molecules, respectively, were not reduced to the 2-amino-2-deoxyhexitol, and most of the resistant units were located in oligosaccharides, the compositions of which were reported.

The biosynthesis of the carbohydrate moieties of sheep thyroid microsomal glycoproteins, particularly thyroglobulin, has been investigated.[235] The presence of a polysaccharide-containing cell coat on the surfaces of keratinizing cells of the Romney wool follicle has been revealed.[236]

α-Fucosidase removed 80—90% of the L-fucose residues from porcine and canine submaxillary mucins.[41]

The composition and biosynthesis of glycoproteins in isolated synaptosomes and synaptosomal intraneural mitochondria from guinea-pig cerebral cortex have been investigated by the incorporation of radioactive

232 P. A. Roukema and J. Heijlman, *J. Neurochem.*, 1970, **17**, 773.
233 I. Holovatsky and A. M. Stadnik, *Ukrain. biokhim. Zhur.*, 1970, **42**, 439.
234 M. Bertolini and W. Pigman, *Carbohydrate Res.*, 1970, **14**, 53.
235 S. Bouchilloux, O. Chabaud, M. Michel-Béchet, M. Ferrand, and A. M. Athouel-Haon, *Biochem. Biophys. Res. Comm.*, 1970, **40**, 314.
236 D. F. G. Orwin, *Austral. J. Biol. Sci.*, 1970, **23**, 623.

monosaccharides.[237] The presence of L-fucose, hexose, 2-amino-2-deoxyhexose, and sialic acid was demonstrated in this way. Glycoproteins containing D-galactose, D-mannose, L-fucose, 2-amino-2-deoxyhexose, and sialic acid in mouse fibroblast L cells and surface membranes have been investigated.[238]

A comparative study of the glycoproteins from the surface of control and Rous sarcoma virus-transformed hamster cells showed that pronasic digestion yielded glycopeptides of higher molecular weight in the transformed cells.[239] The glycoproteins contained sialic acid and incorporated radioactive L-fucose and 2-amino-2-deoxy-D-glucose when the cells were cultured in media containing the labelled monosaccharides. Glycoproteins containing 2-amino-2-deoxyhexoses were released from baby hamster cells by treatment with trypsin and edta.[240]

The 2-amino-2-deoxy-D-galactose, -glucose, and -mannose contents of rat brain glycoproteins have been determined using differential colorimetry.[13] Glycopeptides containing D-galactose, D-mannose, L-fucose, 2-amino-2-deoxyhexose, sialic acid, and sulphate have been obtained from rat brain glycoproteins following pronasic digestion and removal of the glycosaminoglycans by precipitation with cetyl pyridinium chloride.[241] Preliminary studies indicated the presence of D-galactose 6-sulphate and 2-acetamido-2-deoxy-D-glucose 6-sulphate. Also present were non-sulphated glycopeptides which were similar to the sulphated glycoproteins in their molar ratios of 2-amino-2-deoxyhexoses and neutral sugars. The sialoglycoprotein content of rat brain remained constant during development;[242] the biosynthesis of these glycoproteins has also been studied.[243]

Rat small intestinal mucins have been shown to be glycoproteins containing hexose, L-fucose, 2-amino-2-deoxyhexose, and sialic acid.[244] Three distinct classes of glycoprotein were identified by use of 2-amino-2-deoxy-D-[1-^{14}C]glucose as precursor. A glycopeptide isolated from the same source contained D-galactose, L-fucose, 2-amino-2-deoxy-D-glucose, and sialic acid (molar ratios 12 : 4 : 1·2 : 1, respectively).[245] The biosynthesis of the glycopeptide in endoplasmic reticulum from normal and vitamin-A deficient animals was studied *in vitro* using UDP-2-acetamido-2-deoxy-D-[1-^{14}C]glucose and [^{3}H]serine.

Glycoproteins of rat kidney-cell cytoplasm have been studied immunoelectrophoretically.[246] Glomerular basement membrane contained D-galac-

[237] H. B. Bosmann and B. A. Hemsworth, *J. Biol. Chem.*, 1970, **245**, 363.
[238] M. C. Glick, C. Comstock, and L. Warren, *Biochim. Biophys. Acta*, 1970, **219**, 290.
[239] C. A. Buck, M. C. Glick, and L. Warren, *Biochemistry*, 1970, **9**, 4567.
[240] C. Snow and A. Allen, *Biochem. J.*, 1970, **119**, 707.
[241] R. K. Margolis and R. U. Margolis, *Biochemistry*, 1970, **9**, 4389.
[242] R. H. Quarles and R. O. Brady, *J. Neurochem.*, 1970, **17**, 801.
[243] P. A. Roukema, D. N. Van den Eijnden, J. Heijlman, and G. Van der Berg, *F.E.B.S. Letters*, 1970, **9**, 267.
[244] G. G. Forstner, *J. Biol. Chem.*, 1970, **245**, 3584.
[245] L. DeLuca, M. Schumacher, and G. Wolf, *J. Biol. Chem.*, 1970, **245**, 4551.
[246] K. P. Kashkin, D. N. Bochkova, and B. P. Surinov, *Biokhimiya*, 1970, **35**, 1170.

tose, D-glucose, D-mannose, L-fucose, 2-amino-2-deoxyhexose, and sialic acid.[247] A glycopeptide isolated after alkaline treatment contained equimolar proportions of D-glucose, D-galactose, and hydroxylysine, the only amino-acid detected. Twice as much D-glucose was present in the glycopeptide isolated from aminonucleoside nephritic rats, and it was suggested that the difference accounted for the functional alterations of nephritic glomerular basement membrane. A macromolecular fraction isolated from rat liver endoplasmic reticulum contained neutral sugar.[248] A stage in the biosynthesis of rat liver glycoproteins has been investigated.[249]

Glycoprotein synthesis by chick fibroblast cells uninfected and infected with Sindbis virus has been investigated.[250]

Five components of (N-β-L-aspartyl)amino-carbohydrate prepared by pronasic digestion of crystalline ovalbumin were purified by gel filtration, ion-exchange chromatography, and column electrophoresis.[251, 252] The molar ratios of D-mannose to 2-acetamido-2-deoxy-D-glucose ranged from 6 : 5—5 : 2. The inter-relationships of the carbohydrate groups in the five components were studied by hydrolysis with α-mannosidase and β-acetamidodeoxyglucosidase, and it was concluded that the carbohydrate sequences could be represented by a general structure (16). This homology was

(D-G*N*Ac)$_{0\ \text{or}\ 1\ \text{or}\ 2}$—D-Man—D-G*N*Ac—D-G*N*Ac—Asn
|
D-Man$_{0\ \text{or}\ 1}$—(D-Man)$_3$
|
D-Man(D-G*N*Ac)$_{0\ \text{or}\ 1}$

(16)

supported by the demonstration of a common glycopeptide nucleus by o.r.d. and c.d. Differences in the linkages of the D-mannose residues were indicated by ^{1}H n.m.r. measurements. Ovalbumin glycopeptides have been used as standards in an improved technique for the determination of monosaccharides in glycoproteins, *etc.*[9]

The utility of heteropolyacids in the purification of ovalbumin and ovomucoid has been demonstrated.[196] A heterogeneous preparation of ovomucin from chicken egg white contained hexose (7·4%), 2-amino-2-deoxyhexose (7·2%), sialic acid (4·0%), and sulphate (0·44%).[253] The amino-acid composition was similar to that of ovomucoid, and the preparation contained little α-helical structure. The partitioning of ovomucin in a dextran–poly(ethylene glycol)–water system has been investigated.[194]

[247] N. A. Kefalides and L. Forsell-Knott, *Biochim. Biophys. Acta*, 1970, **203**, 62.
[248] D. Kaplan, *Biochem. Biophys. Res. Comm.*, 1970, **211**, 396.
[249] L. DeLuca, G. Rosso, and G. Wolf, *Biochem. Biophys. Res. Comm.*, 1970, **41**, 615.
[250] B. W. Burge and J. H. Strauss, *J. Mol. Biol.*, 1970, **47**, 449.
[251] C.-C. Huang, H. E. Mayer, and R. Montgomery, *Carbohydrate Res.*, 1970, **13**, 127.
[252] R. Montgomery and C. C. Huang, *Proc. Austral. Biochem. Soc.*, 1970, **3**, 8.
[253] J. W. Donovan, J. G. Davis, and L. M. White, *Biochim. Biophys. Acta*, 1970, **207**, 190.

A study has been made of cytoplasmic sites of glycoprotein biosynthesis in chick embryo fibroblasts using cellular fractionation after double labelling of the macromolecules with 2-amino-2-deoxy-D-[^{3}H-6]glucose and ^{14}C-labelled amino-acids.[254] The results demonstrated the participation of many subcellular sites in the biosynthesis of the carbohydrate moieties of glycoproteins and confirmed the independence of each. The presence of radioactive 2-amino-2-deoxy-D-glucose on polysomes indicated that the carbohydrate moiety became attached to the polypeptide while the glycoprotein was still linked to the polysome.

Structural studies have been carried out on glycopeptides isolated from pronasic digests of the Lorenzini jelly of two species of elasmobranch fish *Raja clavata* and *Cetorhinus maximus*.[255] The glycopeptide linkage in all of these glycopeptides was *O*-(2-acetamido-2-deoxy-D-galactosyl)threonine, and the majority of the 2-acetamido-2-deoxy-D-galactosyl residues were substituted at C-3 and/or C-6 by oligosaccharides of various sizes.

Studies on glycoproteins isolated from the jelly coat of sea urchin eggs showed them to contain rhamnose, fucose, mannose, galactose, *N*-acetylneuraminic acid, and *N*-glycolylneuraminic acid.[256] A glycoprotein from the jelly coat of *Pseudocentrotus depressus* eggs has been purified until homogeneous by electrophoresis and ultracentrifugation.[257] The glycoprotein carried a strong negative charge, and contained sialic acid (70·0%), hexose (3·7%), fucose (9·2%), 2-amino-2-deoxyhexose (2·4%), and sulphate (7·6%). Although the major sialic acid was identified as *N*-glycolylneuraminic acid, hydrolysis of the glycoprotein with acid and enzymes suggested that several unidentified sialic acids were also present.

The operculum of the gastropod mollusc *Buccinum undatum* contained a class of glycoproteins ('operculins') which contained mannose, xylose, rhamnose, galactose, glucose, fucose, 2-amino-2-deoxyglucose, and 2-amino-2-deoxygalactose.[258]

Spongin A from the connective tissue of the invertebrate *Hippospongia gossypina* contained a complex glycoprotein–polysaccharide containing arabinose (8·2%), fucose (12%), galactose (8%), mannose (1·1%), xylose (0·6%), 2-amino-2-deoxyhexose (6·3%), hexuronic acid (12%), sulphate (22%), and amino-acid (3%); the arabinose had the D-configuration and the distribution of arabinose within the animal kingdom was discussed.[56] A biological role was proposed for the spongonucleotides.

The single glycoprotein from the membrane of Sindbis virus, an arbovirus grown in chick fibroblast cells, contained D-galactose and D-mannose (5·1% combined), L-fucose (0·5%), 2-amino-2-deoxy-D-glucose (7·3%), and

254 M. Pradal, D. Lebre, P. Louisot, and R. Got, *Comparative Biochem. Physiol.*, 1970, **35**, 31.
255 M. J. How, J. V. S. Jones, and M. Stacey, *Carbohydrate Res.*, 1970, **12**, 171.
256 K. Hotta, H. Hamazaki, M. Horikawa, M. Kurokawa, and S. Isaka, *Seikagaku*, 1970, **42**, 291.
257 K. Hotta, H. Hamazaki, and M. Kurokawa, *J. Biol. Chem.*, 1970, **245**, 5434.
258 S. Hunt, *Biochim. Biophys. Acta*, 1970, **207**, 347.

sialic acid (0·7%).[259] The glycoprotein from virus grown in hamster kidney cells contained a different amount of sialic acid (1·3%). Pronasic digestion of the 'chick' glycoprotein yielded a set of glycopeptides, the smallest of which was estimated to contain a total of nine residues of D-mannose and 2-amino-2-deoxy-D-glucose.[250] The larger glycopeptides contained, in addition, D-galactose, L-fucose, and, presumably, sialic acid. Glycopeptides from the virus grown in hamster cells had a different size distribution but were grossly similar.

[259] J. H. Strauss, B. W. Burge, and J. E. Darnell, *J. Mol. Biol.*, 1970, **47**, 437.

3
Enzymes

A review has appeared on the quaternary structures of biological macromolecules which included data on the subunit formation of some carbohydrases.[2] Circular dichroism studies of a number of carbohydrases have indicated significant proportions of α-helix in the structures of some.[3] The use of enzymes in analytical chemistry has been the subject of a review which includes a section on the determination of carbohydrates.[260] Further information on methods for the determination of carbohydrases and details of inhibitors of carbohydrases has been published.[261]

Acetamidodeoxyhexosidases

The rates of hydrolysis of (2-acetamido-2-deoxy-β-D-glucopyranosyl)-(1 → 3)-*N*-benzyloxycarbonyl-L-serinamide, (2-acetamido-2-deoxy-β-D-galactopyranosyl)-(1 → 3)-*N*-benzyloxycarbonyl-L-serinamide, (2-acetamido-2-deoxy-β-D-glucopyranosyl)-(1 → 3)-L-serine, and other related glycopeptides by β-acetamidodeoxyglucosidase have been investigated and compared with those of the conventional aryl β-glycosides.[262]

Studies on the inhibition of β-acetamidodeoxyglucosidase activity with 2-acetamido-2-deoxy-D-glucono-1,4- and -1,5-lactones showed that only the latter acted as an inhibitor.[263]

The presence of two isoenzymes (A and B) of β-acetamidodeoxyglucosidase in the leukocytes of normal individuals has been confirmed.[264] A quantitative method, based on acrylamide gel electrophoresis, for the determination of the ratio of the isoenzymes was established. The method was proposed as a screening technique for the identification of Tay–Sachs disease since significant differences between the ratios for normals and carriers were demonstrated. Separation on DEAE-cellulose has also been used for the differential assay of the two isoenzymes.[265] The activity of β-acetamidodeoxyglucosidase was elevated in the serum and liver of patients with Hurler and Sanfilippo syndromes.[266]

260 G. G. Guilbault, *Analyt. Chem.*, 1970, 334R.
261 G. G. Guilbault, 'Enzymatic Methods of Analysis,' Pergamon Press, Oxford, 1970.
262 E. Werries and E. Buddecke, *Z. physiol. Chem.*, 1970, **351**, 1089.
263 T. E. Couling and R. Goodrey, *Biochem. J.*, 1970, **119**, 303.
264 J. Friedland, L. Schneck, A. Saifer, M. Pourfar, and B. W. Volk, *Clin. Chim. Acta*, 1970, **28**, 397.
265 N. Dance, R. G. Price, and D. Robinson, *Biochem. J.*, 1970, **119**, 5P.
266 B. A. Gordon and V. Feleki, *Clin. Biochem.*, 1970, **3**, 193.

The specific activity of β-acetamidodeoxyglucosidase in developing rat brain remained constant,[242] and changes in its specific activity in regenerating rat brain have been investigated.[267]

The capability of a multienzyme system from the marine gastropod *Charonia lampas* to degrade keratan sulphate to its constituent monosaccharides and sulphate was attributed to the presence of a β-acetamidodeoxyglucosidase, among other enzymes.[87]

A method has been described for the isolation of β-acetamidodeoxyhexosidase from jack bean meal in electrophoretically homogeneous and crystalline form.[268] Its molecular weight was estimated to be 10^5 by gel filtration. The enzyme liberated terminal β-linked 2-acetamido-2-deoxy-D-glucose and -galactose from various natural and synthetic substrates, and the data indicated that both activities were catalysed by the same enzyme at the same site.

β-Fructofuranosidases

Both isoelectric focusing and Sephadex gel filtration of yeast β-fructofuranosidase (invertase) gave several active fractions.[269] The K_m values for the hydrolysis of raffinose and methyl β-D-fructofuranoside varied according to the fraction used, whereas the value for the hydrolysis of sucrose showed little variation from fraction to fraction. Other studies employed starch gel electrophoresis for the isoenzyme separation, and showed that the K_m values increased with increasing isoelectric points.[270-272]

Low concentrations of some lower aliphatic alcohols stimulated the induction of β-fructofuranosidase in dormant candida of *Aspergillus oryzae*.[273] Yeast β-fructofuranosidase has been prepared in an insolubilised form,[274] and chicken intestinal β-fructofuranosidase has been investigated.[275]

Fucosidases

The α-fucosidase activity in the liver of cases of Hunter and Sanfilippo syndromes was greater than normal.[266]

Porcine kidney α-L-fucosidase has been obtained free from α-D-mannosidase activity, and its high substrate specificity was demonstrated using a number of synthetic substrates.[38] It was shown that L-fucono-1,4-lactone is a specific inhibitor, and that the enzyme split off terminal L-fucose units from fragments of blood group (A + H) substance of different molecular

267 T. Suematsu, N. Iwabori, and T. Koizumi, *Biochim. Biophys. Acta*, 1970, **201**, 378.
268 S. Li and Y. Li, *J. Biol. Chem.*, 1970, **245**, 5153.
269 B. Berggren, *Arkiv. Kemi*, 1970, **31**, 555.
270 B. Berggren, *Arkiv. Kemi*, 1970, **32**, 143.
271 B. Berggren, *Arkiv. Kemi*, 1970, **32**, 161.
272 B. Berggren, *Arkiv. Kemi*, 1970, **32**, 167.
273 H. Sinohara, *Biochim. Biophys. Acta*, 1970, **224**, 541.
274 H. Filippusson and W. E. Hornby, *Biochem. J.*, 1970, **120**, 215.
275 J. Dautlick and C. F. Strittmatter, *Biochim. Biophys. Acta*, 1970, **222**, 447.

weights. However, it was also shown that porcine kidney α-L-fucosidase was inactive against fragments of molecular weight greater than 4000.[39]

The specific activity of α-fucosidase of rat brain has been shown to decrease as the brain develops.[242] The exofucoidanase activity of the α-L-fucosidase of an abalone (*Haliotis gigantea*) liver has been studied.[276]

An α-L-fucosidase was purified from *Clostridium welchii*.[41a] The enzyme occurred in several isoenzyme forms with molecular weights greater than 2×10^5, and was specific for α-(1 → 2)-linkages in oligosaccharides and glycoproteins, but had no action on simple methyl or nitrophenyl L-fucosides.

A purified preparation of 1,2-α-L-fucosidase, free from other carbohydrase activities, has been obtained from *Aspergillus niger*.[41] Detailed specificity studies showed that the enzyme was highly specific for non-reducing, terminal L-fucose residues linked α-(1 → 2) to D-galactose residues. It hydrolysed L-fucose quantitatively from a series of 2-*O*-α-L-fucosyl saccharides, but was inactive against *p*-nitrophenyl α-L-fucoside.

Galactosidases

o-Nitrophenyl 6-deoxy-α-L-*arabino*-hex-5-enopyranoside acted as a substrate for β-galactosidase.[277] In the presence of glycerol as acceptor, the product of enzymic action was 1′-D-glyceryl 6-deoxy-α-L-*arabino*-hex-5-enopyranoside. It was proposed that the mechanism of enzymic reaction involved protonation of the glycosidic oxygen and transferance of the sugar component as a cyclic unit.

In cases of Morquio–Ullrich's disease, dermal biopsies showed a decreased, and sera an increased, content of β-galactosidase compared with normals.[99] The β-galactosidase activity of liver samples from cases of G_{M1}-gangliosidosis was very low.[141] The normal liver enzyme was stimulated by chloride ions.[278]

Human and rat tissues were shown to contain an enzyme that liberated D-galactose from highly purified keratan sulphate.[88] The enzyme had its maximum activity at pH 4·0, was preferentially localised in the lysosomes, and released D-galactose from the non-reducing end of the keratan sulphate molecule. Rat-kidney lysosomal β-galactosidase has been resolved into four active components by DEAE-cellulose chromatography.[279] Different forms of rat small-intestinal 'acid' β-galactosidase have been separated and their properties investigated.[280]

One of the enzymes of a multienzyme system from *Charonia lampas*, which degraded keratan sulphate, was considered to be a β-galactosidase.[87]

[276] K. Tanaka and S. Sorai, *F.E.B.S. Letters*, 1970, **9**, 45.
[277] J. Lehmann and H. Reinshagen, *Annalen*, 1970, **732**, 112.
[278] M. W. Ho and J. S. O'Brien, *Clin. Chim. Acta*, 1970, **30**, 531.
[279] V. Patel and A. L. Tappel, *Biochim. Biophys. Acta*, 1970, **220**, 622.
[280] N. Asp, *Biochem. J.*, 1970, **117**, 369.

The β-galactosidase activity of microbial fermentation filtrates has been initially purified by precipitation with tungstophosphoric, tungstosilicic, or molybdosilicic acids.[196] Precipitates formed when the pH of the polyacid–enzyme solutions were close to the isoelectric point of the enzyme. The groups combining with the tungstophosphoric acid were the electropositive guanidoyl-, imidazol-, and ε-amino-groups, no precipitation being given by the α-amino-groups.

α- and β-Galactopyranosidases from *Aspergillus niger* were purified from a commercial preparation and their properties investigated.[281] The β-galactosidase activity in extracts of *Trichomonas foetus* was separable into two fractions by gel filtration;[282] one form was found to be an exoglycosidase that also released β-linked D-galactose joined to various aglycones. Its action on terminal, non-reducing β-D-galactosyl residues in oligosaccharides was, however, prevented by a substituent on either the terminal or subterminal residue. β-Galactosidase has been identified in *Kluyoeromyces aestuarii* and *K. wikenii*.[283] *Aerobacter cloacae* β-galactosidase was purified and characterised; it was non-isoenzyme-forming.[219] Its properties, including subunit formation, were compared with those of the β-galactosidase of *Escherichia coli*.

Values in the region of $8{\cdot}5 \times 10^4$ have been obtained for the molecular weight of *E. coli* β-galactosidase in concentrated solutions of guanidine hydrochloride.[284] The same enzyme dissociated on treatment with β-mercaptoethanol with rapid loss of activity.[285] Inactivation and dissociation were prevented by prior or simultaneous addition of magnesium ions, indicating that the latter stabilised the enzyme; treatment with magnesium ions alone caused only a slight initial loss of activity. In the presence of urea, the loss of activity and dissociation of the β-galactosidase into subunits were not parallel reactions, the former being more rapid than loss of the tetrameric structure.[286] A mechanism of conformational changes, forming on one hand a stable, inactive tetramer and inactive subunits on the other, gave the best fit with the experimental data when programmed and simulated by an analog computer.

Studies on the interaction of the inhibitor *o*-mercuriphenyl β-D-galactoside chloride with *E. coli* β-galactosidase demonstrated that, although the compound modified some sulphydryl groups with inactivation, substrate binding in the tetramer was not significantly reduced and no sulphydryl group was involved in the actual catalytic mechanism.[287]

Enhanced stability against heat denaturation of *E. coli* wild type and mutant β-galactosidases was found in the presence of specific antibodies.[288]

[281] Y. C. Lee and V. Wacek, *Arch. Biochem. Biophys.*, 1970, **138**, 264.
[282] G. J. Harrap and W. M. Watkins, *Biochem. J.*, 1970, **117**, 667.
[283] J. B. Fiol. *Ann. Inst. Pasteur*, 1970, **118**, 697.
[284] R. P. Erickson, *Biochem. J.*, 1970, **120**, 255.
[285] S. Shifrin, B. J. Grochowski, and S. W. Luborsky, *Nature*, 1970, **227**, 608.
[286] V. M. Wickson and R. E. Huber, *Biochim. Biophys. Acta*, 1970, **207**, 150.
[287] F. G. Loontiens, K. Wallenfels, and R. Weil, *European J. Biochem.*, 1970, **14**, 138.
[288] F. Melchers and W. Messer, *Biochem. Biophys. Res. Comm.*, 1970, **40**, 570.

Complementation reactions have been carried out on *E. coli* mutant peptides produced by cleavage of the aminoethyl-β-galactosidase with cyanogen bromide.[289] The interaction between metabolic intermediates and β-galactosidase from the organism has been investigated.[290] The biosynthesis of the enzyme in a cell-free system has been studied, and the characteristics of the enzyme defined.[291]

Although the β-glucosidase activity could not be removed from sweet almond emulsin β-galactosidase, it was concluded that the activities belonged to two different enzyme molecules since they were not coincident in their response to inhibitors.[292] α- and β-Galactosidases were extracted from spinach leaves, and characterised after separation by ion-exchange chromatography.[293]

Glucosidases

Nojirimycin (5-amino-5-deoxy-D-glucopyranose) was found to be a potent inhibitor of β-glucosidase.[294]

Raised urinary α-glucosidase output provided an index of human kidney tubular damage arising from various kidney diseases.[295] β-Glucosidase activities in human tissues and in cases of Gaucher's diseases have been investigated.[296]

Cattle-liver lysosomal α-glucosidase was shown to be composed of subunits of similar molecular weight ($2{\cdot}5 \times 10^4$) which were bound non-covalently.[297] Chicken intestinal α-glucosidase has been identified.[275]

α- and β-Glucosidases were found in *Kluyoeromyces aestuarii* and *K. wikenii*,[283] and α-glucosidase synthesis has been induced in *Saccharomyces carlsbergensis*[298] and in *Mucor rouxii*.[299] In the latter case, maltose was a suitable inducer and most of the enzyme was bound to the cell wall.

Extracts of *Neurospora crassa* also contained β-galactosidases strongly bound to the cell wall, but the two enzymes had different pH optima of 7·5 and 4·2.[300] The latter enzyme was purified by precipitation with ammonium sulphate and cation-exchange chromatography and, when the organism was grown on lactose, it was released into the culture medium in two electrophoretically distinguishable forms.[301]

[289] S. Lin, M. Villarejo, and I. Zabin, *Biochem. Biophys. Res. Comm.*, 1970, **40**, 249.
[290] V. Moses and P. B. Sharp, *Biochem. J.*, 1970, **118**, 491.
[291] G. Zubay and D. A. Chambers, *Cold Spring Harbor Symposia on Quantitative Biology*, 1969, **34**, 753.
[292] J. Schwartz, J. Sloan, and Y. C. Lee, *Arch. Biochem. Biophys.*, 1970, **137**, 122.
[293] S. Gatt and E. A. Baker, *Biochim. Biophys. Acta*, 1970, **206**, 125.
[294] T. Niwa, S. Inouye, T. Tsuruoka, Y. Koaze, and T. Niida, *Agric. and Biol. Chem. (Japan)*, 1970, **34**, 966.
[295] P. A. Bonini, G. Ceriotti, and C. Franzini, *Clin. Chem. Acta*, 1970, **27**, 415.
[296] B. Hultberg and P. A. Öckerman, *Clin. Chim. Acta*, 1970, **28**, 169.
[297] C. B. Bruni, V. Sica, F. Auricchio, and I. Covelli, *Biochim. Biophys. Acta*, 1970, **212**, 470.
[298] R. A. de Kroon and V. V. Koningsberger, *Biochim. Biophys. Acta*, 1970, **204**, 590.
[299] A. Flores-Carreon, E. Reyes, and J. Ruiz-Herrera, *Biochim. Biophys. Acta*, 1970, **222**, 354.
[300] H. N. Johnson and A. G. DeBusk, *Arch. Biochem. Biophys.*, 1970, **138**, 408.
[301] H. N. Johnson and A. G. DeBusk, *Arch. Biochem. Biophys.*, 1970, **138**, 412.

Several β-glucosidase-active components from sweet almond emulsin contained β-galactosidase activity, but the different activities were not coincident in their response to inhibitors.[292] Emulsin β-glucopyranosidase has been dissociated into subunits, and subsequent recombination achieved with regeneration of activity.[302] The effect of ferrous ions and organic reducing agents on the thioglucosidase activity of *Crambe abyssinica* seed has been studied.[303]

Glucuronidases

A fluorometric method for the assay of β-glucuronidase was based on the use of harmol glucuronide as substrate.[304] The method was used to assay the β-glucuronidase present in human serum and in the organs and sera of various animals.

The effect of various substituents on the hydrolysis of mono-substituted phenyl β-D-glucosiduronic acids by β-glucuronidase has been investigated.[305] The values obtained for the various rate constants indicated that hydrolysis was facilitated by electron-withdrawing substituents and that its rate was determined by their strength. These facts suggested that enzymic hydrolysis was initiated by the attack of a nucleophile, such as a hydroxy- or amino-group of the enzyme protein (17), and not by protonation of the glycosidic oxygen. The various K_m values indicated that

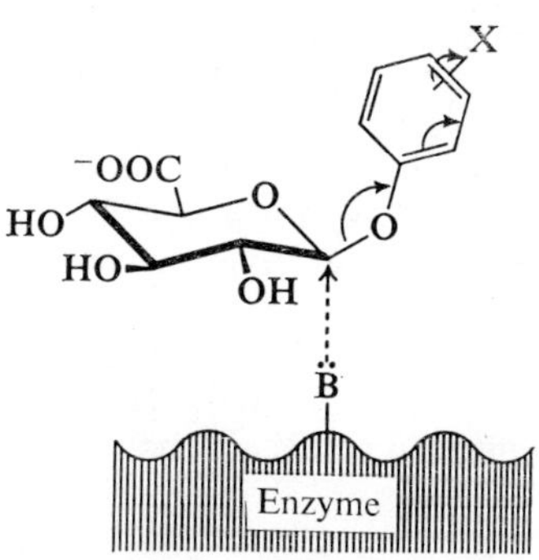

(17)

formation of the enzyme–substrate complex was facilitated by the presence of electron-withdrawing groups on the aromatic ring, implying that the site of interaction in the substrate molecule was located at C-1.

The results of a study on the reversible inhibition of β-glucuronidase showed that the relationship of cholesterol to atherosclerosis is not

302 T. Kleinschmidt, H. Glossman, and J. Horst, *Z. physiol. Chem.*, 1970, **351**, 349.
303 H. L. Tookey and I. A. Wolff, *Canad. J. Biochem.*, 1970, **48**, 1024.
304 K. P. Wong, *Analyt. Biochem.*, 1970, **35**, 35.
305 T. Watanabe and K. Suzuki, *Chem. and Pharm. Bull.* (*Japan*), 1970, **18**, 414.

explained by its action on the enzyme, as had been previously suggested.[306] The β-glucuronidase activity of serum and liver was found to be greater than normal in cases of Hurler and Sanfilippo syndromes.[266]

β-Glucuronidase has been detected in a fractionated homogenate of porcine blood platelets[307] and in caecal contents from germ-free rats.[308] The distribution of the granule-associated enzyme in guinea-pig polymorphonuclear leucocytes has been assessed.[309]

Rat-liver lysosomal β-glucuronidase was fractionated into two active components, both of which had a molecular weight of $2{\cdot}5 \times 10^5$ after purification.[310] Use of synthetic 1-thio-β-D-glucosiduronic acids as substrates for rat-liver β-glucuronidase showed that the enzyme preparation hydrolysed 2-benzothiazolethiol and *p*-nitrothiophenyl β-D-glucuronides but not thiophenyl β-D-glucuronide.[311]

The incorporation of 2-acetamido-2-deoxy-D-[1-^{14}C]glucose into mouse kidney β-glucuronidase by gonadotrophin indicated that the enzyme was a glycoprotein.[312]

Mannosidases

The liver α-mannosidase activity of patients with Hurler and Sanfilippo syndromes was increased above normal.[266] The specific activity of rat-brain α-mannosidase decreased as the brain developed.[242]

α-Mannosidase from the limpet, *Patella vulgata*, was purified 150-fold with 40% recovery.[313] Studies on the purified form showed that the enzyme was a metalloenzyme or enzyme–metal ion complex dissociable at the pH of activity, and that it required zinc ions specifically.

The properties and biosynthesis of *Streptomyces* α-mannosidase (mannosidostreptomycinase) have been reviewed.[314] The specificity of an α-mannosidase isolated from *Streptomyces WI*6 has been defined using a series of mannosides as substrates; the enzyme appears to be an exo-mannanase.[315]

Sweet almond emulsin α-mannosidase was purified by carboxymethyl-cellulose chromatography and isoelectric focusing, and its properties, along with those of the purified emulsin β-mannosidase, were studied.[292]

Sialidases

The intracellular distribution of the sialidase of adult grey matter was determined, and the enzyme found to be concentrated in the synaptosome

[306] J. E. Christner, S. Nand, and N. S. Mhatre, *Biochem. Biophys. Res. Comm.*, 1970, **38**, 1098.
[307] E. Weber, E. Walter, E. Morgenstern, H. Mondt, U. Rose, H. Towliati, and K. Knaudt, *Biochem. Pharmacol.*, 1970, **19**, 1893.
[308] H. Eriksson and J. A. Gustafsson, *European J. Biochemistry*, 1970, **13**, 198.
[309] R. H. Michell, M. J. Karnovsky, and M. L. Karnovsky, *Biochem. J.*, 1970, **116**, 207.
[310] E. Delvin and R. Gianetto, *Biochim. Biophys. Acta*, 1970, **220**, 93.
[311] C. Hétu and R. Gianetto, *Canad. J. Biochem.*, 1970, **48**, 799.
[312] K. Kato, H. Ide, T. Shirahama, and W. H. Fishman, *Biochem. J.*, 1970, **117**, 161.
[313] S. M. Snaith, G. A. Levvy, and A. J. Hay, *Biochem. J.*, 1970, **117**, 129.
[314] A. L. Demain and E. Inamine, *Bacteriol. Rev.*, 1970, **34**, 1.
[315] J. K. Baird and W. L. Cunningham, *Biochem. J.*, 1970, **120**, 18P.

fraction, particularly in the synaptosomal membrane.[316] The sialidase showed activity towards an intrinsic endogenous substrate, as well as towards added disialo- and trisialo-gangliosides and sialyllactose.

The major sialidase of human brain cortex has been obtained by a multistage procedure; its nature and mode of action on sialosyl substrates, particularly the brain gangliosides, were investigated in detail.[317] Sialosyl-(2 → 3)-galactosyl linkages were hydrolysed more readily by the enzyme than were sialosyl-(2 → 8)-sialosyl linkages.

The distribution of sialidase in various regions of bovine brain has been assessed.[232] The presence in rat-liver lysosomes of two neuraminidases having pH optima of 4·4 and 5·8 has been demonstrated.[318] Both enzymes were active against a sialoglycopeptide fraction from a pronase digest of ovine submaxillary glycoprotein, but one of them was inactive against the intact glycoprotein and brain gangliosides. The specific activity of rat-brain sialidase increased as the brain developed.[242]

Neuraminidase from a rough strain of *Diplococcus pneumoniae* has been separated into a number of isoenzymes by column chromatography and electrophoresis.[319]

The action of *Vibrio cholerae* neuraminidase was inhibited by 5-acetamido-6-formyl-4-hydroxy-2,3-dehydropyran-2-carboxylic acid (a periodate oxidation product of methyl 2-deoxy-2,3-dehydro-*N*-acetylneuraminic acid), by its *N*-isonicotinoylhydrazone, and by the reduced form, 5-acetamido-4-hydroxy-6-hydroxymethyl-2,3-dehydropyran-2-carboxylic acid.[320]

Neuraminidase activity was not found in highly concentrated preparations of measles virus, the enzymic activity previously reported being attributed to a latent virus.[321]

α-Amylases

An assay for the specific measurement of α-amylase in the presence of glucoamylase was based on the inability of the latter enzyme to release soluble coloured products from Cibachron Blue amylose.[322] α-Amylase was able to 'by-pass' the substituted D-glucose units with consequent release of coloured material into solution.

A study of the action of α-amylase on maltodextrins, either reduced or oxidised at C-1 of the reducing-end D-glucose unit, showed that such modifications did not alter the ability of dilute solutions of the enzyme to cleave the tetra-, penta-, and hexa-saccharides more rapidly than malto-

[316] C. Schengrund and A. Rosenberg, *J. Biol. Chem.*, 1970, **245**, 6196.
[317] R. Ohman, A. Rosenberg, and L. Svennerholm, *Biochemistry*, 1970, **9**, 3774.
[318] D. R. P. Tulsiani and R. Carubelli, *J. Biol. Chem.*, 1970, **245**, 1821.
[319] S. W. Tanenbaum, J. Gulbinsky, M. Katz, and S. C. Sun, *Biochim. Biophys. Acta*, 1970, **198**, 242.
[320] P. Meinol and H. Tuppy, *Monatsh.*, 1970, **101**, 639.
[321] C. Howe and A. Schluederberg, *Biochem. Biophys. Res. Comm.*, 1970, **40**, 606.
[322] J. J. Marshall, *Analyt. Biochem.*, 1970, **37**, 466.

triose.[323] Such modification of maltotriose, however, rendered the molecule resistant to enzymic attack.

The α-amylase of human pancreatic juice comprised six isoenzymes, the properties of which were compared with those of human parotid saliva α-amylase.[324] The latter enzyme was shown to consist of at least five separable isoenzymes, which could be grouped into two families.[325] The major difference lay in the carbohydrate contents; the members of one family contained 8 and 4 mol of neutral monosaccharide and 2-amino-2-deoxy-D-glucose, respectively, per mol of enzyme, whereas the other contained no detectable carbohydrate.

α-Amylase has been identified in, and extracted from, C_3H mouse submaxillary gland.[326] The extract was purified by Sephadex gel filtration based on the knowledge that α-amylase is retarded when eluted through dextran gels. Refiltration indicated a molecular weight of 6×10^4 for the enzyme, which sedimented as a single peak.

The action pattern on amylose of the α-amylases from porcine pancreas, human saliva, *Bacillus subtilis*, and malted rye has been studied in the initial stages.[327] Porcine pancreatic α-amylase was unique in showing evidence of multiple attack at certain pH values. The presence of glycerol changed the mode of action in the initial stages to an essentially random form.

An investigation of the multiple attack and polarity of action of porcine pancreatic α-amylase established that the optimum pH for multiple attack was 6·9, and that the direction of multiple attack was towards the non-reducing end of the substrate molecule.[328] The information indicated that, after initial attack by the enzyme, the right-hand fragment of the substrate dissociated from the enzyme surface, whereas the left-hand fragment remained at the active site long enough for it to become repositioned and to undergo further attack.

Malto-oligosaccharides, terminated at the reducing ends by [U-^{14}C]-sucrose, were used to investigate the action patterns of pancreatic, *B. subtilis*, and *Endomycopsis* α-amylases.[329] The α-amylases had similar action patterns on oligosaccharides with or without D-fructose at the reducing end, but saliva, malt, *B. stearothermophilus*, *Rhizopus niveus*, *Aspergillus niger*, and Taka α-amylases had somewhat different action patterns.

B. subtilis α-amylase was inactivated by treatment with *N*-acetylimidazole and tetranitromethane,[330] but full catalytic activity was restored

[323] F. W. Parrish, E. E. Smith, and W. J. Whelan, *Arch. Biochem. Biophys.*, 1970, **137**, 185.
[324] B. J. Allan, N. I. Zager, and P. J. Keller, *Arch. Biochem. Biophys.*, 1970, **136**, 529.
[325] D. L. Kauffman, N. I. Zager, E. Cohen, and P. J. Keller, *Arch. Biochem. Biophys.*, 1970, **137**, 325.
[326] L. Frati and A. Caputo, *J. Chromatog.*, 1970, **48**, 547.
[327] W. Banks, C. T. Greenwood, and K. M. Kahn, *Carbohydrate Res.*, 1970, **12**, 79.
[328] J. F. Robyt and D. French, *Arch. Biochem. Biophys.*, 1970, **138**, 662.
[329] S. Okada, S. Kitahata, M. Higashihara, and J. Fukumoto, *Agric. and Biol. Chem. (Japan)*, 1970, **34**, 1407.
[330] J. M. Connellan and D. C. Shaw, *J. Biol. Chem.*, 1970, **245**, 2845.

when the acetylated molecule was treated with hydroxylamine, indicating that tyrosyl residues were required for enzymic activity. Other data supported the evidence that the active physical entity of molecular weight $4{\cdot}8 \times 10^4$ was composed of two equivalent subunits. *B. subtilis* α-amylase was also inactivated by treatment with monoiodoacetate; the reaction with methionyl residues proceeded stoicheiometrically.[331] The decrease of methionyl residues coincided with loss of activity, and the enzyme was completely inactivated when three of the five methionyl residues had reacted.

The characteristics of the α-amylase purified from culture filtrates of *B. stearothermophilus* showed remarkable differences according to the temperature of growth.[332] The distribution of α-amylase-forming ability between membrane and soluble fractions of a cell-free preparation of *B. amyloliquefaciens* was assessed.[333] The contaminant protease of crude bacterial α-amylase, which cleaved the β-chains of tropocollagen to α-chains, could be specifically inhibited with phenylmethylsulphonylfluoride.[58] The α-amylase was then entirely satisfactory for the preparation of polymerised collagen bundles.

Wheat α-amylase was irreversibly inactivated by treatment at pH 2·5 for short periods.[334] Active, insolubilised derivatives of α-amylase have been prepared and their properties investigated.[335, 336]

β-Amylases

The ability of β-amylase to cleave maltotetraose, maltopentaose, and maltohexaose more rapidly than maltotriose was unimpaired by reduction or oxidation of C-1 of the tetra-, penta- and hexa-saccharides.[323] Such modification of maltotriose, however, rendered the molecule immune to attack by β-amylase.

From a study of the kinetics of inactivation of β-amylase, it was proposed that the thiol sites were regulatory entities *in vivo*; this was supported by demonstrating that the enzyme could be reversibly inactivated *via* disulphide interchanges.[337]

The β-amylases of rye and wheat and an alien genome combinant have been compared; the properties of the hybrid enzyme were intermediate between those of the parent species, but the hybrid enzyme had a discrete proteinaceous nature.[338] The relationship between free and latent wheat β-amylases has been investigated by ion-exchange chromatography,[339] and

331 M. Ohnishi and H. Hatano, *Biochim. Biophys. Acta*, 1970, **220**, 350.
332 K. Isono, *Biochem. Biophys. Res. Comm.*, 1970, **41**, 852.
333 G. Coleman, *Biochem. J.*, 1970, **116**, 763.
334 P. Meredith, *Cereal Chem.*, 1970, **47**, 492.
335 S. A. Barker, P. J. Somers, R. Epton, and J. V. McLaren, *Carbohydrate Res.*, 1970, **14**, 287.
336 S. A. Barker, P. J. Somers, and R. Epton, *Carbohydrate Res.*, 1970, **14**, 323.
337 J. Spradlin and J. A. Thoma, *J. Biol. Chem.*, 1970, **245**, 117.
338 W. Y. Lee and A. M. Unrau, *Cereal Chem.*, 1970, **47**, 351.
339 J. E. Kruger, *Cereal Chem.*, 1970, **47**, 79.

the enzyme was inactivated irreversibly by treatment at pH 2·5 for short periods.[334] Active, insolubilised derivatives of β-amylase have been prepared and their properties investigated.[335, 336]

Glucoamylases

Glucoamylase (γ-amylase) has been used for direct analysis of starch.[340] Quantitative enzymic hydrolysis was performed in a DMSO–citrate buffer, and the liberated D-glucose was measured either colorimetrically or by glucose oxidase. Glucoamylase was unable to release soluble coloured products from Cibachron Blue amylose, thus allowing the specific assay of α-amylase in the presence of glucoamylase.[322]

An enzyme purified from rat-liver lysosomal fraction had a molecular weight of $1{\cdot}14 \times 10^5$, and apparently possessed inseparable α-1,6- and α-1,4-glucosidase and transglucosylase activities.[341] Kinetic studies of its action suggested that it had multiple binding sites.[342] Although it was able to catalyse the total hydrolysis of glycogen to D-glucose, debranching of the polysaccharide in the course of hydrolysis was the rate-limiting step. Indirect evidence for the identity of both α-1,6- and α-1,4-activities with a single enzyme was provided by the finding that both α-1,6- and α-1,4-glucosidase activities were simultaneously absent in tissues of children with Type II glycogen storage disease.[343]

Highly purified preparations of glucoamylase obtained from the liver, spleen, and intestine of monkeys have been characterised.[344]

The glucoamylase activity of microbial fermentation filtrates has been initially purified by precipitation with tungstophosphoric, tungstosilicic, or molybdosilicic acids.[196]

Alkaline borohydride treatment of *Aspergillus niger* glucoamylase I showed that the carbohydrate moieties, which contained galactose, glucose, and mannose, were linked by *O*-glycosidic bonds to approximately 45 serine and threonine residues, presumably on the surface of the enzyme molecule.[345] Since the molecule (molecular weight $1{\cdot}1 \times 10^5$) contained 100 monosaccharide units, it appeared that the maximum length of the carbohydrate chains was three units. Oxidation of approximately one-third of the carbohydrate residues with periodate did not affect enzymic activity, but markedly reduced the stability on storage at low temperature. It was suggested that the carbohydrate moieties functioned as stabilisers of the tridimensional structure of the enzyme and, in turn, of the catalytic property of the enzyme.

[340] R. A. Libby, *Cereal Chem.*, 1970, **47**, 273.
[341] P. L. Jeffrey, D. H. Brown, and B. I. Brown, *Biochemistry*, 1970, **9**, 1403.
[342] P. L. Jeffrey, D. H. Brown, and B. I. Brown, *Biochemistry*, 1970, **9**, 1416.
[343] B. I. Brown, D. H. Brown, and P. L. Jeffrey, *Biochemistry*, 1970, **9**, 1423.
[344] B. Seetharam, N. Swaminathan, and A. N. Radhakrishnan, *Biochem. J.*, 1970, **117**, 939.
[345] J. H. Pazur, H. R. Knull, D.-L. Simpson, *Biochem. Biophys. Res. Comm.*, 1970, **40**, 110.

Radioactively-labelled *A. niger* glucoamylases, prepared by growing the organism in media containing either D-[1-^{14}C]glucose- or D-[1-^{14}C]-mannose on purification and periodate oxidation gave products which indicated that the majority of the radioactivity was located in C-1 of the hexose units.[346] Some radioactivity was located in C-6, but from these results it was deduced that the hexoses had largely been incorporated without C—C fission, consistent with a nucleotide–hexose pathway of biosynthesis.

The extracellular glucoamylase of *A. phoenicis*, as purified by DEAE-cellulose chromatography, consisted of two forms, one of which was isolated in a high degree of purity.[347] The optimum conditions for its action were determined. The enzyme had a molecular weight of $6{\cdot}3 \times 10^4$, and contained a total of 17% carbohydrate consisting in part of D-galactose (1·2%), D-glucose (2·8%), and D-mannose (8·0%). The enzyme did not form subunits on treatment with urea.

Data obtained for the mechanism of action of *A. awamori* glucoamylase on maltodextrins, and the inhibition of its action by shortening of the substrate chain-length, have been compared with mathematical models.[348]

An active, insolubilised form of glucoamylase has been used for the continuous degradation of starch.[349]

Dextrin-1,6-glucosidases

A micro-method developed for the determination of dextrin-1,6-glucosidase (amylo-1,6-glucosidase) utilised the ability of the enzyme to incorporate radioactive D-glucose into glycogen, the polymer subsequently being precipitated and counted.[350] The method was specific for dextrin-1,6-glucosidase and was applicable to its detection in tissue extracts and erythrocyte hemolysates. Using this type of assay, it was possible to show the presence of dextrin-1,6-glucosidase in human fibroblasts derived from skin biopsies.[351] The marked decrease of this enzyme activity in cultured fibroblasts from a child with Type III glycogen storage disease indicated the suitability of the method for diagnoses and for prenatal detection of debranching-enzyme deficiency.

According to the results of studies on the action of dextrin-1,6-glucosidase, the enzyme proceeded in both the synthetic and degradative directions with retention of the anomeric configuration.[352] The enzyme synthesised branched products from D-glucose, maltotetraose, and

346 I. J. Russell and D. R. Lineback, *Carbohydrate Res.*, 1970, **15**, 123.
347 D. R. Lineback and W. E. Baumann, *Carbohydrate Res.*, 1970, **14**, 341.
348 K. M. Bendetsky, V. L. Yarovenko, L. N. Ermakova, and L. N. Luk'yanova, *Biokhimiya*, 1970, **35**, 1099.
349 M. J. Bachler, G. W. Strandberg, and K. L. Smiley, *Biotechnol. and Bioeng.*, 1970, **12**, 85.
350 T. E. Nelson and J. Larner, *Analyt. Biochem.*, 1970, **33**, 87.
351 P. Justice, C. Ryan, D. Y. Hsia, and E. Krmpotik, *Biochem. Biophys. Res. Comm.*, 1970, **39**, 301.
352 T. E. Nelson and J. Larner, *Biochim. Biophys. Acta*, 1970, **198**, 538.

cyclohexa-amylose. The reverse action of dextrin-1,6-glucosidase has been further studied using a purified rabbit muscle enzyme, which, on incubation with radioactive D-glucose and maltosaccharides, formed radioactive, branched oligosaccharides.[353] Maltotetraose was the smallest oligomer to act as an acceptor, and the variation of extent of incorporation with pH was determined using glycogen, cycloamyloses, and malto-oligosaccharides of DP up to 13. The linkage of the incorporated D-glucose to the maltosaccharides was shown to be (1 → 6) by partial acid hydrolysis and methylation analysis of the branched heptasaccharide formed from maltohexaose and radioactive D-glucose. The products of incorporation were resistant to pullulanase, indicating that the labelled D-glucose residue was present as a single stub.

Partial amino-acid sequences of porcine pancreatic dextrin-1,6-glucosidase isoenzymes have been determined and compared.[354]

An enzyme system, which debranched glycogen, has been isolated and purified from *Saccharomyces cerevisiae*, and was characterised as a dextrin-1,6-glucosidase.[355] This and the transferase activities of the enzyme were similar to those found previously only in mammalian systems, and were associated with a single species of molecular weight $2{\cdot}8 \times 10^5$, which dissociated into two or three subunits on gel electrophoresis in the presence of sodium dodecyl sulphate.

Dextrin-1,6-glucosidase (molecular weight $9{\cdot}5 \times 10^4$), purified from the culture fluid of *Pseudomonas* sp., hydrolysed all the 1,6-α-D-glucosidic inter-chain linkages in glycogen, amylopectin, and their phosphorylase limit dextrins.[356] The branch points of β-amylase limit dextrins were not hydrolysed completely by the enzyme.

Lysozymes

Lysozyme (muramidase) has been found in human leukocyte extracts.[357] The activity could be separated into two different fractions which had the same pH optima but were activated by different amino-acids.

The levels of lysozyme in human serum and synovial fluid have been determined in order to assess the significance of the enzyme in rheumatoid arthritis.[358] The serum levels of the enzyme in subjects with polycythemia vera were found to be significantly elevated above normal.[359]

A comparison of the complete primary structures of duck egg-white lysozyme II and hen egg-white lysozyme has been made.[360] Although the

353 J. R. Stark and V. Thambyrajah, *Biochem. J.*, 1970, **120**, 17P.
354 P. Cozzone, L. Pasero, B. Beaupoil, and G. Marchis-Mouren, *Biochim. Biophys. Acta*, 1970, **207**, 490.
355 E. Y. C. Lee, J. H. Carter, L. D. Nielsen, and E. H. Fischer, *Biochemistry*, 1970, **9**, 2347.
356 K. Yokobayashi, A. Misaki, and T. Harada, *Biochim. Biophys. Acta*, 1970, **212**, 458.
357 P. Ö. Söder, C. E. Nord, G. Lundblad, and O. Kjellman, *Acta Chem. Scand.*, 1970, **24**, 129.
358 W. Pruzanski, S. Saito, and M. A. Ogryzlo, *Arthritis Rheumat.*, 1970, **13**, 389.
359 R. A. Binder and H. S. Gilbert, *Blood*, 1970, **36**, 228.
360 J. Hermann and J. Jollès, *Biochim. Biophys. Acta*, 1970, **200**, 178.

amino-acid sequences of hen egg lysozyme and bovine α-lactalbumin exhibited considerable homology, the two molecules had quite different molecular conformations in solution.[361] The lysozyme underwent very little change of conformation on dissolution.

The thermodynamics of the temperature dependence of the denaturation of lysozyme by guanidine hydrochloride,[362] and the self-association of the enzyme[363] have been studied. The kinetics of lysozyme–substrate interactions were investigated using chitin oligosaccharides as substrates.[364] Using *Micrococcus lysodeikticus* cells as substrate, goose and hen egg-white lysozymes were found to have different enzyme–substrate interaction characteristics.[365] The former enzyme was absorbed onto the substrate, whereas the latter was not.

Incubation of the bacterial cell-wall tetrasaccharide (18) with hen egg-white lysozyme in the presence of D-glucose lead to the formation of a trisaccharide (19) and a pentasaccharide (20).[366] The dependence of the formation of the two new oligosaccharides on the incubation time and D-glucose concentration was accounted for by a transglycosylation mechanism in which the cell-wall tetrasaccharide (18), or the corresponding hexasaccharide (21) formed from it, served as sources for the formation of the disaccharide donor (22) and tetrasaccharide donor (23), respectively. The overall scheme proposed for the lysozyme-catalysed reactions is shown (Scheme 6). The lysozyme-catalysed transfer of bacterial cell-wall oligosaccharide to forty other mono- and di-saccharides was investigated, and the findings were in agreement with the three-dimensional lysozyme–substrate model (24). Contacts a—e, h, and i had already been proposed on the basis of model building. The structures of the trisaccharides formed when D-xylose, D-galactose, and 2-acetamido-2-deoxy-D-glucose were used as acceptors in the lysozyme-catalysed, transglycosylation reaction of cell-wall tetrasaccharide (18) were (25), (26), and (27), respectively.[367]

The interaction of lysozyme with a series of low molecular weight inhibitors and modified substrates containing (1 → 4)- and (1 → 6)-*O*-(2-acetamido-2-deoxy-β-D-glucose) bonds, *viz.* 2-acetamido-2-deoxy-D-glucose, *O*-(2-acetamido-2-deoxy-β-D-glucopyranosyl)-(1 → 4)-2-acetamido-2-deoxy-D-glucose (*NN'*-diacetylchitobiose), *O*-(2-acetamido-2-deoxy-β-D-glucopyranosyl)-(1 → 4)-*O*-(2-acetamido-2-deoxy-β-D-glucopyranosyl)-(1 → 4)-2-acetamido-2-deoxy-D-glucose (*NN'N"*-triacetylchitotriose), *p*-nitrophenyl 2-acetamido-2-deoxy-β-D-glucopyranoside, *p*-nitrophenyl *O*-(2-acetamido-2-deoxy-β-D-glucopyranosyl)-(1 → 4)-2-acetamido-2-

361 W. R. Krigbaum and F. R. Kügler, *Biochemistry*, 1970, **9**, 1216.
362 C. Tanford and K. C. Aune, *Biochemistry*, 1970, **9**, 206.
363 R. C. Deonier and J. W. Williams, *Biochemistry*, 1970, **9**, 4260.
364 E. Holler, J. A. Rupley, and G. P. Hess, *Biochem. Biophys. Res. Comm.*, 1970, **40**, 166.
365 A.-C. Dianoux and P. Jollès, *Bull. Soc. Chim. biol.*, 1969, **51**, 1559.
366 J. J. Pollock and N. Sharon, *Biochemistry*, 1970, **9**, 3913.
367 J. J. Pollock and N. Sharon, *Carbohydrate Res.*, 1970, **13**, 211.

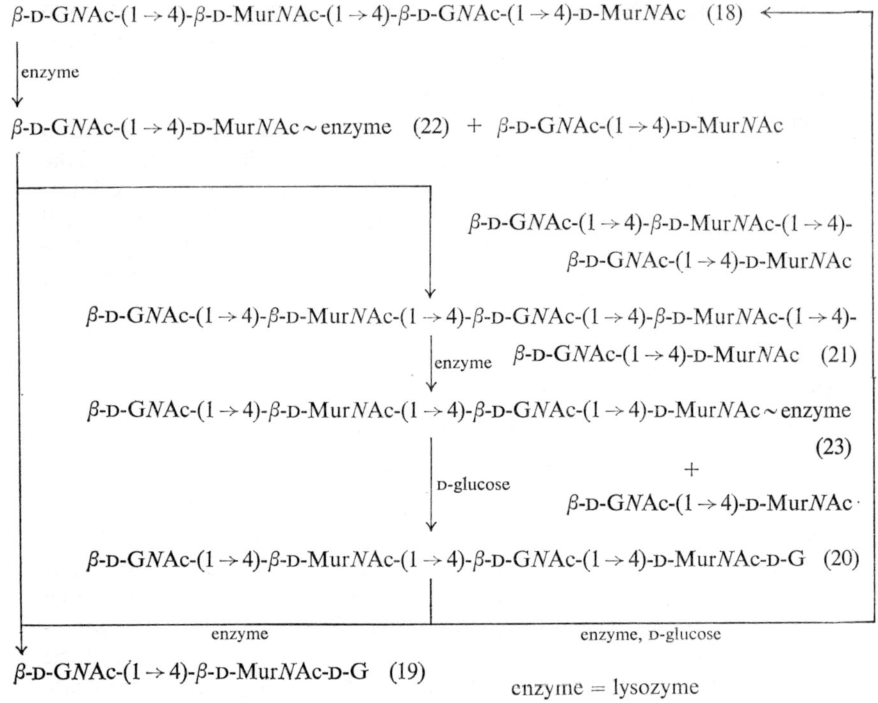

Scheme 6

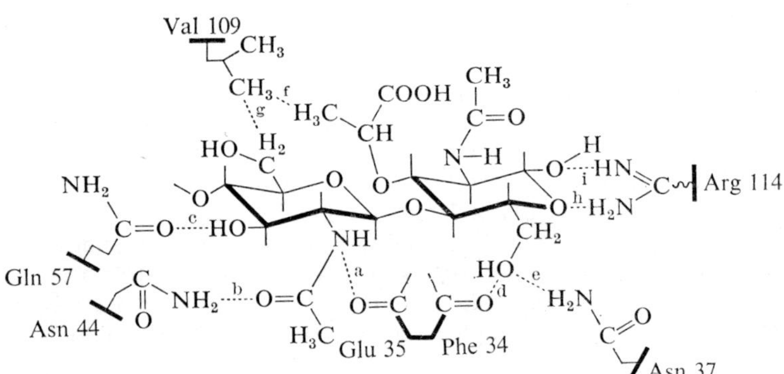

(24)

O-β-D-GNAc-(1 → 4)-O-β-D-Mur$N$Ac-(1 → 4)-D-Xyl (25)

O-β-D-GNAc-(1 → 4)-O-β-D-Mur$N$Ac-(1 → 2)-D-Gal (26)

O-β-D-GNAc-(1 → 4)-O-β-D-Mur$N$Ac-(1 → 4)-D-GNAc (27)

deoxy-β-D-glucopyranoside (A), *p*-nitrophenyl *O*-(2-acetamido-2-deoxy-β-D-glucopyranosyl)-(1 → 6)-2-acetamido-2-deoxy-β-D-glucopyranoside (B), *p*-nitrophenyl *O*-(2-benzamido-2-deoxy-β-D-glucopyranosyl)-(1 → 6)-2-acetamido-2-deoxy-β-D-glucopyranoside, and *p*-nitrophenyl *O*-(2-acetamido-2-deoxy-β-D-glucopyranosyl)-(1 → 4)-*O*-(2-acetamido-2-deoxy-β-D-glucopyranosyl)-(1 → 6)-2-acetamido-2-deoxy-β-D-glucopyranoside (C) has been investigated.[368] The concentrations inducing 50% inhibition of the lysis of *M. lysodeikticus* cell walls were determined, and these along with other parameters allowed quantitative estimation of the affinity of the molecules for the active site of the enzyme. It was shown for the biosides A and B that the position of the glycosidic bonds determined the affinity of the substrate for the active site, whereas trioside C possessed strong affinity for the active site.

The interaction site of the *N*-methylnicotinamide chloride–hen egg-white lysozyme complex has been analysed by means of chemical modifications to the enzyme and by competition for complex formation by carbohydrate inhibitors.[369] The binding sites were identified and compared with those of bovine α-lactalbumin. Acetylated allyl β-glycosides of 2-acetamido-2-deoxy-D-glucose, *O*-(2-acetamido-2-deoxy-β-D-glucopyranosyl)-(1 → 4)-2-acetamido-2-deoxy-D-glucose, and *O*-(2-acetamido-2-deoxy-β-D-glucopyranosyl)-(1 → 4)-*O*-(2-acetamido-2-deoxy-β-D-glucopyranosyl)-(1 → 4)-2-acetamido-2-deoxy-D-glucose were specific inhibitors of lysozyme, their action being irreversible.[370]

The retention of enzymic and immunological properties of lysozyme has been studied using a series of reversible blocking reagents for the amino-group.[371] Modified lysozyme molecules have also been used in a partial determination of the immunological determinants of the enzyme.[372] Antibodies from rabbits sensitized with crystalline lysozyme reacted with the reduced and *S*-carboxymethylated forms as well as with the natural form.[373] Inhibition of these cross-reactions was investigated using tryptic peptides of *S*-carboxymethyl-lysozyme.

A method involving Sephadex and Bio-gel chromatography has been developed for the isolation of the most heavily damaged part of γ-irradiated lysozyme.[374] The method minimised the secondary changes of protein structure occurring after irradiation.

Other Polysaccharidases and Oligosaccharidases

Cellulases.—The cellulase production of the fungus *Myrothecium verrucaria* was regulated by the carbohydrate uptake of the organism, but neither

368 E. A. Shashkova, I. V. Vikha, A. A. Vichutinsky, and A. Y. Khorlin, *Biokhimiya*, 1970, **35**, 124.
369 R. A. Bradshaw and D. A. Deranleau, *Biochemistry*, 1970, **9**, 3310.
370 E. W. Thomas, *Carbohydrate Res.*, 1970, **13**, 225.
371 A. F. S. A. Habeeb and M. Z. Atassi, *Biochemistry*, 1970, **9**, 4939.
372 A. D. Strosberg and L. Kanarek, *European J. Biochem.*, 1970, **14**, 161.
373 J. D. Young and C. Y. Leung, *Biochemistry*, 1970, **9**, 2755.
374 V. I. Maksimov and V. A. Mosin, *J. Chromatog.*, 1970, **47**, 361.

cellulose nor other carbohydrate was required *per se* for enzyme production.[375] Cellulolytic fungi grown on D-glucose as the sole carbon source effected a small but definite depolymerisation of carboxymethylcellulose; depolymerase activity increased considerably at the time of exhaustion of D-glucose from the medium. It was concluded from the various studies that the cellulases were produced in response to a strict regulation of metabolism rather than by the inductive influence of a specific medium additive. Hence, they seemed to be constitutive enzymes, the production of which could be repressed by an abundance of any carbon source capable of being utilised rapidly.

Chitinases.—Melanin was an inhibitor of the chitinase activity of a multi-polysaccharidase system from *Streptomyces* sp., which degraded hyaline and melanised cell walls of *Aspergillus nidulans*.[376] The role of chitinase and other lysozymal enzymes of *Coprinus lagopus* in the autolysis of the fruiting bodies has been investigated.[377]

The extracellular, bacteriolytic activity found in the cultures of *Staphylococcus aureus* was purified by isoelectric focusing, and the enzyme had a molecular weight of 7×10^4 by gel filtration.[378] The properties and mode of action of this enzyme, which was produced by several strains of *S.aureus*, were investigated and the results suggested that it was not a lysozyme, but an endo-acetamidodeoxyglucosidase.[379, 380]

Dextranases.—Dextranase activity has been found associated with the coleoptiles of *Avena sativa*; previously the enzyme had been found only in micro-organisms and animals.[381] The enzyme was sensitive to the plant-growth hormone auxin, and it was considered to be an important factor in the plasticisation of the cell wall in auxin-regulated cell elongation in coleoptiles.

Galactanases.—The galactanase produced by *Phytophthora infestans* degraded a 1,4-linked β-D-galactose polymer present in potato and lupin pectins. The enzyme rapidly reduced the viscosity of solutions of the polysaccharide with concomitant release of reducing groups and D-galactose in monomeric or oligomeric form.[382] The enzyme released total carbohydrate more rapidly than D-galactose from potato cell walls, and appeared on the basis of the results to be an endo-glycanohydrolase having negligible β-galactosidase activity.

Heparin Lyases and Heparan Sulphate Lyases.—A crude enzyme preparation obtained from heparin-induced *Flavobacteria* has been fractionated into

375 M. A. Hulme and D. W. Stranks, *Nature*, 1970, **226**, 469.
376 A. T. Bull, *Arch. Biochem. Biophys.*, 1970, **137**, 345.
377 W. Iten and P. Matile, *J. Gen. Microbiol.*, 1970, **61**, 301.
378 T. Wadström and K. Hisatsune, *Biochem. J.*, 1970, **120**, 725.
379 T. Wadström and K. Hisatsune, *Biochem. J.*, 1970, **120**, 735.
380 T. Wadström, *Biochem. J.*, 1970, **120**, 745.
381 A. N. J. Heyn, *Biochem. Biophys. Res. Comm.*, 1970, **38**, 831.
382 M. Knee and J. Friend, *J. Gen. Microbiol.*, 1970, **60**, 23.

a heparan sulphate lyase, which acted on heparan sulphate and related compounds, and a heparin lyase, which acted mainly on heparin.[86] Both enzymes were purified to one hundred times the purity of previous preparations, and again acted as eliminases rather than as true hydrolases yielding products containing 4-deoxy-L-*threo*-hex-4-enosyluronic acid residues. The specificity of the heparan sulphate lyase appeared to be such that it required the presence of acetamido- or sulphamido-groups and the absence of *O*-sulphate groups. The specificity of the heparin lyase required the presence of *O*-sulphate and sulphamido-groups, but derivatives containing acetamido- or free amino-groups were not substrates. The heparin lyase degraded heparan sulphate to a limited extent apparently by acting on the heparin-like portions of the structure.

Hyaluronidases and Hyaluronate Lyases.—The hyaluronidase activity in the livers of patients with Hurler and Sanfilippo syndromes was increased above the normal value.[266]

A novel hyaluronate lyase, an eliminase, was isolated from *Streptomyces hyalurolyticus* nov. sp.[383] The enzyme had unusual substrate specificities, including its inability to act on other glycosaminoglycuronans.

Laminarinases.—Laminarinase has been isolated from the crystalline style of a member of the bivalvia, *Spisula sachalinensis*, by gel filtration and sulphoethyl–Sephadex chromatography.[384] Tests with a range of substrates showed that the enzyme had specific laminarinase activity and hydrolysed laminaribiose at a very slow rate. Melanin was an inhibitor of the laminarinase activity of a multi-enzyme system prepared from *Streptomyces* sp., which degraded hyaline and melanised cell walls of *Aspergillus nidulans*.[376]

Oligo-1,6-glucosidases.—A study of the oligo-1,6-glucosidases (limit dextrinases) of cereals showed their ability to release (1 → 6)-linked α-maltosyl or α-maltotriosyl units from oligosaccharides (*e.g.* α-maltotriosyl-(1 → 6)-maltotriose and maltosyl-cycloamyloses) and certain polysaccharides (*e.g.* pullulan and amylopectin β-limit dextrin).[385] In this respect, the enzymes were similar to the bacterial pullulanases; however, they differed significantly from the latter in having no action on amylopectin, glycogen, or glycogen β-limit dextrin. The development of an oligo-1,6-glucosidase during cereal germination has been studied.[386]

Polygalacturonases and Polygalacturonate Lyases.—Polygalacturonase formation in *Erwinia aroidae* was stimulated by an active factor in carrot

383 T. Ohya and Y. Kaneko, *Biochim. Biophys. Acta*, 1970, **198**, 607.
384 V. V. Sova, L. A. Elyakova, and V. E. Vaskovsky, *Biochim. Biophys. Acta*, 1970, **212**, 111.
385 D. J. Manners, J. J. Marshall, and D. Yellowlees, *Biochem. J.*, 1970, **116**, 539.
386 D. J. Manners and D. Yellowlees, *Biochem. J.*, 1970, **117**, 22P.
387 H. Tomizawa, K. Izaki, and H. Takahashi, *Agric. and Biol. Chem.* (*Japan*), 1970, **34**, 1064.

extracts;[387] the partial purification of this factor and the stimulating effect of other compounds have been described.

A polygalacturonate lyase isolated from cultures of *Bacillus sphaericus* released, *via* an eliminative mechanism, oligomers containing terminal, non-reducing 4-deoxy-L-*threo*-hex-4-enosyluronic acid residues from Vi antigen, a surface polysaccharide formed by some members of the Enterobacteriaceae.[388] The enzyme was also able to degrade de-*O*-acetylated Vi antigen, pectin, and polygalacturonic acid, the end products being analogous to those formed from the Vi antigen. The enzyme appeared to be a single species. The Vi antigen was also attacked by an apparently true hydrolase produced by a strain of *B. polymyxa* induced with pectin.

Pullulanases.—The specificity of a pullulanase from *Aerobacter aerogenes* has been tested using a number of oligosaccharides.[389] Hydrolysis was confined to cleavage of α-(1 → 6)-interchain links between oligosaccharide chains containing a minimum of two D-glucose units per chain. Starch oligosaccharides and dextrins containing two or more α-(1 → 6)-links were also cleaved, but in no case was D-glucose observed as a product of enzymic action. A single (1 → 6)-linked α-D-glucose 'stub' did not interfere with pullulanase action, and oligosaccharides containing such residues were found as products. Under specific conditions, pullulanase preparations exhibited condensation and transferase properties.

Trehalases.—Plasma trehalase activity is elevated above control levels in cases of diabetes mellitus.[390]

Xylanases.—The production of xylanase activity by *Streptomyces xylophagus* has been investigated.[391]

Carbohydrate Oxidases

Glucose Oxidases.—The anomeric specificity of the glucose oxidase reaction has been studied by n.m.r. spectroscopy using D-glucose, 2-deoxy-D-*arabino*-hexose, and D-*arabino*-hexosulose as substrates.[392] Although the presence of the enzyme caused substantial line broadening in the spectra of the carbohydrates, the results indicated that, in all three cases, the β-pyranose form was preferentially oxidised during the reaction. That the signal diminution was a result of enzyme catalysis was evident from the fact that β-signals reappeared at the expense of α-signals on heat-inactivation of the enzyme. The experiments also indicated that the reactions of 2-deoxy-D-*arabino*-hexose and D-*arabino*-hexosulose were stereochemically equivalent to that of D-glucose. Chromatographic analysis of the products of enzymic

388 L. A. McNicol and E. E. Baker, *Biochemistry*, 1970, **9**, 1017.
389 M. Abdullah and D. French, *Arch. Biochem. Biophys.*, 1970, **137**, 483.
390 L. C. Eze and D. A. P. Evans, *Clin. Chim. Acta*, 1970, **28**, 153.
391 T. Kawaminami and H. Iizuka, *Agric. and Biol. Chem.* (*Japan*), 1969, **33**, 1787.
392 M. S. Feather, *Biochim. Biophys. Acta*, 1970, **220**, 127.

action upon 2-deoxy-D-*arabino*-hexose and D-*arabino*-hexosulose showed them to be 2-deoxy-D-*arabino*-hexonic acid and 2-keto-D-gluconic acid, respectively; *i.e.* compounds analogous to D-glucono-1,5-lactone, which was the product obtained from D-glucose.

Insolubilised derivatives of glucose oxidase have been prepared using nickel–nickel oxide [393] and polystyrene [394] as the support.

Galactose Oxidases.—It emerged from an investigation of a number of compounds as substrates for galactose oxidase, that dihydroxyacetone had an activity relative to D-glucose of 1·1.[395]

Peptidases

Four proteolytic enzymes isolated from autolysing *Saccharomyces carlsbergensis* were found to be glycoproteins, the carbohydrate moieties containing glucose and mannose residues in the ratios 1 : 2, 1 : 1, 2 : 1, and 2 : 1, respectively. The same yeast secreted four glycoprotein proteases into the growth medium, one of which contained mannose as the only carbohydrate. The other three contained glucose and mannose (3 : 1, 1 : 3, and 1 : 3).[396]

The composition and structure of the carbohydrate prosthetic group of the proteolytic enzyme pineapple-stem bromelain have been investigated using the glycopeptide obtained from it by pronase digestion.[397] This glycopeptide contained D-mannose, L-fucose, D-xylose, and 2-acetamido-2-deoxy-D-glucose (3 : 1 : 1 : 2). Two of the D-mannose residues, the L-fucose residue and the D-xylose residue, were specifically liberated by the action of α-D-mannosidase, α-L-fucosidase, and β-D-xylosidase, respectively. β-Acetamidodeoxyglucosidase released 2-acetamido-2-deoxy-D-glucose from the (neutral sugar)–free glycopeptide. The results of partial acid hydrolysis, periodate oxidation, Smith degradation, and other studies were consistent with the structure (28) for the glycopeptide.

α-L-Fuc β-D-Xyl
1 1
↓ ↓
6 ?
α-D-M*p*-(1 → 2)-α-D-M*p*-(1 → 2 or 6)-α-D-M*p*-(1 → ?)-β-D-G*pN*Ac-
(1 → 3 or 4)-β-D-G*pN*Ac-Asn-peptide

(28)

Guinea-pig serum and liver L-asparaginases appeared to be glycoproteins, and immunological studies were carried out on them.[398, 399]

[393] H. H. Weetall and L. S. Hersh, *Biochim. Biophys. Acta*, 1970, **206**, 54.
[394] W. E. Hornby, H. Filippusson, and A. McDonald, *F.E.B.S. Letters*, 1970, **9**, 8.
[395] G. T. Zancan and D. Amaral, *Biochim. Biophys. Acta*, 1970, **198**, 146.
[396] I. S. Maddox and J. S. Hough, *Biochem. J.*, 1970, **117**, 843.
[397] Y. Yasuda, N. Takahashi, and T. Murachi, *Biochemistry*, 1970, **9**, 25.
[398] H. M. Suld and P. A. Herbut, *J. Biol. Chem.*, 1970, **245**, 2797.
[399] H. M. Suld and P. A. Herbut, *J. Biol. Chem.*, 1970, **245**, 2802.

Ribo- and Deoxyribo-nucleases

α-Mannosidase treatment of the asparaginyl-oligosaccharide derived by trypsin and carboxypeptidase treatment of bovine pancreatic juice ribonuclease B yielded a product containing asparagine, 2-acetamido-2-deoxy-D-glucose, and mannose (1 : 2 : 1).[400] The remaining mannose residue was enzyme resistant, but could be removed by Smith degradation, a procedure which did not affect the 2-acetamido-2-deoxy-D-glucose residues. The product was shown to be *O*-(2-acetamido-2-deoxy-β-D-glucopyranosyl)-(1 → 4)-2-acetamido-1-[(*N*-β-L-aspartyl)amino]-2-deoxy-β-D-glucopyranosylamine. The mannose residue appeared to be linked to the 3-position of the terminal, non-reducing 2-acetamido-2-deoxy-β-D-glucopyranosyl residue, rather than to the 4-position.

From studies carried out on a series of glycopeptide fractions from porcine pancreatic ribonuclease, the three sites of polysaccharide attachment were recognised to be at positions 21, 34, and 76 of the 124 amino-acid chain, and were designated sites I, II, and III.[401] Aspartic acid occupied each of the three positions, and it was presumed that the carbohydrate–peptide bonds were 2-acetamido-1-[(*N*-β-L-aspartyl)amino]-2-deoxy-D-glucosylamine linkages. The oligosaccharides attached to the three sites had notably different compositions. The side-chain at site II contained 2-acetamido-2-deoxyglucose (2) and mannose (6 residues), and was homologous with the single oligosaccharide side-chain in bovine ribonuclease B with respect to composition and site of attachment. The side-chains at sites I and III were about twice as large, more complex, and contained galactose, fucose, and *N*-glycolylneuraminic acid. The three sites were recognised to be at residues having external side-chains in regions of the molecule remote from the active site. All the sites were in sequences of the type (29), where at

```
   carbohydrate
   |
-Asn-A-(Ser or Thr)-
```

(29)

sites I and III, A = serine, and at site II, A = methionine. From these data it was concluded that the nature of A exerted a determinative influence on the manner and complexity in which the oligosaccharide chain was elaborated during biosynthesis of the glycoprotein.

Porcine pancreatic ribonuclease P contained mannose (7·2%), galactose (2·2%), fucose (1·2%), and 2-acetamido-2-deoxyglucose (10·5%).[402] Pronase and trypsin digestion gave several glycopeptides, and it was concluded that the ribonuclease structure had at least two carbohydrate chains branching

400 A. Tarentino, T. H. Plummer, and F. Maley, *J. Biol. Chem.*, 1970, **245**, 4150.
401 R. L. Jackson and C. H. W. Hirs, *J. Biol. Chem.*, 1970, **245**, 624.
402 T. Tsuruo, S. Yamashita, T. Terao, and T. Ukita, *Biochim. Biophys. Acta*, 1970, **200**, 544.

from the protein backbone. The glycopeptide linkages were isolated and identified as 2-acetamido-1-[(*N*-β-L-aspartyl)amino]-2-deoxy-β-D-glucosylamine.

Whale pancreatic ribonuclease W_2 contained mannose (1·4%), galactose (1·8%), fucose (1·0%), 2-amino-2-deoxyglucose (3·1%), and 2-amino-2-deoxygalactose (0·7%), and had a molecular weight of $1{\cdot}5 \times 10^4$.[403] A glycopeptide isolated from a pronase digest of the nuclease contained aspartic acid, serine, and threonine (1 : 1 : 1), and had aspartic acid as the *N*-terminus. 2-Acetamido-1-[(*N*-β-L-aspartyl)amino]-2-deoxy-β-D-glucosylamine was isolated from the glycopeptide after hydrolysis.

The extracellular nuclease of *Micrococcus sodonensis* had a molecular weight of 5×10^5 and consisted of 21% carbohydrate.[404] Glucose, galactose, rhamnose, and 2-amino-2-deoxyglucose were identified as the component monosaccharides (2 : 1 : 1 : 4). A 2-amino-2-deoxyglucosyl–serine glycopeptide linkage, involving at least 80% of the serine residues, was demonstrated by β-carbonyl elimination. The carbohydrate moiety of the purified enzyme was related both chemically and immuno-logically to the cell-wall carbohydrate.

The earlier resolution of preparations of bovine pancreatic deoxyribonuclease into active fractions by sulphoethyl–Sephadex chromatography has been improved using columns of phosphocellulose.[405] Gradient elution with sodium acetate (pH 4·7) showed the presence of at least four isoenzymes which all possessed comparable specific activities. Pancreatic juice contained deoxyribonucleases which were chromatographically indistinguishable from the three major forms isolated from tissue. The three main fractions (A—C) present in the pancreatic juice, molar proportions 4 : 1 : 1, were glycoproteins having *N*-terminal leucine, *C*-terminal threonine, and carbohydrate attached at a single position. Deoxyribonucleases A and B had the following sugar composition: A, mannose (6 residues), 2-acetamido-2-deoxyglucose (2); B, galactose (1), mannose (5), 2-acetamido-2-deoxyglucose (3), and sialic acid (1). Enzyme C was similar to A in that it possessed a neutral carbohydrate moiety, mannose (5), 2-acetamido 2-deoxyglucose (2), but contained one less histidine residue and one more proline residue than enzymes A and B.

Miscellaneous Enzymes

N-Acetylneuraminate lyase isolated from beef-kidney cortex was not inhibited by 3-deoxy-3-fluoro-*N*-acetylneuraminic acid and, in contrast to the *N*-acetylneuraminate lyase from *Clostridium welchii*, the compound was not a substrate.[406]

[403] T. Tsuruo, K. Sudo, T. Terao, and T. Ukita, *Biochim. Biophys. Acta*, 1970, **200**, 560.
[404] S. A. Berry, K. G. Johnson, and J. N. Campbell, *Biochim. Biophys. Acta*, 1970, **220**, 269.
[405] J. Salnikow, S. Moore, and W. H. Stein, *J. Biol. Chem.*, 1970, **245**, 5685.
[406] D. A. Sirbasku and S. B. Binkley, *Biochim. Biophys. Acta*, 1970, **206**, 479.

A glucosylamidase, isolated from the snail *Limnaea stagnalis*, was highly specific, cleaving the amide bond in glycopeptides containing 2-acetamido-1-[(*N*-β-L-aspartyl)amino]-2-deoxy-D-glucosylamine units in which the amino- and carboxy-groups of the aspartic acid residue were free. The activity was separated into two fractions having different pH optima.[407]

Pectinesterase was extracted from the pulp of tomato fruit, *Lycopersicum esculentum*, and obtained in ultracentrifugally and disc electrophoretically pure form.[408] Compositional studies showed that several amino-acids were absent.[409] The mode of action of the pectinesterase was studied using poly-(methyl galacturonate) methyl glycoside as substrate.[410] Determination of the amount of enzymic de-esterification occurring at the 'reducing-ends' of the substrate chains, using a polygalacturonase produced by *Clostridium multifermentans*, showed that half of the pectinesterase activity was initiated near the reducing-ends of highly esterified pectin molecules.

It was possible to replace the *N*-acetylneuraminic acid units of neuraminidase-treated sheep-brain alkaline phosphatases using CMP–*N*-acetylneuraminic acid.[411] The original, desialylated, and re-sialylated enzymes had the same activities and kinetics.

A sulphamidase has been isolated from lymphoid tissues of man, dogs, and rats, and purified by gel electrophoresis and filtration.[85] It catalysed the hydrolysis of a single *N*-sulphate residue from heparin and from hexa- and tetra-saccharides originating from the degradation of heparin by bacterial enzymes. In contrast to bacterial sulphamidase, the enzyme had no action upon 2-deoxy-2-sulphamido-D-glucose 6-sulphate, and was also inactive against *N*-sulphated disaccharides and several sulphated glycosamino-glycuronans.

A multi-enzyme system capable of degrading keratan sulphates to D-galactose, 2-acetamido-2-deoxy-D-glucose, and inorganic sulphate was found in the liver of *Charonia lampas*.[87] The concerted action of the enzyme system was attributed in part to a sulphatase, designated as 'keratosulphatase'. The enzyme could be separated into two forms by DEAE–Sephadex chromatography; both forms released all the sulphate from keratan sulphate, but neither was identical with the glycosulphatase or chondrosulphatase present in *Ch. lampas*.

Bovine-plasma monoamine oxidase contained galactose (0·85%), mannose (1·51%), 2-amino-2-deoxyglucose (1·09%), and sialic acid (0·84%).[412] A number of carbohydrate fragments containing carbohydrate, obtained by

[407] E. D. Kaverzneva, A. I. Tschuchrova, and V. V. Kisseleva, *Annalen*, 1970, **738**, 130.
[408] H. Nakagawa, Y. Yanagawa, and H. Takehana, *Agric. and Biol. Chem.* (*Japan*), 1970, **34**, 991.
[409] H. Nakagawa, Y. Yanagawa, and H. Takehana, *Agric. and Biol. Chem.* (*Japan*), 1970, **34**, 998.
[410] M. Lee and J. D. Macmillan, *Biochemistry*, 1970, **9**, 1930.
[411] S. Sarawathi and B. K. Bachhawat, *Biochim. Biophys. Acta*, 1970, **212**, 170.
[412] K. Watanabe and K. T. Yasunobu, *J. Biol. Chem.*, 1970, **245**, 4612.

pronase digestion of the phenylhydrazone derivative of the oxidase, were separated by ion-exchange chromatography and gel filtration to give a major component comprising asparagine (1), D-mannose (5), and 2-acetamido-2-deoxy-D-glucose (2 residues). The susceptibility of this component to glucosylamidase indicated a glycosylamine-type linkage between asparagine and 2-acetamido-2-deoxy-D-glucose. Three residues of D-mannose, one residue of 2-acetamido-2-deoxy-D-glucose, and the two remaining residues of D-mannose, in this order, were released on successive treatment with Jack bean α-mannosidase, β-acetamidodeoxyglucosidase, and α-mannosidase.

α-Lactalbumin, one of the two component enzymes of lactose synthetase, had a similar amino-acid sequence to hen egg muramidase, but assumed a different molecular conformation in solution.[361] The binding sites of the macromolecule from a bovine source have been compared with those of lysozyme.[369] G.l.c. of the trimethylsilyl derivatives of the methyl glycosides released on methanolysis of bovine α-lactalbumin showed that all preparations contained D-galactose, D-mannose, 2-acetamido-2-deoxy-D-glucose, and 2-acetamido-2-deoxy-D-galactose;[413] the total carbohydrate content ranged from 0·25 to 1·5%.

Pig-heart citrate synthetase contained hexose (2%), but it was doubted if this represented a functional part of the enzyme due to the failure to detect any 2-amino-2-deoxyhexose.[414]

Modification of the visible spectrum of the amylopectin–iodine complex by dextrin glycosyltransferase has been used to assay the enzyme.[415]

The monosaccharides obtained from heparin by treatment with heparin-induced *Flavobacterium heparinum* enzymes did not include D-glucuronic acid.[83] Accordingly, a uronic acid isomerase, a constitutive enzyme of *F. heparinum*, was isolated and characterised. The enzyme converted D-glucuronic acid and D-galacturonic acid into D-fructuronic and D-tagaturonic acids, respectively. Monosaccharides derived from heparin and chondroitin sulphates by incubation with *F. heparinum* enzymes did not affect the activity of the isomerase, whereas the monosaccharides from hyaluronic acid markedly inhibited the enzyme.

An enzyme fraction from the seaweed *Gigantina stellata* effected formation of 3,6-anhydride units in κ-carrageenan.[416] This enzyme appeared to be capable of reducing the 'kinks' in the polysaccharide chains, thereby increasing chain rigidity, and of effecting metabolic control of the polysaccharide's conformation and function.

[413] E. J. Hindle and J. V. Wheelock, *Biochem. J.*, 1970, **119**, 14P.
[414] M. Singh, G. C. Brooks, and P. A. Srere, *J. Biol. Chem.*, 1970, **245**, 4636.
[415] T. E. Nelson, D. H. Palmer, and J. Larner, *Biochim. Biophys. Acta*, 1970, **212**, 269.
[416] C. J. Lawson and D. A. Rees, *Nature*, 1970, **227**, 392.

Index of Enzymes Referred to in Chapter 3*

Trivial Name	*Systematic Name*	*EC No.*	*Page*
β-Acetamidodeoxy-glucosidase	β-2-acetamido-2-deoxy-D-glucoside acetamidodeoxy-glucohydrolase	3.2.1.30	222
β-Acetamidodeoxy-galactosidase	β-2-acetamido-2-deoxy-D-galactoside acetamidodeoxy-galactohydrolase		
N-Acetylneuraminate lyase	*N*-acetylneuraminate pyruvate-lyase	4.1.3.3	243
Alkaline phosphatase	orthophosphoric monoester phosphohydrolase	3.1.3.1	244
α-Amylase	α-1,4-glucan 4-glucanohydrolase	3.2.1.1	229
β-Amylase	α-1,4-glucan maltohydrolase	3.2.1.2	231
L-Aspariginase	L-asparagine amidohydrolase	3.5.1.1	241
Bromelain		3.4.4.24	241
Cellulase	β-1,4-glucan 4-glucanohydrolase	3.2.1.4	237
Chitinase	poly-β-1,4-(2-acetamido-2-deoxy)-D-glucoside glycanohydrolase	3.2.1.14	238
Chondrosulphatase	chondroitin-sulphate sulpho-hydrolase	3.1.6.4	244
Citrate synthetase	citrate oxaloacetate-lyase (CoA-acetylating)	4.1.3.7	245
Deoxyribonuclease	deoxyribonucleate oligonucleo-tidohydrolase	3.1.4.5	242
Dextranase	α-1,6-glucan 6-glucanohydrolase	3.2.1.11	238
Dextrin-1,6-glucosidase	dextrin 6-glucanohydrolase	3.2.1.33	238
Dextrin glycosyltrans-ferase	α-1,4-glucan:α-1,4-glucan 4-glycosyltransferase		245
β-Fructofuranosidase	β-D-fructofuranoside fructohydrolase	3.2.1.26	223
α-Fucosidase	α-L-fucoside fucohydrolase		223
Galactanase			238
Galactose oxidase	D-galactose:oxygen oxidoreductase	1.1.3.9	241
α-Galactosidase	α-D-galactoside galactohydrolase	3.2.1.22	224
β-Galactosidase	β-D-galactoside galactohydrolase	3.2.1.23	224
Glucoamylase	α-1,4-glucan glucohydrolase	3.2.1.3	232
Glucose oxidase	β-D-glucose:oxygen oxidoreductase	1.1.3.4	240
α-Glucosidase	α-D-glucoside glucohydrolase	3.2.1.20	226
β-Glucosidase	β-D-glucoside glucohydrolase	3.2.1.21	226
Glucosylamidase	2-acetamido-1-[(*N*-β-L-aspartyl)amino]-2-deoxy-β-D-glucosylamine amidohydrolase		244
β-Glucuronidase	β-D-glucuronide glucurono-hydrolase	3.2.1.31	227
Glycosulphatase	sugar-sulphate sulphohydrolase	3.1.6.3	244
Heparan sulphate lyase			238
Heparin lyase	heparin glycanohydrolase	3.2.1.19	238
Hyaluronate lyase	hyaluronate lyase	4.2.99.1	239

* See Introduction. (Part II, Chapter 1)

Index of Enzymes Referred to in Chapter 3 (*cont.*)

Trivial Name	*Systematic Name*	*EC No.*	*Page*
Hyaluronidase	hyaluronate glycanohydrolase	3.2.1.35	239
Keratosulphatase			244
α-Lactalbumin			245
Laminarinase	β-1,3(4)-glucan glucanohydrolase	3.2.1.6	239
Lysozyme	mucopeptide *N*-acetylmuramyl-hydrolase	3.2.1.17	234
α-Mannosidase	α-D-mannoside mannohydrolase	3.2.1.24	228
β-Mannosidase	β-D-mannoside mannohydrolase	3.2.1.25	228
Monoamine oxidase	monoamine:oxygen oxidoreductase (deaminating)	1.4.3.4	244
Neuraminidase	mucopolysaccharide *N*-acetyl-neuraminyl hydrolase	3.2.1.18	228
Oligo-1,6-glucosidase	oligodextrin 6-glucanohydrolase	3.2.1.10	239
Pectinesterase	pectin pectyl-hydrolase	3.1.1.11	244
Peptidases			241
Polygalacturonase	poly-α-1,4-galacturonide glycanohydrolase	3.2.1.15	239
Polygalacturonate lyase	poly-α-1,4-D-galacturonide lyase	4.2.99.3	239
Pullulanase			240
Ribonuclease	ribonucleate pyrimidinenucleotido-2′-transferase (cyclising)	2.7.7.16	242
Sialidase†			228
Sulphamidase			244
Trehalase	trehalose 1-glucohydrolase	3.2.1.28	240
Uronic acid isomerase			245
Xylanase	β-1,4-xylan xylanohydrolase	3.2.1.8	240

† Sialidase is used generally to include neuraminidase and unspecified neuraminate glycohydrolases.

4
Glycolipids and Gangliosides

A recent review[417] on glycosphingolipids (sugar sphingosine conjugates) included sections on glycosphingolipid nomenclature, stereochemical aspects of sphingosines and phytosphingosines, natural and synthetic sphingosines, dihydrosphingosines and phytosphingosines of different chain lengths, the biosynthesis of sphingosines and glycosphingolipids, ceramide mono- and oligo-saccharides, sulphatides, glycosphingolipids of plant materials and micro-organisms, and gangliosides.

T.l.c. has been used for the separation of glycolipids from lipid extracts of animal tissues (*e.g.* rat spleen) into different classes *viz.* ceramide glucose and galactose mono-, di-, tri-, and tetra-hexosides, sulphatides, and mono-galactosyl diglycerides.[418] Ceramide monogalactosides (galactocerebrosides), ceramide monoglucosides (glucocerebrosides), and ceramide dihexosides (lactosides) from bovine brain sphingolipids, and ceramide tri- and tetra-hexosides from human erythrocyte stroma were also investigated using t.l.c.

It has been demonstrated that the losses of monosaccharides in hydrolyses of glycolipids with hydrochloric acid are due to transglycosylation (Maillard) reactions occurring during evaporation.[10] Detection of such undesirable reactions was achieved by inclusion of D-[^{14}C]glucose in the hydrolysate, which was then investigated by paper chromatography and autoradiography.

Data have been presented which indicate that the Lewis Le^{a} and Le^{b} antigens in human plasma are glycosphingolipids.[419] Active purified preparations contained a mixture of glycosphingolipids, acid hydrolysis of which gave fucose, galactose, glucose, and 2-amino-2-deoxyglucose.

Monosialoganglioside, the end-product of neuraminidase action on the intrinsic ganglioside substrate of adult grey matter, remained associated with the membrane fragments.[316]

Two major groups of glycolipids have been isolated from human tumour tissue: group 1 contained fucose and group 2 contained sialic acid.[420] Fractionation of group 1 on Biosil A gave two major fractions A and B, which were subfractionated on silica gel plates into A_{1-5} and B_{1-2}, A_1 and

[417] J. Kiss, *Adv. Carbohydrate Chem.*, 1970, **24**, 382.
[418] N. M. Neskovic, J. L. Nussbaum, and P. Mandel, *J. Chromatog.*, 1970, **49**, 255.
[419] D. M. Marcus, *Ann. New York Acad. Sci.*, 1970, **169**, 161.
[420] S. I. Hakomori and H. D. Andrews, *Biochim. Biophys. Acta*, 1970, **202**, 225.

B_1 having the highest mobilities. A_1, the major component of A, did not exhibit H, Le^a, or Le^b activities, whereas A_3 and A_4 were highly Le^a-active in precipitin and agglutination tests; the latter fractions had similar carbohydrate compositions and released a lacto-*N*-fucopentaose on cleavage of the glycoside–lipid bond. Fraction A_2 had a similar carbohydrate composition to that of A_3 and A_4, while fraction A_5 was highly active in Le^b precipitin tests and hemagglutination inhibition. Methylation studies suggested that A_5 had the structure (30). Fraction B_2 yielded another Le^b-

Fuc-(1 → 2)-Gal-(1 → 3)-G*N*Ac-(1 → 3)-Gal-(1 → 4)-G-ceramide
4
↑
1
Fuc

(30)

active glycolipid, and showed a high inhibition of Le^b hemagglutination but only a faint line with anti-Le^b serum. The carbohydrate sequence and partial structure (31) were indicated by degradation studies, methylation,

Fuc-(1 → 2)-Gal-(1 → 3)-G*N*Ac-(1 → 3)-Gal-(1 → ?)-G*N*Ac-(1 → 3)-Gal-
4 (1 → 4)-G-ceramide
↑
1
Fuc

(31)

and cross-reactions. Blood-group A and B activities were not detected in these glycolipid fractions, whereas those from normal glandular tissues (intestinal tract and pancreas) contained significant quantities of blood-group A or B glycolipids. These results established that both Le^a- and Le^b-glycolipids are present in human adrenocarcinoma tissue regardless of the Lewis blood type of the tumour donor. Liver samples from cases of G_{M1}-gangliosidosis Type 1 contained increased amounts of G_{M1}-ganglioside.[141]

The gangliosides from human spleen and bovine spleen and kidney have been extracted and their carbohydrate moieties characterised.[421] Those from both bovine locations contained the same sialo-oligosaccharide moieties, and, in all, eleven bovine and five human gangliosides were identified. They were derived from *O*-β-D-galactosyl-(1 → 4)-D-glucose (lactose), *O*-(2-acetamido-2-deoxy-β-D-galactosyl)-(1 → 4)-*O*-β-D-galacto-syl-(1 → 4)-D-glucose, *O*-β-D-galactosyl-(1 → 3)-*O*-(2-acetamido-2-deoxy-β-D-galactosyl)-(1 → 4)-*O*-β-D-galactosyl-(1 → 4)-D-glucose, *O*-β-D-galactosyl-(1 → 3)-*O*-(2-acetamido-2-deoxy-β-D-glucosyl)-(1 → 3)-*O*-β-D-galactosyl-(1 → 4)-D-glucose, and *O*-β-D-galactosyl-(1 → 4)-*O*-(2-acetamido-2-deoxy-β-D-glucosyl)-(1 → 3)-*O*-β-D-galactosyl-(1 → 4)-D-glucose, and differed in their sialic acid contents. Gangliosides from human tissue

[421] H. Weigandt and H. W. Bücking, *European J. Biochem.*, 1970, **15**, 287.

contained *N*-acetylneuraminic acid only, whereas those from bovine tissue contained *N*-acetylneuraminic acid or *N*-glycolylneuraminic acid or both. Human normal and cataractous lens, and bovine lens and iris contained a ganglioside composed of sialic acid, glucose, galactose, and ceramide (1 : 1 : 1 : 1).[422] The human lens contained, in addition, a second component comprising sialic acid, glucose, galactose, 2-amino-2-deoxygalactose, and ceramide (1 : 1 : 2 : 1 : 1). No differences were detected between the ganglioside compositions of normal and cataractous human lens.

Ceramide mono- and di-hexosides have been isolated from cows' milk.[423] The structures of these glycolipids were established by methylation and partial hydrolysis as β-D-glucosyl-(1 → 1)-*N*-acylsphingosine [ceramide glucoside, (32)] and β-D-galactosyl-(1 → 4)-β-D-glucosyl-(1 → 1)-*N*-acylsphingosine [ceramide lactoside, (33)]; the D-configurations of the hexoses

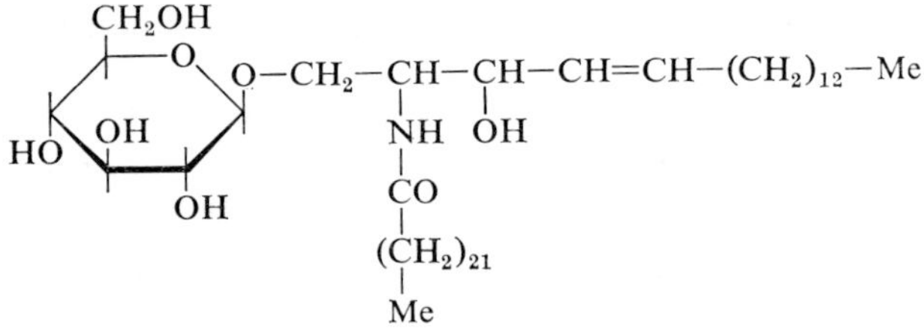

(32)

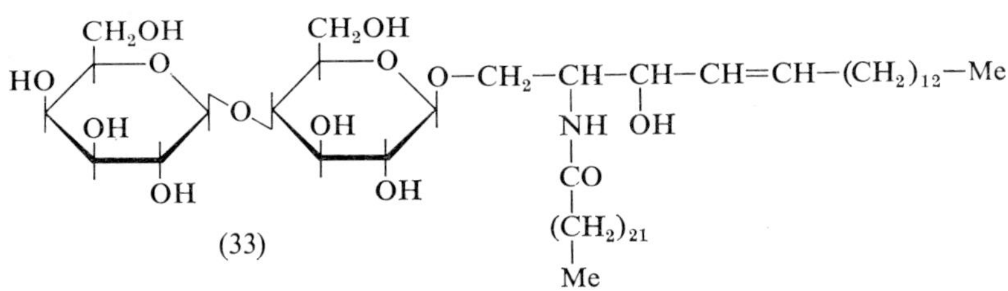

(33)

were assumed. Studies on the J blood-group substance of cattle suggested that the determinant group is a glycosphingolipid.[424] Gangliosides from various bovine brain regions have been determined, and their distribution was assessed by measurement of the sialic acid.[232] A controlled procedure for the quantitative extraction and isolation of five pure gangliosides from bovine kidney has been described.[425] Two of these gangliosides (72% of the total) were hematosides, a third (22%) was a dineuraminyl-dihexoside, whereas the other two (together 6%) were more complex

[422] A. S. Windeler and G. L. Feldman, *Biochim. Biophys. Acta*, 1970, **202**, 361.
[423] Y. Fujino, M. Nakano, and T. Saeki, *Agric. and Biol. Chem. (Japan)*, 1970, **34**, 442.
[424] O. W. Thiele and J. Koch, *European J. Biochem.*, 1970, **14**, 379.
[425] K. Puro, *Acta Chem. Scand.*, 1970, **24**, 13.

polysialogangliosides. Calf retinal gangliosides have been separated, and were shown to contain *N*-acetylneuraminic acid.[426]

Porcine plasma and erythrocytes were found to contain, in similar concentrations, the same glycosphingolipids as human sources, namely glucosyl ceramide, lactosyl ceramide, and galactosyl-galactosyl-glucosyl-ceramide.[427] *In vivo* labelling indicated the site of synthesis of these molecules and showed that the release or catabolism of erythrocyte glycosphingolipids was the major source of all four plasma glycosphingolipids.

The composition and synthesis of glycolipids and glycoproteins in isolated synaptosomes and synaptosomal intraneural mitochondria from guinea-pig cerebral cortex have been determined by the incorporation of labelled monosaccharides.[237] Sialic acid, fucose, hexose, and 2-amino-2-deoxyhexose were identified as components.

Glycolipids, containing fucose, galactose, mannose, 2-amino-2-deoxy-hexose, and sialic acid, from mouse fibroblast (L cell) and its surface membrane have been investigated.[238] The glycolipids of the fibroblast (L cell) were isolated and separated into four classes: ceramide lactoside (20% of total), mono- and di-sialogangliosides (38%), and hematosides (42%).[428] The sequences, determined after partial hydrolysis with acid, were galactosyl-(2-amino-2-deoxy-galactosyl)-galactosyl-glucosyl-ceramide for the mono- and di-sialogangliosides, and galactosyl-glucosyl-ceramide for the hematosides, which also contained *N*-acetylneuraminic acid and *N*-glycolylneuraminic acid. Only the disialogangliosides and hematosides were identified in the cell-surface membrane. The effect of testosterone on the composition and biosynthesis of mouse-kidney glycosphingolipids has been investigated.[429]

An investigation of the biosynthesis of rat-liver mannolipid showed that retinol (vitamin A) stimulated the incorporation of [^{14}C]mannose from GDP-[^{14}C]mannose.[249] Enzymes involved in the synthesis of gangliosides in developing rat brain have been studied;[243] *N*-acetylgalactosaminyl-(*N*-acetylneuraminyl)-galactosyl-glucosyl-ceramide galactosyltransferase has been located in rat brain.[430]

About half of the gangliosides from various fish brains consisted of the ceramide tetra- and penta-*N*-acetylneuraminylganglio-*N*-tetraosides.[431] They differed from the brain gangliosides of warm-blooded animals in the occurrence of 8-*ON*-diacetylneuraminic acid and the preponderance of tetra- and penta-sialogangliosides.

[426] D. Kostic, P. F. Urban, B. Lemieux, and P. Mandel, *Bull. Soc. Chim. biol.*, 1969, **51**, 1632.

[427] G. Dawson and C. C. Sweeley, *J. Biol. Chem.*, 1970, **245**, 410.

[428] D. B. Weinstein, J. B. Marsh, M. C. Glick, and L. Warren, *J. Biol. Chem.*, 1970, **245**, 3928.

[429] J. B. Hay and G. M. Gray, *Biochim. Biophys. Acta*, 1970, **202**, 566.

[430] G. B. Yip and J. A. Dain, *Biochim. Biophys. Acta*, 1970, **206**, 252.

[431] I. Ishizuka, M. Kloppenburg, and H. Wiegandt, *Biochim. Biophys. Acta*, 1970, **210**, 299.

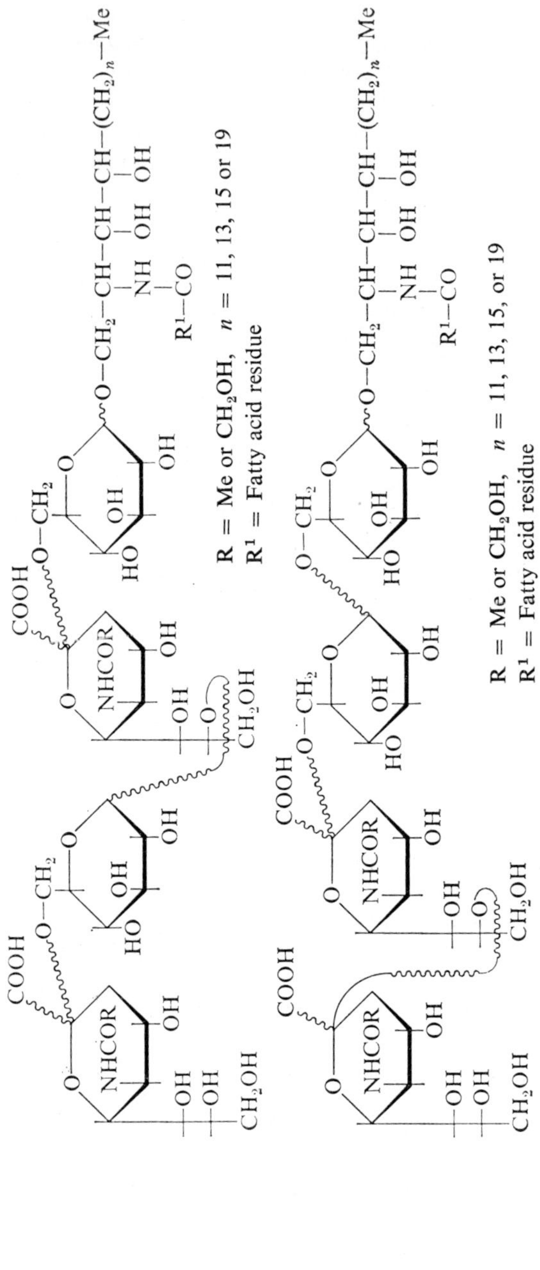

R = Me or CH_2OH, n = 11, 13, 15 or 19
R^1 = Fatty acid residue

R = Me or CH_2OH, n = 11, 13, 15, or 19
R^1 = Fatty acid residue

(35)

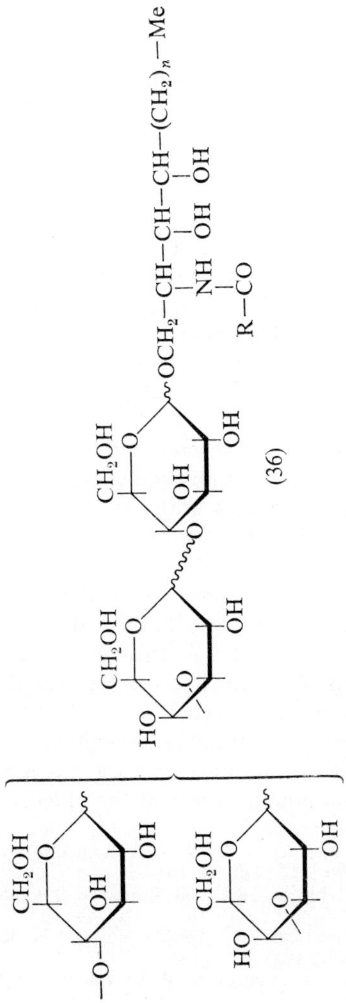

(36)

R = Fatty acid residue, n = 11, 13, 15, or 19

The monosaccharide contents of lipid extracts of fifty species of marine invertebrates have been determined.[432] Glycolipids were found in all of the species, for which Spongia and Echinodermata had the highest and Coelenterata and Arthropoda the lowest contents. The glycolipids were characterised by t.l.c. as acylated cerebroside-like lipids, cerebroside-like lipids, and glycolipids of middle and high polarity, and were shown to contain hexose, 6-deoxyhexose, pentose, and unidentified monosaccharides. Gangliosides were found only in Echinodermata. A new sialoglycolipid, containing ceramide, *N*-acetylneuraminic acid, and D-glucose, has been isolated from the sea urchin *Strongylocentrotus intermedius*, and was suggested from hydrolysis, periodate oxidation, and methylation studies to have either structure (34) or (35).[433] The long-chain bases of the sialoglycolipid were shown to contain phytosphingosine and dehydrophytosphingosine, and the compositions of the fatty acids acylating the amino-group of the sphingosine base were determined.[434] A sialoglycolipid isolated from the digestive gland of the starfish *Distolasterias nippon*, by preparative t.l.c. of the lipid extract, contained phytosphingosine, D-glucose, D-galactose, and sialic acids (1 : 2 : 2 : 2·6), the latter being a mixture of *N*-acetylneuraminic acid and an unidentified sialic acid.[435] Partial hydrolysis yielded lactosyl ceramide, while methylation studies showed that the glycolipid had linear oligosaccharide chains with both D-glucose residues being substituted at C-4 and both D-galactose residues being substituted at C-3 (36).[436]

The positions of the acyl groups in two partly acylated glycolipids from strains of *Ustilago* (smut fungi) were located using a series of reactions which involved replacement of the acyl groups with methyl groups, and subsequent identification of the positions of the methyl groups in the sugar residues after acid hydrolysis.[437]

Investigations of a glucose-containing phosphoglyceride of *Mycoplasma laidlawii* indicated a structural similarity to a diglycosyl-diglyceride containing glycerophosphate bound to one of the glucose units.[438]

[432] V. E. Vaskovsky, E. I. Kostetsky, V. I. Svetashev, I. G. Zhukova, and G. P. Smirnova, *Comparative Biochem. Physiol.*, 1970, **34**, 163.

[433] N. K. Kochetkov, G. P. Smirnova, and I. G. Zhukova, *Doklady Akad. Nauk S.S.S.R.*, 1970, **193**, 344.

[434] I. G. Zhukova, G. P. Smirnova, I. S. Glukhoded, and N. K. Kochetkov, *Doklady Akad. Nauk S.S.S.R.*, 1970, **192**, 563.

[435] I. G. Zhukova, T. A. Bogdanovskaya, G. P. Smirnova, and N. K. Kochetkov, *Biokhimiya*, 1970, **35**, 775.

[436] N. K. Kochetkov, I. G. Zhukova, G. P. Smirnova, and T. A. Bogdanovskaya, *Doklady Akad. Nauk S.S.S.R.*, 1970, **191**, 358.

[437] S. S. Bhattacharjee, R. H. Haskins, and P. A. J. Gorin, *Carbohydrate Res.*, 1970, **13**, 235.

[438] H. M. Verheij, P. F. Smith, P. P. M. Bonsen, and L. L. M. Van Deenen, *Biochim. Biophys. Acta*, 1970, **218**, 97.

5

Chemical Synthesis and Modification of Polysaccharides *etc.*

Synthesis of Polysaccharides, Oligosaccharides, and Glycopeptides

Polymerisation of 1,6-anhydro-β-D-mannopyranose in the presence of chloroacetic acid gave a high molecular weight, ethanol-insoluble fraction which, on the basis of periodate oxidation studies, contained (1 → 6)-like linkages (42%), (1 → 2)- or (1 → 4)-like linkages (36%), and (1 → 3)-like linkages (22%). End-group analysis gave $\overline{DP} = 113$. The polymer reacted vigorously with concanavalin A, indicating that it was a highly branched polymer containing multiple α-D-mannopyranosyl residues at chain-ends.[439]

The Lewis acid-catalysed polymerisation of 1,6-anhydro-2,3-di-*O*-benzyl-4-*O*-(2,3,4,6-tetra-*O*-benzyl-α-D-glucopyranosyl)-β-D-glucopyranose (hexabenzyl-1,6-anhydromaltose) led to products of number-average molecular weight up to $1{\cdot}4 \times 10^4$ and specific rotations up to + 97°; these parameters were considerably affected by variations in the reaction conditions.[440] Debenzylation of the products having the highest molecular weights and optical rotations yielded hydrated, comb-shaped polysaccharides in which the rotational values corresponded to a high degree of stereoregularity in the main chain and predominantly α-D-linkages in both the main and side-chains. Similar Lewis acid-catalysed polymerisation of 1,6-anhydro-2,3-di-*O*-benzyl-4-*O*-(2,3,4,6-tetra-*O*-benzyl-β-D-glucopyranosyl)-β-D-glucopyranose (hexabenzyl-1,6-anhydrocellobiose) gave products of number-average molecular weight up to 7×10^3 and specific rotation up to + 80°.[441] The effect of variation of reaction conditions on the characteristics of the polymer was investigated, and the conditions giving maximum stereoregularity and molecular weight were similar to those observed for the corresponding maltose derivative. Debenzylation of the products yielded non-dialysable, hydrated, comb-shaped polysaccharides, analysis of which indicated the presence of 1·5 moles of water per repeating disaccharide unit; the same value was obtained for the water-content of the maltose derivative.

[439] R. Robinson and I. J. Goldstein, *Carbohydrate Res.*, 1970, **13**, 425.
[440] O. Veruovic and C. Schuerch, *Carbohydrate Res.*, 1970, **14**, 199.
[441] V. Masura and C. Schuerch, *Carbohydrate Res.*, 1970, **15**, 65.

Methods for the preparation of β-1,4-linked oligosaccharides of 2-acetamido-2-deoxy-D-glucose (chitin oligosaccharides) with DP 2—6 have been described.[442] Comparable products were obtained with the two routes employed: *viz.* acetolysis of chitin followed by de-*O*-acetylation, and hydrolysis of chitin with 2N-HCl, but the latter route was recommended on account of its greater convenience. Good separations of the oligosaccharide products were obtained using Sephadex LH20 columns, whilst use of a previously recommended medium (charcoal–Celite) proved unsatisfactory due to slow flow rates and the reverse order of elution of the oligosaccharides.

Malto-oligosaccharides, terminated by [U-^{14}C]sucrose, have been prepared by incubation of cyclohexa-amylose (α-cyclodextrin) and [U-^{14}C]sucrose with *Bacillus macerans* transferase.[329]

A di-*N*-acetylchitobiose-asparagine derivative, 2-acetamido-4-*O*-(2-acetamido-2-deoxy-β-D-glucopyranosyl)-1-[(*N*-β-L-aspartyl)amino]2-deoxy-β-D-glucopyranosylamine (38) has been synthesised from α-chitobiose octa-acetate (37) (Scheme 7).[443]

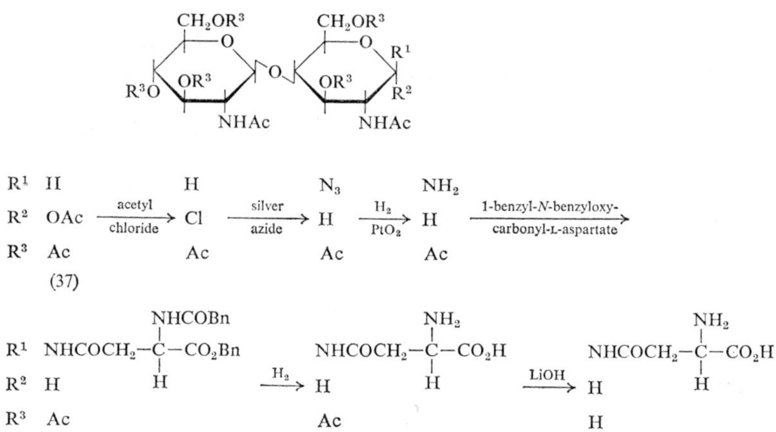

(38)

Scheme 7

Modification of Polysaccharides and Uses of Modified Polysaccharides

A review of graft polymerisation onto polysaccharides covers methods of initiation, copolymerisation reaction methods, solid- and solution-state grafting, and characterisation.[444] Considerable attention is given to grafting onto starch and cellulose and their derivatives.

[442] B. Capon and R. L. Foster, *J. Chem. Soc.* (*C*), 1970, 1654.
[443] M. Spinola and R. W. Jeanloz, *J. Biol. Chem.*, 1970, **245**, 4158.
[444] J. C. Arthur, *Adv. Macromol. Chem.*, 1970, **2**, 1.

A dichlorotriazine dye, CCAB10B, has been coupled to laminarin, amylopectin, dextran, sargassan, pelvetian, pectin, zosterine, and a galacturonan from zosterine pre-treated with diazomethane.[445] This process of polysaccharide modification was used for following these polysaccharides during gel electrophoresis and gel filtration.

Ivory nut mannan B and cellulose have been oxidised with nitrogen tetroxide to transform the C-6 hydroxy-groups into carboxy-groups. Both products contained keto-groups which were formed as a result of side-reactions.[446] Reduction of the oxidised polysaccharides with sodium borotritide and determination of the incorporated radioactivity indicated that the ratio of keto-groups in the oxycellulose to those in the oxymannan was 2 : 1. The oxymannan contained one keto-group for every ten anhydro-D-mannose units.

Cotton cellulose, modified with *N*-methylbis-(2-chloroethyl)amine, incorporated 0·034 mol of reagent per D-glucopyranosyl unit. Hydrolysis of the product gave D-glucose, a series of substituted D-glucoses {*i.e.* 2-*O*-, 3-*O*-, and 6-*O*-[(*N*-methyl-*N*-2-hydroxyethylamino)ethyl]-D-glucopyranoses}, and a series of simple cross-linked D-glucoses {*i.e.* *N*-methylaminobis-[*O*-(*N*-2-ethyl)-D-glucopyranoses]} with D-glucose linkages at 2,2′-, 6,6′-, and 2,6′-positions only.[447] These structures were confirmed by synthesis, and the relative distribution of (*N*-methyl-*N*-2-hydroxyethylamino)ethyl groups among the 2-, 3-, and 6-positions was 2·30 : 0·28 : 1·00. The relative distribution of cross-linkages among the 2,2′-, 6,6′-, and 2,6′-positions was 2·10 : 3·72 : 1·00. The monofunctionally-substituted fraction, having a single reagent unit in the substituent group, accounted for 42·6% of the nitrogen, whereas the cross-linked fraction, having a single reagent unit in the cross-link, accounted for 25·8%; complex products accounted for the remainder of the nitrogen of the cross-linked cellulose.

Cellulose furoates (39) were prepared by the reaction of purified cotton cellulose with 2-furoyl chloride, 5-bromo-2-furoyl chloride, 5-methyl-2-furoyl chloride, and 2-furanacryloyl chloride in pyridine as the acid

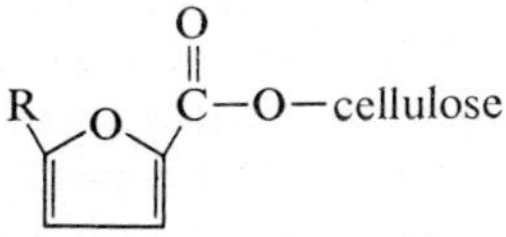

(39)

scavenger and DMF as the diluent.[448] The bromo-substituted cellulose furoates underwent nucleophilic displacement reactions with pyridine, *NN*-dimethylcyclohexylamine, and triethylamine to give the corresponding salts; possible mechanisms for the displacement reactions were discussed.

[445] A. F. Pavlenko and Y. S. Ovodov, *J. Chromatog.*, 1970, **52**, 165.
[446] C. Mercer and H. I. Bolker, *Carbohydrate Res.*, 1970, **14**, 109.
[447] E. J. Roberts and S. P. Rowland, *Canad. J. Chem.*, 1970, **48**, 1383.
[448] S. Singh and J. C. Arthur, *Carbohydrate Res.*, 1970, **14**, 73.

The temperature dependence of radiation-induced grafting of styrene to cellulose, the monomer consumption, and the replacement of styrene with sixteen different styryl monomers have been studied.[449] For the system cellulose–styrene–methanol irradiated in air, additives (*e.g.* pyridine, acetone, and stilbene) showed inhibition of styrene grafting at low concentrations but sensitisation at high concentrations.[450]

The swelling of cellulose in aqueous zinc chloride solutions depends on a mechanism involving the formation of a complex with the vicinal hydroxy-groups on C-2 and C-3 of the D-glucopyranose repeating unit.[451]

Modified cellulose acetates have been formed by acetylating *O*-(carboxymethyl)-, *O*-nitro-, and *O*-(cyanoethyl)-cellulose.[452] Elastic films produced from *O*-(cyanoethyl)cellulose acetate were tough, and had better desalination characteristics than the standard cellulose acetate membrane. 6-Azido-6-deoxycellulose acetate (40) has been prepared and converted into 6-aldehydocellulose (41) by photolysis (Scheme 8).[453] The degree of substitution of the 6-aldehydocellulose was determined by formation of the 6-(2,4-dinitrophenylhydrazone) (42) or by reduction with sodium borodeuteride followed by hydrolysis, acetonation, and mass spectrometric determination of the extent and position of deuterium incorporation into the resultant 1,2:5,6-di-*O*-isopropylidene-α-D-glucofuranose (43) (Scheme 8). The gel chromatographic properties of cellulose nitrate in acetone [454] and factors affecting the dissolution rate of cellulose acetate phthalate in aqueous solution in relation to drug availability of enteric-coated preparations [455] have been studied.

Carboxymethylcellulose has been converted into *N*-(*m*-dihydroxyborylphenyl)carbamylmethylcellulose (44) by reaction of its azide with an aqueous solution of *m*-aminobenzeneboronic acid (Scheme 9).[456] Aminoethylcellulose reacted with an aqueous solution of *N*-(*m*-dihydroxyborylphenyl)succinamic acid (45) in the presence of *N*-cyclohexyl-*N*′-β-(4-methylmorpholinium)ethylcarbodi-imide *p*-tolylsulphonate to yield *N*-[*N*′-(*m*-dihydroxyborylphenyl)succinamyl]aminoethylcellulose (46) (Scheme 9). The two cellulose derivatives were shown to form specific complexes of variable stabilities with nucleic acid components, sugars, and other polyols. In chromatographic columns of these cellulose derivatives, the retention volume of a particular polyol depended on the availability of a glycol group with the appropriate configuration and conformation, the pH of the elution solvent, the ionic strength and nature of the cations in the

[449] S. Dilli and J. L. Garnett, *Austral. J. Chem.*, 1970, **23**, 1163.
[450] S. Dilli and J. L. Garnett, *Austral. J. Chem.*, 1970, **23**, 1767.
[451] N. J. Richards and D. G. Williams, *Carbohydrate Res.*, 1970, **12**, 409.
[452] M. A. El-Taraboulsi, M. A. Mandil, and H. E. M. Ali, *Carbohydrate Res.*, 1970, **13**, 83.
[453] D. M. Clode and D. Horton, *Carbohydrate Res.*, 1970, **12**, 477.
[454] G. Meyerhoff, *Makromol. Chem.*, 1970, **134**, 129.
[455] M. Hayashi, T. Nagai, and H. Nogami, *Chem. and Pharm. Bull.* (*Japan*), 1970, **18**, 2350.
[456] H. L. Weith, J. L. Wiebers, and P. T. Gilham, *Biochemistry*, 1970, **9**, 4396.

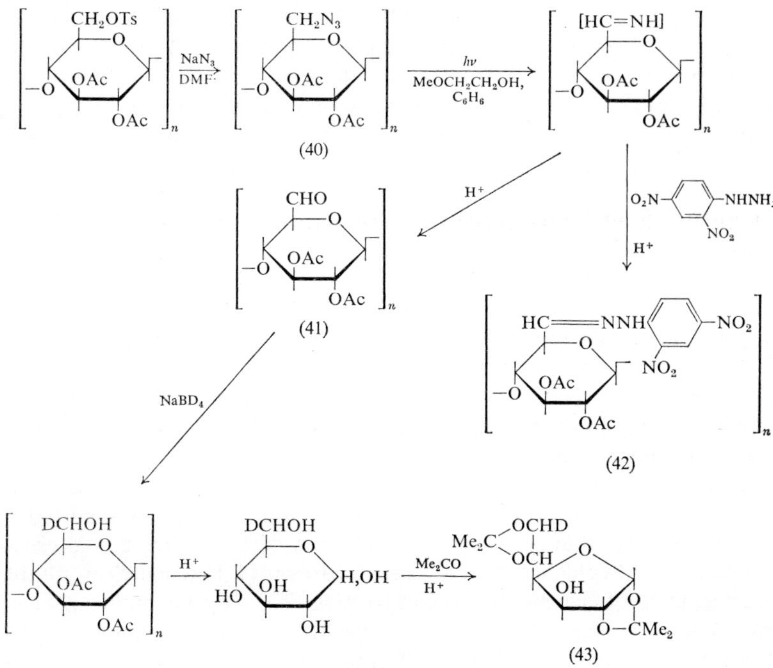

Scheme 8

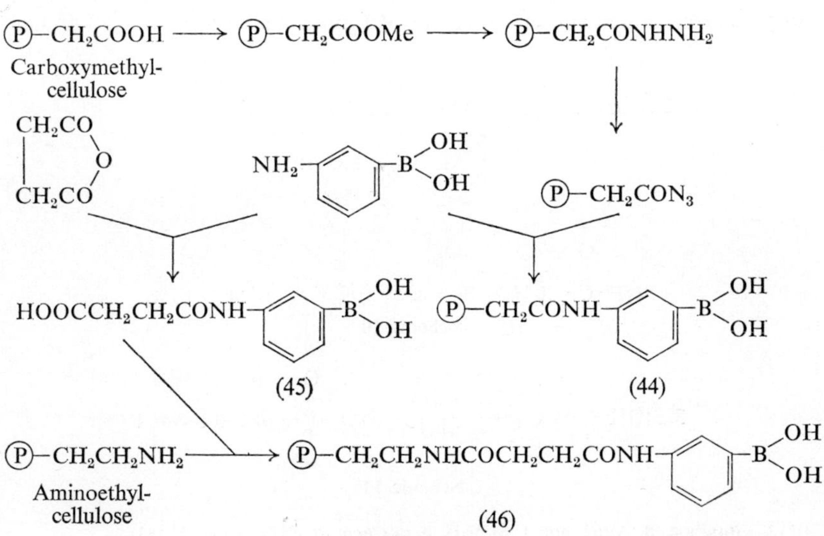

Scheme 9

elution solvent, and, in the case of nucleotides, the nature of the base attached to the glycol group. A study of the complex formation of two carboxymethylcelluloses, substituted to different extents, at acid pH with α-, αs-, β-, and κ-caseins revealed that complex formation between different casein components and carboxymethylcellulose commenced at different pHs, and that selective precipitation could be achieved for mixtures of caseins.[179]

A method has been outlined whereby chlorohydroxypropyl derivatives of cross-linked hydroxypropylcellulose and hydroxypropyl-Sephadex can be prepared with a controlled DS.[457] The introduction of halogen into the insoluble polymer facilitated further substitution reactions exemplified by the preparation of aminohydroxypropyl, aminoethoxypropyl, *N*-hydroxyethylaminohydroxypropyl, lithocholamidohydroxypropyl, and mercaptohydroxypropyl derivatives. These Sephadexes proved useful for gel chromatography in organic solvents.

DEAE-cellulose has been used as the support in an insolubilised form of glucoamylase,[349] and the further use of diazo-, isothiocyanato-, and azido-derivatives for the preparation of insolubilised amylases has been reported.[336] The action of organic cyanates on cellulose Avicel, Sepharose, and cross-linked dextran (Sephadex) to give activated forms of these matrices suitable for attachment of enzymes has been investigated.[220] Cyclic imidocarbonate structures were suggested to constitute the activated groups on the polymers. Aminoethylcellulose, bromoacetylcellulose, carboxymethylcellulose, and Sepharose 4B have been used to form myoglobin derivatives, which could then be used as immunoabsorbents.[458]

The preparation and properties of modified starches, particularly adipate cross-linked starches (Scheme 10), phosphate cross-linked starches

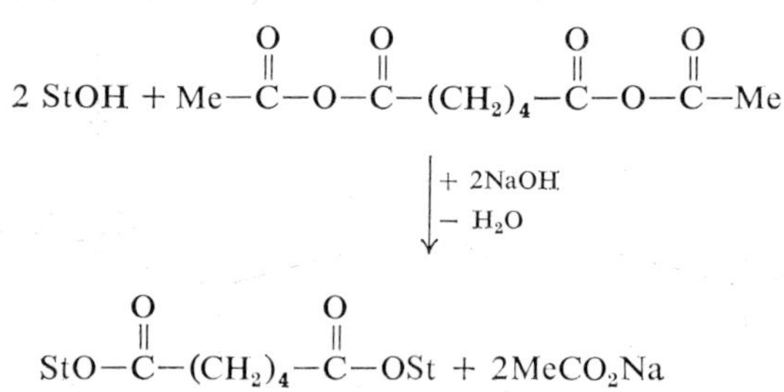

Scheme 10

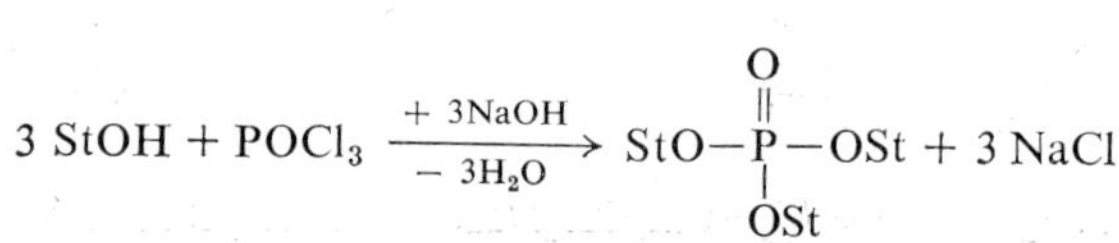

Scheme 11

[457] J. Ellingboe, B. Almé, and J. Sjövall, *Acta Chem. Scand.*, 1970, **24**, 463.
[458] R. J. Boegman and M. J. Crumpton, *Biochem. J.*, 1970, **120**, 373.

$$\mathrm{StOH + Me{-}\overset{\overset{O}{\|}}{C}{-}O{-}\overset{\overset{O}{\|}}{C}{-}Me \xrightarrow[-\,H_2O]{+\,NaOH} Me{-}\overset{\overset{O}{\|}}{C}{-}OSt + MeCO_2Na}$$

Scheme 12

$$\mathrm{StOH + Me{-}\overset{\overset{O}{\|}}{C}{-}O{-}CH{=}CH_2 \xrightarrow{base} Me{-}\overset{\overset{O}{\|}}{C}{-}OSt + MeCHO}$$

Scheme 13

(Scheme 11), and starch stabilised to acid by acetylation (Schemes 12 and 13) have been briefly reviewed.[459] Carboxymethylation of potato starch with monochloroacetic acid in aqueous methanol gave products having DS of 0·20, 0·50, 0·97, and 1·23.[460] The material of DS 0·50 had the highest viscosity, and it was concluded that the quantity and distribution of the carboxymethyl groups played a dominant rôle in determining the viscosity of aqueous solutions of the carboxymethyl-starches.

Treatment of insoluble wheat starch with acidified ceric ammonium nitrate or ferrous ammonium sulphate–hydrogen peroxide under conditions used for graft polymerisation gave products only partially dispersible in aqueous DMSO.[461] Treatment with dilute nitric acid alone gave a completely dispersible product, as did electron-beam irradiation. Possible structures for the modified starches were discussed.

A new type of highly insoluble, polyfunctional, diazotisable resin (47) has been prepared by the condensation of dialdehyde starch with *p,p′*-diaminodiphenylmethane and subsequent reduction of the Schiff-base-type polymeric product (Scheme 14). After diazotisation, the resins were suitable for the attachment of enzymes.[462]

Acid hydrolysis of *O*-(2-hydroxyethyl)starch (DS 0·1) gave, in addition to the expected 2-, 3-, and 6-*O*-(2-hydroxyethyl)-D-glucoses, 1,2-*O*-ethylene-α-D-glucofuranose, 1,2-*O*-ethylene-β-D-glucopyranose, and 1,2-*O*-ethylene-α-D-glucopyranose.[463] A series of methyl α-maltodextrins[464] and maltodextrins reduced or oxidised at C-1 of the terminal reducing D-glucose unit have been prepared.[323]

6-*O*-Tritylamylose has been oxidised with DMSO–acetic anhydride and the product oximated.[465] Subsequent reduction, detritylation, and dialysis yielded a compound bearing an amino-group at C-2, which was oriented

459 O. B. Wurzburg and C. D. Szymanski, *J. Agric. Food Chem.*, 1970, **18**, 997.
460 T. Shimizu and M. Araki, *J. Agric. Chem. Soc. Japan*, 1970, **44**, 107.
461 G. F. Fanta, R. C. Burr, C. R. Russell, and C. E. Rist, *Cereal Chem.*, 1970, **47**, 85.
462 L. Goldstein, M. Pecht, S. Blumberg, D. Atlas, and Y. Levin, *Biochemistry*, 1970, **9**, 2322.
463 H. C. Srivastava, K. V. Ramalingham, N. M. Doshi, and A. S. Chaudhari, *Carbohydrate Res.*, 1970, **12**, 23.
464 J. A. Thoma, C. Brothers, and J. Spradlin, *Biochemistry*, 1970, **9**, 1768.
465 M. L. Wolfrom and P. Y. Wang, *Carbohydrate Res.*, 1970, **12**, 109.

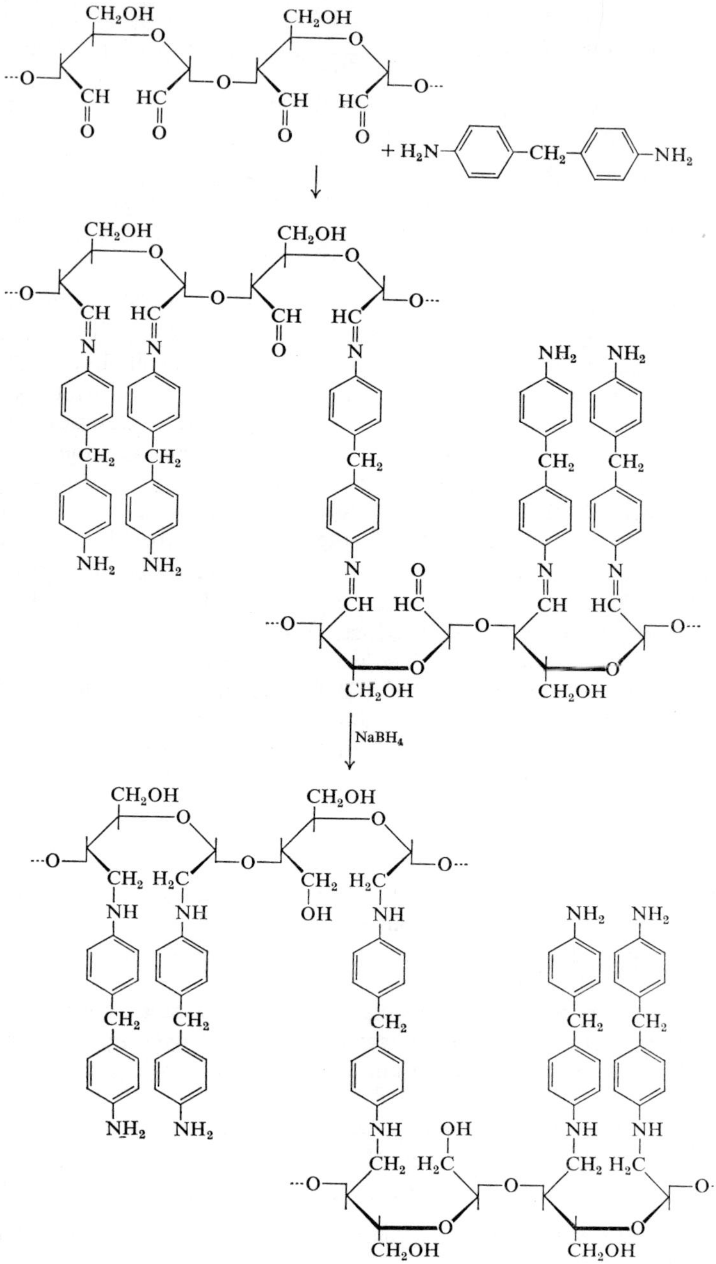

(47)

Scheme 14

almost entirely in the D-*gluco* configuration. The reaction of amylose with iodine has been used in the detection of carboxylic acids on thin-layer chromatograms.[466] A coloured derivative of amylose (Cibachron Blue amylose) has proved useful for the specific assay of α-amylase in the presence of glucoamylase.[322]

The preparation of a number of agarose bead (Sepharose) derivatives containing ω-aminoalkyl, bromoacetamidoethyl, succinylaminoethyl, *p*-aminobenzamidoethyl, tyrosyl, and sulphydryl groups has been described.[467] The use of these derivatives in the purification of proteins by affinity chromatography was investigated, and Sepharose was used in a similar way for the purification of a carbohydrase.[219] Cyanogen bromide derivatives of agarose have been employed in the preparation of insolubilised concanavalin A [24, 25] and enzymes.[468]

Periodate-oxidised amylose reacted with aqueous ammonia to give imidazole and 4(5)-(2-hydroxyethyl)imidazole, whereas periodate-oxidised dextran gave imidazole and 4(5)-methylimidazole; the reaction with periodate-oxidised laminarin gave only traces of imidazoles.[469] Mechanisms were proposed for these reactions, which appear to be specific for the linkage in the parent polysaccharide.

The peracetates of amylose, cyclohexa- and cyclohepta-amylose have been investigated by i.r. and n.m.r. spectroscopy to establish the conformation of the polymer chains.[470] Gel-forming materials have been prepared by cross-linking of cyclohepta-amylose with epichlorhydrin, and their use in chromatographic separations of nucleic acid derivatives, nucleotides, and nucleosides was described.[471] Separation was believed to rely on inclusion complex formation similar to that of cycloamylose in solution.

The preparation of benzoylated DEAE-Sephadex and its application in the synthesis of deoxyribonucleotides has been reported.[472]

Xylo- and manno-dextrans, containing 10% D-xylose and 5% D-mannose, respectively, have been prepared by transglycosylation reactions in alkali. Methylation analyses of the product polysaccharides showed that α-D-xylopyranosyl and α-D-mannopyranosyl residues had been introduced as end-groups mainly at the C-2 positions of the dextran.[473]

Pectic acid propionate was reduced with gaseous diborane to yield a series of polysaccharides having different carboxy-group contents.[474] Sulphation of the reduced polymers with various molar ratios of the

[466] L. Chafetz and M. H. Penner, *J. Chromatog.*, 1970, **49**, 340.
[467] P. Cuatrecasas, *J. Biol. Chem.*, 1970, **245**, 3059.
[468] K. Mosbach and B. Mattiasson, *Acta Chem. Scand.*, 1970, **24**, 2093.
[469] E. L. Richards, *Austral. J. Chem.*, 1970, **23**, 1033.
[470] B. Casu, M. Reggiani, G. G. Gallo, and A. Vigevani, *Carbohydrate Res.*, 1970, **12**, 157.
[471] J. L. Hoffman, *Analyt. Biochem.*, 1970, **33**, 209.
[472] J. J. Michniewicz, O. S. Bhandt, J. Goodchild, S. K. Dheer, R. H. Wightman, and S. A. Narang, *Biochim. Biophys. Acta*, 1970, **224**, 626.
[473] B. Lindberg and S. Svensson, *Acta Chem. Scand.*, 1970, **24**, 711.
[474] R. L. Whistler and G. A. Towle, *Arch. Biochem. Biophys.*, 1970, **138**, 39.

sulphur trioxide–DMF complex yielded a series of polysaccharides having different contents of *O*-sulphate and carboxy-groups. Treatment of the reduced polymers with ethylenimine prior to sulphation yielded a second series of polysaccharides having different proportions of *N*- and *O*-sulphate and carboxy-groups. All the sulphated polysaccharides showed activity in delaying the clotting time of recalcified human plasma. The anticoagulant activity of the reduced *O*-sulphated polysaccharides increased with increasing degree of sulphation and decreasing carboxy-group content. An increase in anticoagulant activity with decreasing carboxy-group content was also observed for preparations containing *N*- and *O*-sulphate groups, but the nitrogen contents, representing the *N*-sulphate contents, had little relationship to the anticoagulant activities.

Peracetylated and permethylated derivatives of orange pectic acid, and methylene blue and ruthenium red complexes of pectic acid have been formed for use in conformational studies of the D-galactopyranosiduronic acid moiety.[475] Polymethyl-poly-D-galacturonic acid methyl glycoside was used as substrate for studies on a pectinesterase.[410]

Chitin has been degraded, and chitosan formate prepared from the resultant chitosan by formic acid treatment.[476] The derivative was employed to form chitosan-impregnated cellulose thin layers, which were used for ion-exchange t.l.c. of nucleic acid constituents.

A ^{59}Fe-labelled iron–chondroitin sulphate colloid was used in an investigation of the iron distribution in mice.[80] A method has been developed whereby pure tritium-labelled heparin can be prepared.[84]

Escherichia coli lipopolysaccharide has been cross-linked and insolubilised by glutaraldehyde, and was then used to isolate anti-polysaccharide antibodies.[221]

Modification of Glycoproteins and Enzymes and Uses of Modified Glycoproteins and Enzymes

The preparation and use of glutaraldehyde-cross-linked concanavalin A for the isolation of polysaccharides and glycoproteins has been reported.[477] The insoluble derivative was suitable for use in a column, and removed from solution polysaccharides and glycoproteins that normally react with concanavalin A, *e.g.* glycogen, immunoglobulin M, and dextrans. Laminarin and cyclohexa-amylose, which do not normally react, were not absorbed by the matrix. The absorbed macromolecules could be desorbed by using methyl α-D-glucopyranoside or methyl α-D-mannopyranoside. In combination with enzymes, the glutaraldehyde-cross-linked concanavalin A has been used for the detection and measurement of cell interior and surface glycoproteins.[6]

[475] S. Hirano, M. Manabe, N. Miyazaki, and K. Onodera, *Agric. and Biol. Chem.* (*Japan*), 1970, **34**, 1530.
[476] K. Nagasawa, H. Watanabe, and A. Ogamo, *J. Chromatog.*, 1970, **47**, 408.
[477] E. H. Donelly and I. J. Goldstein, *Biochem. J.*, 1970, **118**, 679.

A cyanogen bromide-activated derivative of concanavalin A acted as a group-specific absorbent for serum components of high carbohydrate content.[25] The properties of a similar concanavalin A derivative and those of an insoluble poly-L-leucyl derivative, formed by reaction of concanavalin A with L-leucine-*N*-carboxyanhydride, were found to be identical.[24] They bound highly branched α-D-glucose and α-D-mannose polymers, but not an α-D-glucose polymer with a low degree of branching. The utility of the preparations in the fractionation of carbohydrate polymers was demonstrated by the separation of blood-group (A + H) substance into two fractions.

Dimeric bovine serum albumin was prepared by oxidation of mercaptoalbumin; on treatment with alkali at pH 10·3, the dimer reverted to the monomer.[199] The *S*-carboxymethyl derivative of orosomucoid,[191] the *S*-sulphonyl derivative of luteinizing hormone,[170] and the *S*-alkyl [222] and *O*-succinyl [223] derivatives of Tamm–Horsfall glycoprotein have been prepared.

Immunoadsorbents were prepared by covalently coupling macromolecules such as human IgG and antibodies to porous glass using a silane as coupler.[478] The immunoadsorbents complexed with specific antibodies and enzymes, which could be desorbed in high purity and in high yield. The insolubilised antibodies were very stable and could be used many times.

Comparisons of the properties of aminoethylcellulose, bromoacetylcellulose, carboxymethylcellulose, ethylene–maleic anhydride copolymer, and Sepharose 4B derivatives of myoglobin indicated that Sepharose-myoglobin was superior as an immunoadsorbent.[458] Molar propionic acid eluted 60—100% of the adsorbed antibody, and 60—90% of the eluted protein was precipitable antibody; the yields were higher than those achieved with other dissociating solvents. Immunoadsorbents have also been formed by making insoluble derivatives of insulin antibodies,[220] anti-β-galactosidase antibody,[219] and *Escherichia coli* lipopolysaccharide.[221] In the latter case, bound antibodies could be eluted with 0·2M-glycine buffer (pH 2·2), and the antibody recovered amounted to 10%. The immunoasdorbent was used repeatedly without significant reduction in its binding capacity for antibodies.

Active insolubilised derivatives of α-amylase have been prepared by coupling to diazo-, isothiocyanato-, and azido-groups of modified cross-linked polyacrylamides.[335] Analogous derivatives of β-amylase could be prepared by coupling through diazo- and isothiocyanato-groups only. The stabilities of the enzyme preparations were comparable to those of the corresponding cellulose derivatives of the enzymes. In the case of α-amylase, the polyacrylamide derivatives were more stable than the free enzyme in solution. Comparisons of the retention of activity of the

[478] H. H. Weetall, *Biochem. J.*, 1970, **117**, 257.

cellulose-bound insolubilised enzymes showed that, although considerable desorption of physically-bound enzyme took place on initial re-use, the derivatives could then be re-used with comparatively little loss of activity.[336] The polyacrylamide derivatives were superior in that retention of activity was good on initial use and repeated re-use.

A process involving diazotisation has been described for the chemical attachment of an enzyme to the surface of a polystyrene matrix;[274] for example, β-fructofuranosidase was chemically attached to the surface of polystyrene beads and tubes using this process. The properties of the carrier-fixed enzyme were investigated, and the pH–activity curve showed a marked difference from that of the free enzyme in solution.

Up to 55% of the activity of free glucoamylase was retained on binding to DEAE-cellulose. Derivatisation narrowed the pH optimum and lowered the temperature optimum of the enzyme.[349] The insolubilised enzyme was used in column form for the continuous conversion of starch, and maintained its activity for several weeks.

Glucose oxidase has been covalently attached to polystyrene, which had been nitrated, reduced, and diazotised.[394] The use of the enzyme derivative in automated assays was described. An insoluble derivative of glucose oxidase was also produced by covalently coupling it to nickel oxide on nickel screening.[393] The derivative had kinetic values similar to those of the soluble enzyme, but showed greater thermal and duration stabilities. Water-insoluble derivatives of papain, mercuripapain, subtilopeptidase A, polytyrosyltrypsin,[462] and hexokinase [468] have also been reported.

The aminoethyl derivative of β-galactosidase,[289] the *S*-carboxymethyl [373] and other derivatives [371, 372] of lysozyme, the *O*-acetyl derivative of α-amylase,[330] and the phenylhydrazone of monoamine oxidase [412] have been prepared. De-sialylated and re-sialylated forms of alkaline phosphatase have been produced.[411]

Author Index

Abano, Y., 209
Abdullah, M., 240
Abon El-Khier, A., 169
Acher, A. J., 15, 16
Ackers, J. P., 63
Acton, E. M., 59, 136
Adams, G. A., 6, 24, 168
Adamson, J., 51, 157
Adamyants, K. S., 54, 55
Äyräs, P., 150
Ahluwalia, R., 129
Aida, K., 110
Akazawa, S., 216
Åkerström, S., 215
Akita, E., 131
Akiyama, K., 196
Albrecht, H. P., 105, 109
Alekseev, Yu. E., 91, 115
Alfredsson, G., 22
Ali, H. E. M., 258
Ali, Y., 71
Allalouf, D., 195
Allan, B. J., 230
Allbutt, A. D., 39
Allen, A., 218
Almé, B., 260
Alper, C. A., 215
Al-Radhi, A. K., 45, 64
Amaral, D., 241
Amer, M. M., 169
Aminoff, D., 185
Anastassiadis, P. A., 196
Anderle, D., 168
Andermann, F., 201
Anderson, B., 185
Anderson, L., 212
Anderson, L. E., 123
Anderson, M., 193
Ando, H., 58
Ando, K., 17
Andrews, H. D., 248
Anet, E. F. L. J., 8
Angrist, A. A., 196
Angyal, S. J., 32, 116, 129, 146, 148, 155
Anikeeva, A. N., 171
Anisuzzaman, A. K. M., 72
Anno, K., 43
Anseth, A., 193
Anteunis, M., 111
Antonakis, K., 111, 147
Antonenko, T. S., 15, 41
Antonopoulos, C. A., 193
Araki, M., 261
Araki, Y., 94
Arbatskii, N. P., 179
Arita, H., 24
Arita, M., 17
Aritani, M., 24, 160
Armitage, I., 3
Aro, A., 179
Arora, S. K., 143, 162
Arrambide, E., 205
Arreguin, B., 27, 167
Arthur, J. C., 256, 257
Aseeva, N. N., 90
Ashitaka, Y., 209
Ashtan, J. F., 114
Asp, N., 224
Aspberg, K., 181
Atassi, M. Z., 237
Athonel-Haon, A. M., 217
Atlas, D., 261
Aune, K. C., 235
Auricchio, F., 226
Avigad, G., 8, 165
Avramers, S., 178
Awad, O. M., 5
Ayad, S., 189
Ayer, H. E. M., 179

Bachhawat, B. K., 200, 244
Bachler, M. J., 233
Backinowsky, L. V., 114
Baehler, B., 111
Baer, H. H., 66, 72, 73
Bahl, O. P., 185, 209
Bahurel, Y., 149
Bailey, A. J., 188
Bairamova, N. E., 75
Baird, J. K., 228
Baker, B. G., 9
Baker, D., 105
Baker, D. A., 142
Baker, E. A., 226
Baker, E. E., 240
Baker, E. M., 120
Bakerman, S., 213
Balduini, C., 203
Ball, D. H., 47
Ballard, J. M., 33
Bally, C., 42
Baltes, W., 71
Banks, W., 189, 230
Bardos, T. J., 136
Barford, A. D., 53, 74
Barker, R., 28, 29, 42
Barker, S. A., 201, 206, 231
Barlow, C. B., 145
Barnett, J. E. G., 5, 123
Barrett, J. T., 182
Barsegyan, I. V., 6
Bashey, R. I., 196
Baskevitch, Z., 19
Bateman, D. F., 119
Batstone, G. F., 199
Batyuk, V. S., 20
Bauer, Š., 30
Baugh, P. J., 8
Baumann, W. E., 233
Bayer, E., 157, 165
Bayer, M., 66
Beach, R. L., 73
Beale, R., 198
Beaupoil, B., 234
Beck, E., 171
Bedetti, G., 168
Beeby, R., 210
Beilfuss, W., 58
Beljean, M., 170
Bell, D. J., 169
Bell, K., 210
Bendetsky, K. M., 233
Benkovic, S. J., 141
Bently, J. P., 202
Ber, A., 195
Beran, P., 27
Berenson, G. S., 191
Berger-du Bois, F., 214
Berggren, B., 223
Bergström, J., 189
Berman, H. M., 161
Berrang, B., 80
Berry, S. A., 243
Bertolini, M., 38, 217
Berzate, A., 172
Bettelheim, F. A., 191
Bevill, R. D., 89
Beving, H. F. G., 90
Bewley, T. A., 207
Beychok, S., 185
Bezkorovainy, A., 210
Bhandt, O. S., 263
Bhattacharjee, S. S., 31, 35, 254
Bhatty, R. S., 182
Bierens De Haan, J., 170
Bilik, V., 7
Binder, R. A., 234
Binkley, R. W., 6
Binkley, S. B., 243
Binkley, W. W., 6
Binsch, G., 153
Birkenmeyer, R. D., 133
Birkhäuser, A., 94
Bischof, E., 106
Bistline, R. G. jun., 38
Bitter, T., 201
Bjelle, A., 193
Björndal, H., 160, 185
Blackburn, B. J., 154
Blackford, R. W., 138
Blank, H.-U., 141
Bloch, A., 135
Block, A., 196

Blumberg, S., 261
Blumbergs, P., 57
Blumenfeld, O. O., 213
Bobek, M., 135
Bobrava, L. N., 190
Bochkova, D. N., 218
Bock, K., 96
Boegman, R. J., 260
Boenisch, T., 215
Bogdanov, A. A., 180
Bogdanovskaya, T. A., 254
Bogdashova, L. S., 183
Bogden, A. E., 212
Bognár, R., 12, 50, 69
Bohlmann, F., 130
Bolker, H. I., 257
Bonini, P. A., 226
Bonsen, P. P. M., 254
Borders, D. B., 130
Borén, H. B., 90
Bors, L., 128, 153
Bose, J. L., 5, 17
Bosmann, H. B., 218
Boswell, K. H., 137, 166
Bouchard, J. L., 130
Bouchilloux, S., 217
Bouillant, M. L., 21
Bourgeois, J. M., 107, 111
Bourrillon, R., 211, 216
Bower, R. S., 123
Boyd, W. C., 180
Bradford, B., 167
Bradshaw, R. A., 237
Brady, R. F. jun., 34
Brady, R. O., 218
Brandt, R., 130
Braselton, W. E., 207
Braun, I., 19
Brazhnikova, M. A., 130, 133
Breck, W. G., 145
Breen, M., 193
Breitmaier, E., 156, 157
Brennan, S., 50
Brice, R. E., 123
Brimacombe, J. S., 22, 33, 45, 47, 64, 75, 161
Brodbeck, U., 138
Brooks, G. C., 245
Brossi, A., 115
Brothers, C., 261
Brovelli, A., 203
Brown, B. D., 179
Brown, B. I., 232
Brown, D. H., 232
Brown, R. K., 6, 14, 90, 92
Brüning, J., 86
Bruerton, K. I., 188
Brummel, M. C., 179
Bruni, C. B., 226
Brunngraber, E. G., 179
Bryan, J. G. H., 161
Buck, C. A., 218
Bucking, H. W., 249
Buckingham, J., 67
Buddecke, E., 15, 196, 222
Budzis, M., 75
Budzyński, A. Z., 212
Bueding, E., 189
Bugg, C. E., 163
Buhe, E., 18
Bukhari, S. T. K., 165
Bull, A. T., 238
Buncel, E., 44
Burckard, J., 198
Burge, B. W., 219, 221
Burkhardt, F., 101, 153
Burr, R. C., 261
Burton, R., 151
Butt, W. R., 206
Buys, H. R., 150

Cadenas, R. A., 36, 70
Cagnoli Bellavita, N., 19
Calcagno, M., 205
Calkins, D. F., 60
Callahan, J. W., 201
Callop, P. M., 213
Campbell, J. N., 243
Čápek, K., 35, 36, 58, 91
Capková-Šteffková, J., 36
Capon, B., 57, 256
Caputo, A., 230
Carey, F. A., 39
Carlson, K. D., 72, 151
Carman, R. M., 33
Carmo-sousa, L. Do., 180
Carter, J. H., 234
Carubelli, R., 229
Casagrande, D. J., 168
Castellani, A. A., 203
Castenfors, H., 189
Casu, B., 156, 263
Ceccherelli, P., 19
Cendrowska, H., 131
Cerezo, A. S., 36, 165
Ceriotti, G., 226
Černý, M., 27, 53, 111
Cessi, C., 64
Chabaud, O., 217
Chafetz, L., 263
Chahin, S., 18
Chalet, J. M., 75, 107
Chalk, R. C., 47
Chambers, D. A., 226
Chang, C. J., 15
Chaplin, M. F., 206
Charon, D., 114
Chaudhari, A. S., 26, 261
Chavin, S. I., 214
Chen, C.-Y., 116, 148
Chenery, D. H., 162
Cheng, Y. Y., 136
Chernetskii, V. P., 139
Chernobrovaya, N. V., 20
Chiemprasert, T., 71
Chin, C., 14
Ching-Puente, O. A., 47
Chittenden, G. J. F., 146
Chittenden, R., 164
Chizat, F., 151
Chizhov, O. S., 119, 158, 173, 190
Chládek, S., 135, 141
Chojnowski, W., 137
Chopin, J., 21
Cho Tun, H., 178
Chowdhury, P., 213
Chrispeels, M. J., 182
Christensen, J. J., 8
Christie, P. H., 63
Christner, J. E., 228
Chu, S. Y., 78
Cifonelli, J. A., 193, 197, 199
Cirelli, A. F., 114
Clapp, R. C., 19
Clark, C. R., 18
Clark, D. F., 144
Clausen, J., 195
Clayton, S. J., 135
Cléophax, J., 47, 75
Clode, D. M., 72, 258
Coat, J. P., 139
Coburn, R. A., 19
Cocu, F. G., 128, 153
Coffey, R. G., 169
Cohen, E., 230
Cohen, S., 214
Coleman, G., 231
Collins, P. M., 77
Comstock, C., 218
Conchie, J., 179
Connellon, J. M., 230
Conrad, H. E., 191
Constantopoulos, G., 192
Cook, A. F., 38, 43
Cook, R. E., 161
Cook, S. E., 169
Coppola, J. C., 162
Corina, D. L., 123
Corlett, R. D., 145
Cornwell, G. G., 195
Corse, J., 124
Couling, T. E., 113, 222
Courtois, J. E., 145
Covelli, I., 226
Coxon, B., 152, 154
Cozzone, P., 234
Craigie, J. S., 123
Crawford, J. F., 118
Creasey, S. E., 44
Creeth, J. M., 187
Criddle, W. J., 123
Crowley, G. M., 182
Crumpton, M. J., 260
Csüros, Z., 39, 50
Cuatrecasas, P., 263
Cuervo, L. A., 198
Cunningham, L. W., 187
Cunningham, W. L., 198, 228
Cusack, N. J., 70

D'Agnolo, G., 168
Dain, J. A., 251
Dalferes, E. R., 191
Damodaran, N. P., 109
Dance, N., 222
Danhofer, H. B., 169
Danilov, S. N., 171
Danninger, H., 17
Danyluk, S. S., 155
Darby, G. K., 199
Darnall, D. W., 178
Darnell, J. E., 221
Darzens-Souloumiac, D., 145
Das, I., 190

Dautlick, J., 223
David, S., 58, 138, 139
Davidson, E. A., 44, 47
Davies, D. B., 155
Davies, J. V., 190
Davis, D. D., 164
Davis, J. G., 219
Davis, R. F., 213
Davison, B. E., 161
Davydova, L. P., 63
Dawson, G., 251
Day, W., 156
Dayhoff, M. O., 178
Dea, I. C. M., 17, 18
Deak, G., 39, 50
Dean, C., 164
Dean, M. F., 198
De Bruyn, A., 111
De Bruyne, C. K., 17, 18
DeBusk, A. G., 226
Defaye, J., 29
Deferrari, J. O., 36, 70, 78, 114
Degand, P., 198
Dehmlow, E. V., 130
Dekasan, A., 192
de Kroon, R. A., 226
de Lederkremer, R. M., 78, 114
De Ley, J., 111
Delmau, J., 149
Deluca, L., 218, 219
Delvin, E., 228
Demain, A. L., 228
Demura, H., 170
Denborough, M. A., 187
Denesh, R., 196
Deonier, R. C., 235
DePreux, T., 201
Deranleau, D. A., 237
Derevitskaya, V. A., 179, 183
Desai, P., 8
Descotes, G., 149, 151
Des Gouttes, D., 214
Deshpande, V., 12
Desiderio, D. M., 143
de Urse, C. A., 205
Dheer, S. K., 263
Diana, G. D., 54
Dianoux, A. C., 235
Diehl, H. W., 26
Dietrich, C. P., 192
Dietzel, B., 18
DiFerrante, N., 192, 195
Dijong, I., 79
Dilli, S., 258
Diner, U. E., 148
Dinglinger, F., 19
Di Sanseverino, L. R., 162
Diver-Haser, A., 16
Dmitriev, B. A., 75, 114
Doane, W. M., 25, 26, 40, 82
Dodgson, K. S., 190
Dolak, L. A., 133
Dolhun, J. J., 143
Donaldson, B., 167
Donelly, E. H., 264
Doner, L. W., 110
Donovan, J. W., 219
Dorman, D. E., 155, 156
Doshi, N. M., 26, 261
Dowson, D., 206
Drews, G., 22
Driguez, H., 57
Drygin, Y. F., 180
Dudek, G. A., 212
Duplan, J., 149
Durette, P. L., 152
Durix, A., 21
Durr, G. J., 135
Dutton, G. G. S., 167
Dwek, R. A., 51, 153
Dyggre, H. V., 195

Eckstein, F., 43, 162
Edstrom, R. D., 189
Egami, F., 192
Egan, L. P., 50
Egan, R. S., 133
Eibl, H., 41, 145, 171
Ekblad, M., 207
El Amin, S., 5
El Ashry, S. H., 67
Eliel, E. L., 150, 153
El Khadem, H., 67, 69, 110
El Kholy, I., 67
Ellingboe, J., 260
El-Sayed, M. M., 5
El Sekeki, M., 67
El-Shafei, Z. M., 67, 175
El-Taraboulsi, M. A., 258
Elving, P. J., 143
Elyakov, G. B., 6
Elyakova, L. A., 239
Emanuel, N. M., 123
Emig, P., 139
Emoto, S., 8, 29
Englard, S., 8, 165
Entlicher, G., 181
Epton, R., 231
Erickson, R. P., 215, 225
Eriksson, H., 228
Ermakova, L. N., 233
Esber, H. J., 212
Esknazy, M., 215
Espinosa, F. G., 37
Etzler, M. E., 185
Etzold, G., 138
Eustache, J., 138
Evans, D. A. P., 240
Evstigneeva, R. P., 126
Ezaki, N., 131
Eze, L. C., 240

Fadeeva, T. N., 7
Faivre, L., 75
Falco, E. A., 139
Familusi, J. B., 200
Fanta, G. F., 261
Farber, S. J., 198
Farkas, I., 50
Farkaš, J., 133, 135
Farkas, L., 17
Farkas Szabo, I., 50
Farnsworth, N., 17
Fatiadi, A. J., 123, 126
Fawcett, J. S., 201
Feather, M. S., 8, 112, 240
Fecher, R., 137, 166
Fedorova, G. B., 133
Fedorova, V. I., 179
Feldman, G. L., 250
Feleki, V., 222
Felgenhauer, K., 178
Fenichel, L., 39, 50
Ferens, N. A., 6
Fernández-Bolaños, J., 82
Ferrand, M., 217
Ferretti, A., 210
Ferrier, R. J., 99, 158
Fialkiewicz, Z., 71
Filby, W. G., 160
Filippusson, H., 223, 241
Finan, P. A., 50
Finch, P. R., 170
Finlayson, A. J., 182
Fiol, J. B., 225
Fischer, E. H., 234
Fischer, W., 66
Fisers, G., 172
Fisher, L. V., 133
Fishman, W. H., 190, 228
Flanagan, V. P., 210
Fletcher, A. P., 215
Fletcher, H. G. jun., 26, 41, 113, 118
Fleury, P., 145
Flores-Carreon, A., 226
Florestedt, H., 70
Fodor, G., 37
Follmann, H., 112
Forsell-Knott, L., 219
Forstner, G. G., 218
Foster, A. B., 51, 53, 157
Foster, R. L., 57, 256
Fowler, L. J., 188
Fox, J. J., 54, 60, 133, 135, 136, 139
Frahne, D., 141
Franklin, E. C., 214
Fransson, L.-Å, 193, 195
Franzini, C., 226
Fraser-Reid, B., 94, 95, 96, 98
Frati, L., 230
Freeman, M. A. R., 188
Freidland, J., 222
French, D., 230, 240
Friend, J., 238
Frings, C. S., 170
Fritz, H.-J., 118
Fritz, R., 213
Fromme, J., 90
Fuertes, M., 142
Fujimaki, I., 131
Fujimaki, M., 139
Fujino, M., 208
Fujino, Y., 250
Fujiwara, A. N., 59
Fujiwara, T., 162
Fukumoto, J., 230
Fukushima, D. K., 165, 169
Furukawa, K., 185
Furukawa, Y., 135
Fusaro, R. M., 190

Gaertner, K., 138
Galanos, G., 57
Gale, C. C., 209
Galicka, N., 71
Galkowski, T. T., 113, 146
Gallo, G. G., 263
Gan, J., 207
Ganguly, A. K., 134
Garcia-Munôz, G., 71, 142
Gardell, S., 193
Gardiner, D., 77
Garegg, P. J., 22, 190
Garnett, J. L., 258
Garraway, J. L., 169
Gasser, R., 101, 106
Gastambide-Odier, M., 22
Gateau, A., 86, 172
Gatehouse, B. M., 161
Gatt, S., 226
Gaudemer, A., 45, 86, 100, 172
Gauze, G. F., 130
Gavrilenko, O. A., 108
Gavrilina, G. V., 130
Gehrke, C. W., 209
Gelin, M., 149
Geluk, H. W., 115
Gent, P. A., 33, 45, 124, 126
George, E., 200
Gerber, N. N., 22
Géro, S. D., 45, 47, 57, 75, 77, 86, 100, 172
Gheorghiu, T., 167
Gianetto, R., 228
Gibmeier, H., 18
Gibbs, C. F., 25, 60
Gigg, R., 124, 126
Gilbert, E., 43
Gilbert, H. S., 234
Gilham, P. T., 258
Gillard, G. C., 192
Glaudemans, C. P. J., 38
Glenner, G. G., 178
Glick, M. C., 218, 251
Glick, M. D., 161
Glinski, R. P., 49, 57
Glossman, H., 227
Glukhoded, I. S., 254
Gochnauer, T. A., 5
Goding, L. A., 182
Goebel, D., 70
Goldberg, J. M., 198
Goldstein, I. J., 180, 181, 255, 264
Goldstein, L., 261
Goodchild, J., 263
Goodey, R., 113, 222
Goodman, L., 48, 59, 60, 133, 136, 138
Goodwin, J. C., 34
Goodwin, S. D., 187
Goody, R. S., 60
Gordon, B. A., 200, 222
Gorin, P. A. J., 31, 35, 254
Got, R., 180, 220
Gottikh, B. P., 141
Gottlieb, D., 131
Gough, B. M., 104
Goussault, Y., 216
Goutarel, R., 57
Goyena, H., 205
Graf, R., 60, 107
Grahe, G., 144
Grant, D. M., 156
Grant, M. E., 189
Gray, C. J., 206
Gray, G. M., 251
Gray, G. R., 42
Green, D. P. L., 139
Green, J., 110
Greenwood, C. T., 189, 230
Grey, A. A., 21, 143, 154
Griffiths, W. C., 113, 146
Grigoryan, G. L., 212
Grimek, H. J., 206, 208
Grisebach, H., 104
Grochowski, B. J., 225
Grohlich, D., 210
Gros, E. G., 22
Gross, H., 50
Grossman, A., 188
Gruezo, F., 185
Gruñeiro, E. M., 22
Günther, J., 157
Guilbault, G. G., 222
Gulbinsky, J., 229
Gunner, S. W., 104, 106
Gupta, S. K., 12, 157
Guseinov, E. M., 113
Gustafsson, J. A., 228
Gut, V., 141
Guthrie, R. D., 44, 75, 77, 145, 165
Guzman Fernández-Bolaños, R., 82
Györgydeák, Z., 12

Haas, H. J., 170
Habeeb, A. F. S. A., 237
Haga, M., 33, 87
Haines, A. H., 22
Hakomori, S. I., 248
Hall, D. M., 13
Hall, L. D., 3, 51, 151, 153, 157
Hall, S. R., 143
Ham, J. T., 161
Hamazaki, H., 220
Hammond, S., 135
Hamor, T. A., 161
Hampton, A., 142
Hanessian, S., 98
Hanna, R., 41
Harada, H., 169
Harada, N., 165
Harada, T., 234
Hardegger, E., 20, 82
Hardingham, T. E., 198, 202, 203
Harmon, R. E., 12
Harper, P. J., 142
Harrap, G. J., 225
Harris, J. F., 8
Harris, M. J., 150, 158
Harrison, A. S., 51
Hascall, V. C., 192
Haskins, R. H., 35, 254
Hassing, G. S., 181
Hata, T., 42
Hatanaka, H., 116
Hatano, H., 231
Hauber, R., 168
Hausmann, W. K., 130
Hausser, C., 75
Haust, M. D., 200
Havsmark, B., 195
Hay, A. J., 179, 228
Hay, G. W., 145
Hay, J. B., 251
Hay, R. W., 18
Hayashi, M., 258
Hayden, G. A., 182
Heap, J. M., 44, 81
Heard, D. D., 29
Hearn, V. M., 187
Heddleston, K. H., 199
Heijlman, J., 217, 218
Helferich, B., 13, 14
Heller, D., 70
Heller, J., 216
Hellman, W., 198
Hems, R., 53
Hemsworth, B. A., 218
Henderson, J. F., 78
Hennen, G., 208
Herbut, P. A., 241
Heřmánková, V., 45
Hermann, J., 234
Herold, C.-P., 36, 37
Hersh, L. S., 240
Hess, G. P., 235
Hesse, R. H., 51
Hessler, E. J., 130
Hettler, J., 139
Hétu, C., 228
Hevey, R. C., 141
Hewson, K., 135
Heyn, A. N. J., 238
Heyns, K., 58, 71, 84, 86, 168
Higashihara, M., 230
Hilborn, J. C., 196
Hildesheim, J., 45, 47, 57, 75, 77, 100
Hilz, H., 190
Hindle, E. J., 245
Hineno, H., 45
Hineno, M., 158
Hirano, S., 264
Hirs, C. H. W., 242
Hisatsune, K., 238
Hiss, D., 156
Hitt, W. E., 188
Ho, M. W., 224
Hodge, J. E., 9
Hodges, D. R., 206
Hodgson, K. O., 39
Hörhammer, L., 20
Hoffman, D. J., 135
Hoffman, J. L., 263
Hoffman, P., 179
Hogenkamp, H. P. C., 112
Hoiberg, C. P., 44
Holcomb, G. N., 208
Holler, E., 235
Holmquist, L., 64

Holovatsky, I., 217
Holý, A., 140
Honjo, M., 135
Honma, T., 51, 93
Hopwood, J. J., 192
Horikawa, M., 211, 220
Hornby, W. E., 223, 241
Horner, M. W., 57
Horst, J., 227
Horton, D., 3, 29, 40, 67, 69, 72, 80, 109, 110, 112, 152, 155, 164, 258
Horwitz, J. P., 101, 106
Hotta, K., 211, 220
Hough, J. S., 180, 241
Hough, L., 57, 60, 71
Houminer, Y., 9
Hovingh, P., 192
How, M. J., 220
Howard, J., 114
Howarth, G. B., 132
Howe, C., 213, 229
Howe, M. L., 182
Howell, D. S., 198
Howgate, P., 117
Hoz, S., 9
Hruska, F. E., 21, 143
Hsia, D. Y., 233
Huang, C., 219
Huber, R. E., 225
Hubert-Habert, M., 48
Huckstep, L. L., 156
Hudson, B.-G., 29
Hughes, D. L., 114
Hullar, T. L., 75
Hulme, M. A., 238
Hultberg, B., 226
Hultman, E., 189
Hunedy, F., 47, 75
Hunt, D. J., 162
Hunt, L. T., 178
Hunt, S., 220
Husain, A., 47
Hygstedt, O., 210

Ichimura, K., 75
Ichino, M., 138
Ida, Y., 160
Ide, H., 228
Igarashi, K., 51, 93
Iglesias, J., 71
Igloý, M., 169
Iio, M., 79
Iizuka, H., 240
Ikehara, M., 139, 141, 162
Ikenaka, T., 24
Imagawa, T., 93
Imai, K.-I., 135
Imazawa, M., 139
Inanune, E., 228
Inatsu, Y., 160
Inch, T. D., 27, 32, 107, 152, 164
Ingle, T. R., 17
Inokawa, S., 45, 83, 122
Inouye, S., 130, 226
Irimajiri, T., 45
Irisawa, J., 51
Isaacs, N. W., 29, 162
Isaka, S., 220
Iseki, S., 187
Isemura, M., 24
Ishidate, M., 118
Ishido, Y., 94
Ishikawa, T., 195
Ishizuka, I., 251
Isliker, H., 214
Isono, K., 231
Iten, W., 238
Ito, T., 131
Iwabori, N., 223
Iwasaki, T., 212
Izaki, K., 239
Izatt, R. M., 8

Jackobs, J., 34
Jackson, R. L., 242
Jacobs, H. G., 187
Jahnke, H. K., 130
Jaiswal, P. K., 147
James, K., 32, 116, 146, 148, 155
Jamieson, G. A., 182, 213
Janik, B., 143
Janot, M. M., 57
Jansson, L., 196
Jarý, J., 35, 36, 45, 49, 54, 58, 91
Jasinska, J., 71
Jasinski, T., 71
Jastorff, B., 139
Jeanloz, R. W., 70, 120, 188, 189, 256
Jeffrey, G. A., 3, 19, 123, 161
Jeffrey, P. L., 204, 232
Jennings, H. J., 25, 56
Jensen, F. R., 164
Jewell, J. S., 103
Jirgensons, B., 178
Jochims, J. C., 154
Johansson, G., 211
Johansson, I., 110
Johnson, H. N., 226
Johnson, J. A., 190
Johnson, K. G., 243
Johnson, R. L., 193
Johnson, R. N., 51, 157
Jokić, A., 44
Jollès, J., 234
Jollès, P., 235
Jones, A. J., 156
Jones, A. S., 117, 199
Jones, G. H., 109
Jones, J. K. N., 56, 90, 132
Jones, J. V. S., 220
Jordaan, J. H., 79
Jotterand, A., 75
Judah, J. D., 212
Jung, G., 156
Jurch, G. R. jun., 8
Jurczak, J., 92
Justice, P., 233

Kabat, E. A., 178, 185
Kaboré, I., 57
Kadentsev, V. I., 158, 173
Kadlec, S. B., 124
Kärkkäinen, J., 168
Kagal, S. A., 160
Kågedal, L., 215
Kalmanson, A. E., 212
Kalvoda, L., 133
Kamerling, J. P., 151
Kan, L. S., 143
Kanarek, L., 237
Kaneko, M., 139, 141
Kaneko, Y., 239
Kanke, Y., 196
Kao, K. T., 188
Kaplan, D., 219
Karady, S., 159
Karapally, J. C., 192
Karimoto, Y., 216
Karle, I. L., 163
Karle, J., 163
Karnes, H. A., 133
Karnovsky, M. J., 228
Karnovsky, M. L., 228
Karr, D. B., 120
Kartnig, T., 169
Kashkin, K. P., 218
Kaska, W. C., 46
Kastelan, M., 195
Katchian, H., 124
Kato, K., 58, 160, 228
Kato, N., 36
Kato, T., 42
Katz, M., 229
Katzman, R. L., 188, 189
Kauffman, D. L., 230
Kaverzneva, E. D., 244
Kawai, S., 189
Kawaminami, T., 240
Kawano, N., 116
Kawasaki, T., 6, 24, 58, 160
Kawata, M., 118
Kefalides, N. A., 219
Kefurt, K., 54
Kefurtová, Z., 58
Keglević, D., 38, 151
Keleti, J., 90
Kelleher, P. C., 169
Keller, P. J., 230
Keller, T., 156
Keller-Schierlein, W., 130
Kempson, G. E., 188
Kennard, C. H. L., 29
Kennard, O., 162
Kennedy, J. F., 57, 178, 199, 200, 201, 206
Kent, P. W., 51, 63, 153, 215
Kenyon, G. L., 139
Kerr, K. A., 162
Kershaw, K., 8
Khan, K. M., 230
Khim, H. S., 19
Khorlin, A. Ya., 15, 41, 53, 77, 237
Khuang-Huu, Q., 57
Khwaja, T. A., 41, 137
Kienzle, F., 73
Kikot, G. S., 179, 183
Kikugawa, K., 138
Kilgore, W. W., 119

Kim, H. S., 3, 123, 161
Kim, Y. B., 199
Kimura, M., 118
Kinae, N., 11, 71
Kindel, P. K., 69
King, J., 193, 197
King, J. F., 39
King, R. D., 106
Kinner, L., 46
Kinoshita, M., 63
Kiraly, I. C., 20
Kirby, A. J., 162
Kirby, P. T., 51
Kirichenko, E. A., 171
Kishi, M., 191
Kiss, J., 101, 102, 153, 248
Kisseleva, V. V., 244
Kitahata, S., 230
Kiyozumi, M., 58, 160
Kjaer, A., 19
Kjellman, O., 234
Kleefeldt, L., 18
Klein, R. S., 135
Kleinschmidt, A. K., 198
Kleinschmidt, T., 227
Klemer, A., 18, 30, 98
Klimov, E. M., 15
Kloczewiak, M., 212
Kloppenburg, M., 251
Klotz, I. M., 178
Klyashchitskii, B. A., 123, 124, 126
Knaudt, K., 228
Knee, M., 238
Knull, H. R., 232
Koaze, Y., 130, 226
Kobayashi, T., 131
Koch, H. J., 155, 156
Koch, J., 250
Koch, K. F., 156
Kochetkov, N. K., 9, 15, 18, 54, 55, 75, 119, 123, 173, 179, 183, 190, 254
Kocon, R. W., 113, 146
Kocourek, J., 181
Koeppen, B. H., 36
Kohoutova, J., 27
Koizumi, T., 223
Kolesnikov, D. G., 20
Kolka, S., 71
Kolodynska, Z., 12
Komori, T., 58, 160
Koningsberger, V. V., 226
Konnert, J., 163
Konowal, A., 27, 92, 150
Kopeć, M., 212
Kornhauser, A., 38
Korte, W. D., 46
Kostetsky, E. I., 254
Kostic, D., 251
Koštiř, J. V., 181
Kosuge, T., 119
Kotick, M. P., 133, 136
Kováč, P., 160, 168
Kováč, T., 38
Kováčik, V., 160
Kovalev, U. I., 18
Kovář, J., 7, 72
Kovsharova, I. N., 130
Kownatzki, F., 60
Kowollik, G., 138, 140
Koyama, H., 133, 198
Koyama, M., 170
Kozlov, I. A., 180
Krayevsky, A. A., 141
Kresse, H., 196
Kresse, V. H., 188
Krigbaum, W. R., 235
Krivchuk, P. B., 20
Krmpotik, E., 233
Krol, B., 9
Kroschwitz, J. I., 155
Krüger, W., 13
Kruger, J. E., 231
Kshepinsky, I., 133
Kucar, Š., 7
Kudinova, M. K., 130
Kudryashov, L. I., 9, 18, 123
Kügler, F. R., 235
Kuhfittig, G., 165
Kuhne, H., 79
Kumagai, K., 169
Kumar, S., 77
Kumazawa, S., 72
Kundu, N. G., 118
Kurata, T., 139
Kurdanov, Kh. A., 91, 115
Kurokawa, M., 211, 220
Kuszmann, J., 54
Kutz, R., 18
Kuzuhara, H., 8, 29
Kwan, T., 116, 148

Lamb, D. J., 133
Lambein, F., 71
Lamberson, H. V., 211
Lamkin, W. M., 208
Lancaster, J. E., 130
Lance, D. G., 73
Landge, A. B., 5
Lands, W. E. M., 145
Langen, P., 138, 140
Langerman, N. R., 178
Lapenko, V. L., 26
Lapierre, R. B., 6
Larner, J., 190, 233, 245
Larson, A. C., 162
Laskina, L. P., 108
Latallo, Z. S., 212
Lau, P., 195
Lavrova, M. F., 130
Lawrence, M. A., 216
Lawson, C. J., 245
Lawton, B. T., 56, 90
Laycock, M. V., 123
Lebre, D., 220
Lechevalier, M. P., 22
Leclercq, F., 147
Lee, E. Y. C., 234
Lee, L. T., 213
Lee, M., 244
Lee, W. W., 60, 133, 138
Lee, W. Y., 231
Lee, Y. C., 188, 225, 226
Lefèvre, M., 216
Lehmann, J., 103, 224
Le-Hong, N., 75, 103
Lehrfeld, J., 34
Leibovich, S. J., 189
Lemieux, B., 251
Lemieux, R. U., 3
Leon, M. A., 214
Lerner, L. M., 136
Leslie, G. Q., 214
Leslie, J. G., 188
Lessard, J., 57
Letourneau, F., 37
Leung, C. Y., 237
Leuthardt, F., 42
Leutzinger, E. E., 142
Levin, Y., 261
Levvy, G. A., 179, 228
Lewis, C., 133
Lewis, G. J., 27, 107
Ley, R. V., 32
Li, N. C., 143
Li, S., 211, 223
Li, Y., 211, 223
Liao, T., 208
Libby, R. A., 232
Lichtenthaler, F. W., 66, 139
Lichtman, M. A., 214
Likhosherstov, L. M., 179, 183, 184
Limber, G. K., 213
Lin, S., 226
Lindahl, U., 196, 197
Lindberg, B., 22, 160, 263
Lineback, D. R., 233
Linker, A., 192
Lintner, M. A., 47
Lipscomb, H. S., 195
Lipták, A., 24
Lisowska, E., 188
Lissac-Cahu, M., 149
Listowsky, I., 8, 165
Litter, M. I., 78
Liverstovskaya, T. Ya., 18
Lloyd, K. O., 181, 185
Loewns, F., 124
Lofroth, G., 11
Long, L. jun., 19, 47
Looker, J. H., 160
Loontiens, F. G., 225
Louisot, P., 180, 220
Lowther, D. A., 196
Luborsky, S. W., 225
Lüderitz, O., 57, 90
Luk'yanova, L. N., 233
Lundblad, A., 185
Lundblad, G., 234
Lundie, P. R., 22
Lundin, R. E., 124
Lundt, I., 96
Luttrell, B. M., 129
Lybyer, M. J., 12
Lynch, S. S., 206
Lyons, H., 170
Lyons, T. P., 180

McCauley, J. G., 190
McCloskey, J. A., 143
McDonald, A., 241
McGale, E. H. F., 195
McGavack, T. H., 188

McGinnis, G. D., 28, 151
Machinami, T., 130
Machleidt, H., 15
McIntire, K. R., 214
Mackellar, F. A., 133
McKenzie, H. A., 210
McLaren, J. V., 231
McLaughlin, J. E., 215
McLean, A., 98
Macmillan, J. D., 244
McNicol, L. A., 240
McPhail, A. T., 161
McShan, W. H., 206, 207, 208
Maddox, I. S., 241
Madroñero, R., 71
Madroñero, T., 142
Madruga, J. E., 198
Maeda, S., 94
Maehr, H., 57
Magerlein, B. J., 132
Maghuin-Rogister, G., 208
Mahieu, P., 216
Maichuk, D. T., 38
Maijs, L., 172
Maksimov, V. I., 237
Maley, F., 242
Malysheva, N. N., 173
Mamoir, G., 208
Manase, M., 264
Mandava, N., 37
Mandel, P., 248, 251
Mandil, M. A., 258
Manners, D. J., 239
Marçais, J., 211
Marchis-Mouren, G., 234
Marcus, D. M., 51, 248
Margolis, R. K., 218
Margolis, R. U., 218
Markiw, R. T., 168
Maroteaux, P., 200
Marsh, J. B., 213, 251
Marsh, R. E., 19, 161, 163
Marshall, J. J., 229, 239
Marshall, R. T., 193
Martin, E., 201
Martin, J. C., 3, 149, 151
Martin, J. R., 133
Martinez, A. P., 60
Martinsson, E., 168
Martynova, M. D., 183
Mashburn, T. A., 179
Masojidkova, M., 27
Mason, R. M., 194
Masura, V., 255
Matile, P., 238
Matsuhiro, B., 78
Matsui, M., 165
Matsumoto, T., 139
Matsunaga, I., 116
Matsuo, M., 169
Matsushima, Y., 24
Matsuura, K., 94
Mattiasson, B., 263
Mattsson, K., 215
Matyschok, H., 29
Mayer, H., 22, 36, 90
Mayer, H. E., 219
Mayfield, J. D., 208
Mead, W. L., 160
Meade, H. M., 211
Mehrotra, U. S., 120
Meinol, P., 229
Melchers, F., 225
Melchior, J. C., 195
Meller, A., 18
Mendicino, J., 41
Menninger, F. F., 212
Mercer, C., 257
Mercier, C., 189
Meredith, P., 231
Mersmann, G., 98
Mertes, M. P., 135
Meshreki, M. H., 69, 110
Messer, W., 225
Mester, L., 67, 165
Mester, M., 165
Meyer, D., 211
Meyer, F. A., 196
Meyer, K., 199
Meyerhoff, G., 258
Meyers, E., 130
Meyerson, S., 160
Meyer Zu Reckendorf, W., 15, 58, 64, 65, 66, 106
Mezentsev, A. S., 133
Mhatre, N. S., 228
Micheel, F., 11
Michel-Béchet, M., 217
Michell, R. H., 228
Michniewicz, J. J., 263
Mikhant'ev, B. I., 26
Miljković, M. A., 44, 47
Mills, F. D., 9
Mirelman, D., 169
Miroff, G., 211
Mirzayanova, M. N., 63
Misaki, A., 234
Mishkin, A. R., 123
Mitchell, A. J., 158
Mitscher, L. A., 130
Miyaki, M., 140
Miyata, T., 189
Miyazaki, N., 264
Mizsei, A., 169
Mizutani, Y., 118
Mochizuki, M., 209
Moczar, E., 188
Moczar, M., 188
Mody, N., 106
Moffatt, J. G., 55, 105, 109, 138
Molloy, B. B., 156
Mondt, H., 228
Monneret, C., 57
Montgomery, J. A., 135, 140
Montgomery, R., 179, 219
Montreuil, J., 179, 212
Monsigny, M., 179
Moore, J. S., 190
Moore, S., 243
Moreno, P., 216
Moretti, J., 211
Morgan, P. H., 187
Morgan, W. T. J., 182
Morgenlie, S., 24
Morgenstern, E., 228
Mori, Y., 196
Morita, Y., 182
Morozowich, W., 133
Mosbach, K., 263
Moscarello, M. A., 170
Moses, V., 226
Mosin, V. A., 237
Moskvin, M. D., 6
Motherwell, W. D. S., 162
Mott, M. G., 199
Moudgal, N. R., 207
Mroczkowski, Z., 35, 36
Müller, A., 130
Müller, H., 212
Müller, W. M., 13, 14
Müller-Mulot, W., 120
Muir, H., 188, 198, 203
Muir, L., 188
Mukerjee, H., 63
Muller, F. J., 198
Mullinger, R. N., 180
Mumma, R. O., 44
Munday, K. A., 5
Munns, A. R. I., 163
Murachi, T., 241
Murai, T., 94
Murata, K., 195
Murayama, A., 139
Muro, T., 133
Murofushi, N., 19
Murphy, W. H., 188, 210
Murray, D. H., 135
Mushran, S. P., 120
Myles, A., 141

Nagabhushan, T. L., 14, 63, 126
Nagai, T., 258
Nagasawa, K., 169, 264
Nagy, G., 24
Nair, P. M., 67
Nakada, S., 63
Nakagawa, H., 244
Nakagawa, I., 42
Nakagawa, T., 72
Nakai, T., 170
Nakamura, Y., 120
Nakanishi, K., 165
Nakanishi, Y., 191, 201
Nakano, M., 250
Nánási. P., 24
Nand, S., 228
Namiki, S., 133
Narang, S. A., 263
Nashed, M. A., 69, 110
Nassr, M. A. M., 67
Nayak, U. G., 45, 87
Nechaeva, N. P., 130
Neelakantan, S., 5
Nehring, H., 33
Nelkin, W., 8
Nelson, T. E., 233, 245
Nemes, É. N., 12
Neskovic, N. M., 248
Ness, R. K., 26
Neuberger, A., 215
Neuhahn, H.-J., 130
Neuss, N., 156
Nichols, M. R., 212
Nicot, C., 211
Niedballa, U., 136
Nielsen, L. D., 234

Nienaber, N., 15
Niewoehner, C. B., 124
Nifant'ev, E. E., 171
Niida, T., 130, 131, 226
Nikitin, I. V., 123
Nishida, T., 162
Nishida-Fukuda, M., 192
Nishimura, Y., 14
Niwa, T., 130, 226
Nogami, H., 258
Nogradi, M., 17
Nord, C. E., 234
Novák, P., 49
Nowakowski, A., 35, 36
Nuhn, P., 70
Nussbaum, J. L., 248

Oakenfull, D. G., 42
O'Brien, J. S., 224
Odham, G., 24
O'Donnell, G. W., 29
Öckerman, P. A., 192, 226
Öhman, R., 210
Oesterling, T. O., 133
Oette, K., 167
Ogamo, A., 169, 264
Ogata, T., 45, 122
Ogawa, S., 36, 132
Ogiso, Y., 139
Ogryzlo, M. A., 234
Ogston, A. G., 212
Ohki, E., 132
Ohkuma, S., 214
Ohman, R., 229
Ohnishi, M., 231
Ohrui, H., 8, 29
Ohya, T., 239
Okabe, H., 6
Okada, S., 230
Okigawa, M., 116
Okonenko, Y. U., 20
Okuda, T., 7
Omura, H., 79
Omura, S., 133
Ong, K.-S., 105, 106
Ono, T., 198
Onodera, K., 142, 264
Orgel, L. E., 137
Orrell, S. A., 189
Orwin, D. F. G., 217
Osaki, K., 7
Oshima, Y., 195
Ota, M., 43
Otake, T., 26, 36
Otter, B. A., 139
Overend, W. G., 77, 104, 106
Ovodov, Y. S., 257
Owen, L. N., 44, 81
Ozawa, T., 11, 71, 120

Pacák, J., 27, 53, 111
Paerels, G. B., 115
Paice, J. C., 209
Pal, P. R., 63
Pal, S., 198
Palmer, D. H., 245
Palo J., 215
Pan, S. C., 19
Panke, D., 19
Papkoff, H., 207, 208
Parnaudeau, M. A., 211
Parrish, F. W., 230
Pasero, L., 234
Patel, V., 224
Patterson, E. L., 130
Paulsen, H., 36, 37, 60, 75, 79, 83, 84, 86
Pavlenko, A. F., 257
Pays, M., 170
Pazur, J. H., 232
Pdonsky, J., 19
Peach, C. M., 188
Pecht, M., 261
Pederson, C., 96, 153
Peel, R. P., 107
Peer, H. G., 71
Penner, M. H., 263
Pennock, C. A., 199
Pepper, D. S., 213
Perlin, A. S., 155, 156, 192
Perret, F., 75, 103, 133
Perry, M. B., 168
Perry, M. G., 64
Peseke, K., 71
Pešić, M., 44
Petersonn, G., 159
Petrow, V., 206
Petty, J. D., 33
Peyer, C., 211
Pfleiderer, W., 141
Phelps, C. F., 198, 202
Phillips, G. O., 8, 160, 190
Philpot, C. W., 28, 151
Pickering, W. F., 114
Pierce, J. G., 208
Pigman, W., 217
Pihlaja, K., 150
Pilotti, A., 160
Pines, S. H., 159
Pinter, J. K., 210
Pita, J. C., 198
Planta, R. J., 168
Plummer, T. H., 242
Pocchiari, F., 168
Podešva, J., 53
Pollock, J. J., 235
Polyakov, A. I., 90
Ponomarenko, E. Yu., 26
Ponomareva, E. A., 139
Popa-Luchian, M., 9
Popenoe, E. A., 192
Popoff, T., 118
Poppleton, B. J., 161
Porath, J., 181
Porter, R. D., 180
Posternak, T., 128, 153
Pourfar, M., 222
Powell, K. A., 124
Pozhonji, T., 196
Pradal, M., 220
Prajsnar, B., 118
Praus, R., 197
Pravdić, N., 113, 151
Preobrazhenskii, N. A., 123, 124, 126
Preston, B. N., 196
Prevost, C., 3
Price, R. G., 222
Prior, A. M., 44
Privalova, I. M., 77
Prokop, J., 135
Proshlyakova, V. V., 130
Prox, A., 160
Pruzanski, W., 234
Prystaš, M., 30
Pulkrábek, P., 135
Puro, K., 250
Purygin, P. P., 141
Pusztai, A., 182
Pyler, R. E., 87

Quarles, R. H., 218

Rabinsohn, 7, 15, 16
Rachaman, E. S., 15
Radatus, B., 94, 95, 96
Radhakrishnan, A. N., 232
Radhakrishnamurthy, B., 191
Rahman, A., 162
Rajabalee, F., 66, 73
Ralph, A., 5
Ramalingam, K. V., 26, 52, 261
Rank, W., 73
Rao, A. J., 207
Rao, J. M., 67
Rao, S. T., 143, 162
Ratcliffe, W. A., 215
Rathbone, E. B., 151
Rathnan, P., 207
Ravindrannathan, T., 139
Reardon, K. M., 146
Rebers, P. A., 199
Reed, E. J., 118
Rees, D. A., 245
Reese, C. B., 34, 41, 139
Reggiani, M., 263
Reich, H. J., 155
Reid, P. E., 167
Reinert, H. 144
Reinicke, B., 130
Reinshagen, H., 103, 224
Reist, E. J., 60, 63
Reithel, F. J., 169
Revankar, G. R., 135
Revelskya, L. G., 171
Reyes, E., 226
Ribaldi, M., 19
Richards, E. L., 263
Richards, G. N., 29
Richards, N. J., 165, 258
Richardson, A. C., 57, 71, 74
Richardson, D. I., 42
Richardson, N. G., 185
Richtmyer, N. K., 5, 110, 171
Rico, M., 142
Rieder, H. P., 211
Rienits, K. G., 204
Rinehart, K. L. jun., 131
Rist, C. E., 25, 26, 40, 82, 261
Roberts, E. J., 26, 257
Roberts, J. D., 155, 156
Robertson, J. H., 130
Robins, M. J., 44, 137

Robins, R. K., 44, 135, 136, 137, 142, 156
Robinson, D., 222
Robinson, H. C., 197
Robinson, N., 22
Robinson, R., 255
Robyt, J. F., 230
Roch-Norlund, A. E., 189
Röhle, G., 13
Rogers, G. T., 143, 166
Roglić, G., 38
Rohrer, D. C., 162
Ropuszynski, S., 29
Rosantsev, E. G., 212
Rose, B., 213
Rose, U., 228
Rosenberg, A., 229
Rosenberg, D., 151
Rosenberg, L., 198
Rosenfeld, E. L., 184
Rosenstein, R. D., 161
Rosenthal, A., 104, 105, 106, 107, 142
Rosenthal, S., 118
Rosso, G., 219
Roth, M., 170
Roukema, P. A., 217, 218
Rouzeau, C., 138
Rowe, E. L., 133
Rowe, J. J. M., 167
Rowe, K. L., 167
Rowell, R. M., 110
Rowland, S. P., 26, 257
Rozhdestvenskaya, I. D., 7
Rozmarin, Gh., 9
Rubin, A. L., 189
Ruiz-Herrera, J., 226
Rupley, J. A., 235
Russell, C. R., 40, 82, 261
Russell, I. J., 233
Ruth, J. M., 210
Ryan, C., 233
Ryan, K. J., 136
Rychlik, I., 135
Ryle, M., 206
Rytting, J. H., 8
Rzepka, M., 29

Sable, H. Z., 124
Saeki, H., 132
Saeki, T., 250
Saenger, W., 162
Saffhill, R., 34, 42, 139
Saha, A., 213
Said, F., 169
Saifer, A., 222
Saito, S., 234
Sajdera, S. W., 197
Sakakibara, T., 72
Sakata, M., 81
Saldaña, M. C., 71
Salnikow, J., 243
Salo, W. L., 41
Sambury, S., 213
Samek, Z., 49, 54
Samokhvalov, G. I., 63
Samuelsen, B., 24
Samuelson, O., 168
Samy, T. S. A., 208
Sanchez, R. A., 137
Sandaralingam, M., 34
Sanderson, G. R., 192
Sano, H., 132
Sarawathi, S., 244
Sarfati, R. S., 114
Sarfati, S. R., 145
Sarre, O. Z., 134
Sato, H., 165
Sato, K., 110
Sawada, J., 133
Sawai, H., 182
Sax, K. J., 130
Saxena, B. B., 207
Saxena, O. C., 7
Schachter, H., 170
Schaffner, C. P., 57
Schengrund, C., 229
Scherz, H., 11
Schimmel, S. D., 89
Schimmelpfennig, K., 205
Schluederberg, A., 229
Schmid, K., 212
Schmid, R., 104
Schmidt, R. R., 118
Schneider, G., 165
Schneider, I. A., 9
Schneider, J. J., 17, 169
Schnek, L., 222
Schöllnhammer, G., 104
Schubert, M., 198
Schüep, W., 20, 82
Schuerch, C., 255
Schuller, E., 216
Schumacher, J. N., 37
Schumacher, M., 218
Schuster, A., 19
Schwartz, J., 226
Scott, A. I., 165
Scott, J. E., 63
Scott, R. B., 189
Scott, R. W., 169
Scriver, C. R., 201
Sebastian, S., 75
Secret, D. W., 167
Seetharam, B., 232
Segrest, J. P., 187
Seligmann, O., 17
Senchenkova, S. N., 183, 184
Seno, N., 43, 199
Sepulchre, A. M., 75, 86, 172
Serafini-Cessi, F., 64
Sergienko, I. Z., 114
Sethi, V. S., 170
Shaban, M. A. E., 67
Shafizadeh, F., 28, 151
Shah, R. H., 118, 124
Shapira, J., 6
Shapiro, D., 15, 16
Shaposhnikova, G. I., 6
Shapovalova, S. P., 130
Sharon, N., 235
Sharp, P. B., 226
Sharpatyi, V. A., 18, 123
Sharpless, T. W., 144
Shasha, B. S., 82
Shashkova, E. A., 237
Shaw, D. C., 210, 230
Shaw, D. R. D., 169
Shaw, G., 70
Shaw, S. J., 143
Shekhtman, N. M., 149
Shen, T. Y., 137, 166
Shereshevskii, A. A., 113
Sherman, W. R., 42
Sherwood, O. D., 206, 208
Shestakova, T. G., 171
Shibata, M., 133
Shier, W. T., 131
Shifrin, S., 225
Shileiko, L. V., 7
Shimizu, B., 138, 140
Shimizu, F., 138
Shimizu, S., 201
Shimizu, T., 261
Shinriki, N., 118
Shirahama, T., 228
Shively, J. E., 191
Shoji, T., 36, 65
Shortridge, K. F., 209
Shostakovskii, M. F., 90
Shulman, H. J., 199
Shulman, M. L., 53
Shuman, D. A., 44
Shvets, V. I., 123, 124, 126
Sica, V., 226
Sidbury, J., 189
Siddiqui, I. R., 5
Sie, H., 190
Siegenthaler, P., 201
Sikorski, J., 206
Silpananta, P., 212
Simon, E., 8
Simpson, D.-L., 232
Simpson, R., 44
Sinclair, H. B., 33, 44, 53, 150, 169
Singer, J. A., 170
Singh, M., 245
Singh, M. P., 7
Singh, P. P., 6, 24
Singh, S., 257
Singh, S. V., 7
Singh, U. P., 6
Sinkinson, G., 210
Sinnette, C. H., 200
Sinohara, H., 223
Sinou, D., 149
Sirbasku, D. A., 243
Sirokman, F., 49
Siskin, S. B., 75
Sjövall, J., 260
Sleeter, R. T., 44, 150, 169
Slessor, K. N., 100
Sloan, J., 226
Slusarchyk, D. M., 130
Smart, K., 213
Smiatacz, Z., 71, 152
Smiataczowa, K., 71
Smiley, K. L., 233
Smirnova, G. P., 254
Smith, C. R. jun., 72, 151
Smith, C. W., 75, 77
Smith, E. E., 230
Smith, F. D., 38
Smith, I. C. P., 21, 143, 154
Smith, P. F., 254
Snaith, S. M., 228
Snow, C., 218

Snyder, R. L., 161
Sobell, H. M., 143
Sobiczewski, W., 131
Sobitzkat, H., 11
Söder, P. Ö, 234
Sohař, P., 54
Sokolov, S. D., 124
Sokolowski, J., 71, 152
Somers, P. J., 231
Somogyi, L., 12
Sonnenbichler, J., 135
Sonohara, H., 201
Sorai, S., 224
Šorm, F., 30, 133, 135
Sotsky, S. M., 192
Sova, V. V., 239
Sozmen, M., 160
Spiegel, R. S., 190
Spik, G., 179, 212
Spinola, M., 70, 256
Spiro, R. G., 178
Spradlin, J., 231, 261
Sprinzl, M., 107, 142
Squires, T. G. S., 50
Srere, P. A., 245
Srinivasan, S. R., 191
Srivastava, H. C., 26, 52, 261
Srivastava, R. M., 14, 90, 92
Stacey, B. E., 33
Stacey, M., 75, 78, 220
Stadnik, A. M., 217
Stamm, O. A., 13
Staňek, J., 111
Stanek, J. jun., 11
Stark, J. H., 87
Stark, J. R., 234
Staub, A. P. A., 98
Stavitskii, V. V., 139
St. Cyr, M. J., 169
Steers, E., 215
Stefanovich, V., 196
Šteffková, J., 35
Stein, W. H., 243
Steiner, P.-R., 151, 153, 157
Steinert, G., 83
Steinert, K., 83, 84
Stellner, K., 36
Stenzel, K. H., 189
Stepanenko, B. N., 114, 190
Stephen, A. M., 151
Stern, K. F., 133
Sternberg, M. Z., 212
Steven, F. S., 189
Stevens, C. L., 49, 57
Stevenson, F. K., 215
Stewart, J. C. M., 41
Still, W. J. S., 189
Stirton, A. J., 38
Stockwell, R. A., 194
Stone, A. L., 192
Strachan, I., 179
Strandberg, G. W., 233
Stranks, D. W., 238
Strauss, J. H., 219, 221
Strecker, G., 179
Strittmatter, C. F., 223
Strobach, D. R., 114
Strosberg, A. D., 237
Strucinski, J., 37
Stud, M., 142
Suami, T., 36, 65, 130, 132
Subramanian, E., 162
Sudo, K., 243
Suematsu, T., 223
Sugihara, J. M., 148
Suld, H. M., 241
Sulston, J. E., 34
Suminoe, K., 148
Sun, S. C., 229
Sun, S. F., 212
Sundaralingam, M., 143, 162
Suo, K., 39
Surinov, B. P., 218
Susott, R. A., 28, 151
Suzuki, K., 227
Suzuki, S., 17, 201
Svennerholm, L., 229
Svensson, S., 160, 263
Sveshnikova, I. N., 130
Svetashev, V. I., 254
Sviridov, A. F., 119, 190
Swaminathan, N., 209, 232
Swanson, F. O., 112
Swanson, S. A. V., 188
Sweeley, C. C., 251
Sykulski, J., 70
Symons, R. H., 43
Szabo, L., 114
Szabó, P., 145
Szafranek, J., 71
Szantay, C., 136
Szarek, W. A., 56, 73, 90, 103, 132
Szejtli, J., 5
Szilágyi, L., 12, 69
Szymanski, C. D., 261

Taboada, J., 27, 167
Tagu, T., 7
Taigel, G., 154
Takagi, Y., 64
Takahashi, H., 239
Takahashi, N., 19, 241
Takano, M., 33
Takehana, H., 244
Takenishi, T., 42
Talukder, M. Q.-K., 169
Tamura, H., 209
Tamura, Z., 116
Tanaka, K., 224
Tane, M., 209
Tanenbaum, S. W., 229
Tanford, C., 235
Tanimura, W., 148
Tappel, A. L., 224
Tarentino, A., 242
Tarnowski, W., 190
Tarussova, N. B., 141
Tate, M. E., 123
Tatsuta, K., 131
Tatum, J. H., 8
Tavale, S. S., 143
Taylor, K. G., 49, 57
Teitel, S., 115
Tejima, S., 33, 81
Terao, T., 242, 243
Thambyrajah, V., 234
Theander, O., 118
Thewalt, U., 163
Thiele, O. W., 250
Thoma, J. A., 231, 261
Thomas, E. W., 15, 237
Thomas, H. J., 135, 140
Thomas, R., 18
Thorndahl-Jaccard, S., 75
Tichá, M., 181
Tihlárik. K., 30
Tindall, C. G. jun., 40, 109
Titova, N. B., 7
Tittensor, J. R., 117
Tkhorevskaya, Z. G., 6
Toda, N., 199
Tömpe, L., 216
Tojo, S., 209
Tokura, Y., 209
Tolbert, B. M., 120
Tollin, P., 163
Tomita, K., 162
Tomizawa, H., 239
Tong, G. L., 138
Tookey, H. L., 227
Towle, G. A., 263
Towliati, H., 228
Townsend, L. B., 135, 142
Tracey, A. S., 100
Trapmann, H., 170
Trimnell, D., 40
Tripathi, K. K., 209
Trnka, T., 27
Tronchet, J., 75, 94, 106, 107, 142
Tronchet, J. M. J., 60, 75, 94, 103, 106, 107, 111, 133, 142
Trotter, J., 114
Trowbridge, D. B., 139
Trummlitz, G., 139
Truter, M. R., 123
Tschuchrova, A. I., 244
Tsilvich, T. L., 141
Tsuboyama, K., 143
Tsuchiya, T., 14, 39, 45, 64, 131
Tsuchiya, Y., 122
Tsuji, K., 130, 131
Tsuji, M., 201
Tsukamoto, H., 58
Tsukuda, Y., 133
Tsuruo, T., 242, 243
Tsuruoka, T., 130, 131, 226
Tucker, L. C. N., 45, 64
Tulsiani, D. R. P., 229
Tuppy, H., 229
Turvey, J. R., 150, 158
Tuskui, A., 148
Tutsumi, M., 79

Ueda, T., 139
Uemura, T., 110
Uesugi, S., 141
Uhlemann, G., 30
Ui, N., 209
Ukhdina, R. S., 130
Ukita, T., 242, 243
Ulbricht, T. L. V., 143, 166

Ulyanova, O. I., 124
Umezawa, S., 14, 39, 45, 63, 64, 131
Unrau, A. M., 231
Unsworth, A., 206
Upton, J. D., 169
Urban, P. F., 251
Usher, A., 42
Usherwood, E. W., 98
Usov, A. I., 54, 55
Uvarova, N. I., 6
Uzlova, L. A., 20, 108

Vaillant, R., 190
Van Beeumen, J., 111
van Chuyen, N., 139
Van Deenen, L. L. M., 254
Van Den Bos, R. C., 168
Van den Eijnden, D. H., 218
Van Den Ouweland, G. A. M., 71
Van der Berg, G., 218
van der Heydt, W., 79
Vandermeer, J. W., 5
Van Kamp, G. J., 168
Van Parijs, R., 71
Van Praag, D., 198
Van Wijnendaele, F., 18
Vasilenko, K. D., 6
Vaskovsky, V. E., 239, 254
Vass, G., 165
Veis, A., 193
Vercellotti, J. R., 15, 50
Verheij, H. M., 254
Verheyden, J. P. H., 55
Vermes, J. P., 57
Veruovic, O., 255
Vervoort, A., 17
Vethaviyasar, N., 99, 158
Veyrières, A., 58, 120
Vicari, G., 185
Vichutinsky, A. A., 237
Vigevani, A., 263
Vikha, I. V., 237
Villarejo, M., 226
Villar-Palasi, C., 190
Villé, C., 22
Virag, S., 196
Vitvitskii, V. N., 212
Vliegenthart, J. F. G., 151
Voelter, W., 156, 157, 165
Volk, B. W., 222
Volk, W. A., 57
Vorbrüggen, H., 136
Voznesenskaya, S. V., 9, 18

Wacek, V., 225
Wadström, T., 238
Wagner, G., 70
Wagner, H., 17, 20
Walczak, E., 77
Walker, P. S., 206
Walker, R. T., 199
Wallenfels, K., 225
Walter, E., 228
Wampler, D. L., 162
Wander, J. D., 3, 80, 155, 164
Wang, P. Y., 261
Ward, D. N., 208
Ward, E., 123
Warren, C. D., 124
Warren, L., 218, 251
Warren, S., 162
Waser, J., 19, 161
Wassiliadou-Micheli, N., 15, 64
Watanabe, H., 264
Watanabe, K., 244
Watanabe, K. A., 60, 133, 135
Watanabe, T., 227
Watkins, W. M., 187, 225
Watson, D. G., 162
Watson, D. W., 199
Watt, W. B., 182
Webber, J. M., 22
Weber, E., 228
Weber, P., 44, 179
Webster, B. R., 209
Weckesser, J., 22
Weed, R. I., 214
Weetall, H. H., 240, 265
Weigandt, H., 249
Weigerding, A., 170
Weigert, F. J., 155
Weil, J. K., 38
Weil, R., 225
Weinstein, D. B., 251
Weinstein, H. G., 193
Weis, A., 178
Weisenborn, F. L., 130
Weisman, R. A., 190
Weiss, A. H., 6
Weiss, J. B., 189
Weiss, R., 144
Weissmann, B., 16
Weith, H. L., 258
Wellman, G., 12
Wempen, I., 135
Werries, E., 15, 222
Westphal, O., 36, 41, 171
Westwood, J. H., 45, 53
Wetzel, E. R., 130
Wheelock, J. V., 210, 245
Whelan, W. J., 189, 230
Whistler, R. L., 45, 72, 87, 110, 135, 160, 263
White, A. C., 106
White, L. M., 219
White, R. J., 63
Wickson, V. M., 225
Wiebers, J. L., 143, 258
Wiederschain, G. Y., 184
Wiegandt, H., 251
Wightman, R. H., 263
Wilhoit, R. C., 8
Williams, D. G., 161, 165, 258
Williams, J. M., 31
Williams J. W., 235
Williams, N., 107, 153
Williams, N. R., 104, 106
Williams, W. J., 195
Willy, W. E., 153
Wilson, D. P., 136
Wilson, H. R., 154, 162
Winand, R. J., 216
Windeler, A. S., 250
Winetrout, M., 211
Wing, R. E., 25, 26
Winkley, M. W., 135, 136, 142, 156
Winzler, R. J., 44, 179
Witkowski, J. T., 135
Witt, I., 212
Wittick, J. J., 137, 166
Woermann, D., 18
Wolczunowicz, G., 128, 153
Wolf, G., 218, 219
Wolfe, L. S., 201
Wolff, I. A., 72, 151, 227
Wolfrom, M. L., 261
Wong, K. P., 227
Wood, P. J., 5
Woodbury, R. R., 178
Woods, D. A., 215
Wright, J. A., 54, 136
Wright, V., 206
Wrixon, A. D., 165
Wu, C. Y., 148
Wulf, G., 50, 89
Wulff, G., 13
Wurzburg, O. B., 261
Wusteman, F. S., 194, 197

Yaguchi, M., 64, 168
Yajima, T., 142
Yakunenkova, M. A., 171
Yamada, Y., 110
Yamamoto, A., 58
Yamamoto, H., 116, 148
Yamamoto, S., 187
Yamashita, S., 242
Yanagawa, Y., 244
Yanagisawa, H., 63
Yarovaya, S. M., 9
Yarovenko, V. L., 233
Yasuda, Y., 241
Yasunobu, K. T., 244
Yellowlees, D., 239
Yip, G. B., 251
Yokobayashi, K., 234
Yokota, T., 19
Yoldikov, V. N., 53
Yoshida, H., 45, 122
Yoshikawa, M., 42
Yoshimura, J., 58
Yoshinaga, H., 158
Yoshioka, Y., 135
Young, J. D., 237
Young, N. M., 214
Young, R. C., 153
Yuen, R., 170
Yuferov, V. P., 212
Yunker, M., 98

Zabin, I., 226
Zager, N. I., 230
Zagrodzki, S., 9
Zamojski, A., 27, 92, 96, 150
Zancan, G. V., 241
Zanlungo, A. B., 36, 70
Zarubinskii, G. M., 171
Zefirov, N. S., 149
Žemlićka, J., 101, 135, 141
Zhdanov, Yu. A., 20, 91, 108, 115

Zhelvakova, E. G., 124, 126
Zhukova, I. G., 254
Zinbo, M., 42
Zinner, H., 33, 50, 71, 89
Zissis, E., 118
Zobáčová, A., 45
Zolotarev, B. M., 158
Zolotarev, N. S., 6
Zoltewicz, J. A., 144
Zorbach, W. W., 135
Zubay, G., 226
Zurabyan, S. E., 41
Zwierzchowska, Z., 96